T0207226

Lecture Notes in Business Information Processing 496

LNBIP reports state-of-the-art results in areas related to business information systems and industrial application software development – timely, at a high level, and in both printed and electronic form.

The type of material published includes

- Proceedings (published in time for the respective event)
- Postproceedings (consisting of thoroughly revised and/or extended final papers)
- Other edited monographs (such as, for example, project reports or invited volumes)
- Tutorials (coherently integrated collections of lectures given at advanced courses, seminars, schools, etc.)
- Award-winning or exceptional theses

LNBIP is abstracted/indexed in DBLP, EI and Scopus. LNBIP volumes are also submitted for the inclusion in ISI Proceedings.

Anna Bernasconi

Model, Integrate, Search... Repeat

A Sound Approach to Building Integrated Repositories of Genomic Data

Anna Bernasconi
Politecnico di Milano
Milan, Italy

ISSN 1865-1348 ISSN 1865-1356 (electronic)
Lecture Notes in Business Information Processing
ISBN 978-3-031-44906-2 ISBN 978-3-031-44907-9 (eBook)
https://doi.org/10.1007/978-3-031-44907-9

This book is a revised version of the Ph.D. dissertation written by the author to receive her Ph.D. from the Department of Electronics, Information and Bioengineering at Politecnico di Milano, Italy. The original Ph.D. dissertation is accessible at https://hdl.handle.net/10589/169559. In 2023, the Ph.D. dissertation won the CAiSE Best Ph.D. Award, granted to outstanding Ph.D. theses in the field of information systems engineering.

This Springer imprint is published by the registered company Springer Nature Switzerland AG
The registered company address is: Gewerbestrasse 11, 6330 Cham, Switzerland

Paper in this product is recyclable.

To our models,
your uncertainties,
and their unification.

Foreword

Anna Bernasconi received her Ph.D. in Information Technology in 2021. The focus of her Ph.D. has been data integration and data search applied to biological and genomic data. She obtained outstanding results both for what concerns data abstractions and methods and for what concerns tangible technological results, in the form of open systems made available to the research community.

Specifically, her work on conceptual modeling for genomics has produced the definition of the Genomic Conceptual Model (GCM), an integrative data model collected by extracting common entities and their description from a very large number of genomic data sources; next, her work on data integration pipelines has produced META-BASE, a structured approach to the integration of heterogeneous datasets, consisting of steps o data extraction, cleaning, normalizing, merging, and semantic enrichment. Finally, collected datasets (currently, over 67 datasets with 560K files and a total size of about 10 Tbytes) have been collected to the system GenoSurf, which provides a user-friendly search interface.

With the outbreak of the COVID-19 pandemic, she has been able to convert her work from human genomics to viral genomics, and to reapply her methods to viral sequences – the "Repeat" in her thesis title hints at the ability to apply the "model-integrate-search" paradigm first to human genomics datasets and next to viral sequences. By capitalizing on her experience and thanks to her very systematic approach to research, she developed the Viral Conceptual Model (VCM), an integrative data model based on the most important databases which are used worldwide for viral sequence deposition. She next developed ViruSurf, an integrated repository that collects millions of sequences of the SAR-CoV-2 virus; next, she started working towards a knowledge base of amino acid changes and data visualization and analysis tools for supporting the study of viral variants.

This work occurred within a group of Ph.D. students and PostDocs who have shared with Anna responsibilities, duties, and merits; Anna has been able to be highly collaborative with her peers and highly supportive of younger fellows, acting very positively in support of the group's cohesion. Two years after her defense, Anna continues her research journey at Politecnico di Milano: she has published 30 international journal papers, won three prizes for her Doctoral Thesis, achieved habilitation to tenure, acted as a reviewer in many PCs, and organized several events, being a leading member of the conceptual modeling, information systems, and bioinformatics communities.

Milano, August 2023 *Stefano Ceri*

Preface

This book is a revised version of the Ph.D. dissertation written and successfully defended by the author to receive her Ph.D. from the School of Information Technology, Politecnico di Milano, Italy. A summary of the work was published in SpringerBriefs in Applied Sciences and Technology in https://doi.org/10.1007/978-3-030-85918-3_8.

The challenge of integrating genomic data and their describing metadata is at the same time important (a wealth of public data repositories is available to drive biological and clinical research), difficult (the domain is complex and there is no agreement among the various data formats and definitions), and well-recognized (in the common practice, repositories are accessed one-by-one, with tedious and error-prone efforts). Within the context of the European Research Council project "data-driven Genomic Computing" (ERC AdG 693174), which supports genomic research by proposing bioinformatics abstractions and tools, the Ph.D. dissertation has addressed one of its central objectives, i.e., building an integrated repository for genomic data. This book reflects the research adventure that starts from *modeling* biological data, goes through the challenges of *integrating* complex data and their describing metadata and finally builds tools for *searching* the data empowered by a semantic layer. The results of this thesis are part of a broad vision: availability of conceptual models, related databases, and search systems for both humans and viruses genomics will provide important opportunities for research, especially if virus data will be connected to its host, the human being, who is the provider of genomic and phenotype information.

This Ph.D. work applies the principles of Information Systems development to a very complex domain. Such complexity results in the need for dealing with non-obvious problems in each of the steps and illustrates the versatility and importance of Information Systems Engineering techniques in dealing with upcoming issues of essence to human endeavors. In 2023, the dissertation won the CAiSE Ph.D. Award, granted to outstanding Ph.D. theses in the field of information systems engineering.

Milano, August 2023 *Anna Bernasconi*

Acknowledgements

Working with genomic data management and integration is feeling incredibly current these days: we are spotting variants in the SARS-CoV-2 viruses even before they are discussed in newspapers. I would have never imagined that my Ph.D. would turn out so incredibly interesting and empowering. It started as a bet with myself; four years later, I feel like I won it.

For this, I want and must thank immensely my supervisor Professor Stefano Ceri, who showed never-ending support, instilled motivation every step of the way, and gave the example by working stubbornly also during weekends, vacations, and nights, always ready to help understanding, reasoning, and writing. Thank you, Stefano! My second thank you goes to Professor Alessandro Campi, who introduced me to the world of research, calling me back from industry, where I did not belong. I also thank Professor Marco Masseroli, who collaborated on many of the topics described in this thesis, showing rigor and care for work. A special mention goes to Dr. Arif Canakoglu, my very generous colleague, who taught me the tricks of the data integration world; our collaboration was very fruitful: I will never forget his help and kindness. My gratitude goes also to Professor Barbara Pernici, who was a very helpful tutor, providing guidance and inspiration, even during missions! I am also profoundly thankful to Professors Carlo Batini and Oscar Pastor for accepting the duty of reading and reviewing this manuscript: your interest in this work and your commitment to join the discussion make me feel proud and honored at the same time.

I would like to continue by thanking the whole Genomic Computing group at Politecnico, all the people who went through these four years with me shoulder to shoulder, sharing many happy moments: Eirini, my best Ph.D. mate who I will miss forever; Andrea, the sweet and fun soul of GeCo; Luca, with whom I enjoyed many chats; Gaia, beautiful food&travel laughs; Pietro, his humor must be discovered and is good a surprise; Sara, kind and supportive; Michele, the rock of our group; Olya, strong girl; as well as adoptive members of the Data Science group: Francesco, Marco, and Giorgia. Last but not least: Carlo and Claudio, who have now left Politecnico but were always an example to follow since my Master's.

Thank you Giuli, for being always at my side, for helping me believe in what I am good at, for putting up with the last year of my Ph.D.: it was a lot of work and not so much fun, but you made it full of good distractions!

Thank you to my family, because they have supported me all the way, since the first days of kindergarten, through high school, university "far away" and finally this long research period. You succeeded. In leaving me always free to choose, to study, to not do anything else because "I needed to study". I value immensely this attitude and I feel gratefully lucky for having the chance of focusing on what I love the most. Thank you, Babbo, Mamma, Chiara (and my whole American family, Rio, Rafael, Rei), Silvia (who also gave super helpful research tips!), and Zia Anna.

Finally, *grazie mille* to my best friends, cheering for me from wherever they are in the world at the moment: Ariane, Caterina, Arianna, Fabio, Gaia, Silvano, Ema, Mariele, Sara, Eleonora, Juju, Viola, Carli, Gallo, Lorentz, Giuse.

A wonderful journey has just ended.
Let the next challenge begin!

Contents

1 **Introduction** 1
1.1 Genomic Data Integration 5
1.2 Thesis Contribution within the GeCo Project 7
1.3 The Recent COVID-19 Pandemic 9
1.4 Thesis Structure 10

Part I Human Genomic Data Integration

2 **Genomic Data Players and Processes** 19
2.1 Technological Pipeline of Genomic Data 20
2.1.1 Production 21
2.1.2 Integration 21
2.1.3 Services and Access 25
2.2 Taxonomy of Players Involved in Data Production and Integration 26
2.3 Main Genomic Data Players 30
2.3.1 Contributors 32
2.3.2 Repository Hosts 32
2.3.3 Consortia 33
2.3.4 Integrators 36

3 **Modeling Genomic Data** 39
3.1 The Genomic Data Model 40
3.2 The Genomic Conceptual Model 42
3.2.1 Analysis of Metadata Attributes 43
3.2.2 Model Design 44
3.2.3 Validation: Source-Specific Views of GCM 49
3.3 Related Works 49

4 **Integrating Genomic Data** 53
4.1 Theoretical Rationale 55
4.2 Approach Overview 57

4.3 Data Download 58
4.4 Data Transformation 62
4.5 Data Cleaning 67
4.6 Data Mapping 71
4.7 Data Normalization and Enrichment 76
4.7.1 Search Service and Ontology Selection 77
4.7.2 Enrichment Process 83
4.8 Data Constraint Checking 86
4.9 Architecture Implementation 88
4.9.1 Data Persistence 90
4.10 Architecture Validation 91
4.10.1 Lossless Integration 91
4.10.2 Semantic Enrichment 94
4.10.3 User Evaluation 95
4.11 Related Works 95

5 Snapshot of the Data Repository 99
5.1 Included Data Sources 102
5.2 OpenGDC: Integrating Cancer Genomic Data and Metadata 106
5.3 Towards Automated Integration of Unstructured Metadata 110
5.3.1 Experiments 111
5.3.2 Related Works 117

6 Searching Genomic Data 119
6.1 Issues in Exploiting Semantic Knowledge in Genomics 120
6.2 Inference Explanation 122
6.3 GeKnowGraph: Exploration-Based Interface 124
6.3.1 Exploration Interaction 125
6.4 GenoSurf: a Web-Based Search Server for Genomic Datasets 130
6.4.1 Data Search 132
6.4.2 Key-Value Search 134
6.4.3 Query Sessions 135
6.4.4 Result Visualization 136
6.4.5 Additional Functionalities 136
6.4.6 Use Cases 138
6.4.7 Validation of GenoSurf 143
6.5 Related Works 153

7 Future Directions of the Data Repository 157
7.1 Including New Data Sources 158
7.2 Improving Genomic Data Quality Dimensions 159
7.3 Towards Better Interoperability 161
7.4 Simplifying Data and Tools for End Users 162
7.5 Monitoring Integration and Search Value 164

Part II Viral Sequence Data Integration

8 Viral Sequences Data Management Resources . . . 167
8.1 Landscape of Data Resources for Viral Sequences . . . 168
8.1.1 Fully Open-Source Resources . . . 170
8.1.2 GISAID and its Resources . . . 172
8.2 Integration of Sources of Viral Sequences . . . 173
8.2.1 Metadata Integration . . . 173
8.2.2 Value Harmonization and Ontological Efforts . . . 173
8.2.3 Replicated Sequences in Multiple Sources . . . 175
8.3 SARS-CoV-2 Search Systems . . . 175
8.3.1 Portals to NCBI and GISAID Resources . . . 176
8.3.2 Integrative Search Systems . . . 177
8.3.3 Comparison . . . 178
8.4 Discussion . . . 180
8.4.1 GISAID Restrictions . . . 180
8.4.2 Metadata Quality . . . 181
8.4.3 (Un)Willingness to Share Sequence Data . . . 181

9 Modeling Viral Sequence Data . . . 183
9.1 Conceptual Modeling for Viral Genomics . . . 184
9.2 Answering Complex Biological Queries . . . 189
9.3 Related Works . . . 191

10 Integrating Viral Sequence Data . . . 193
10.1 Database Content . . . 194
10.2 Relational Schema . . . 195
10.3 Data Import . . . 197
10.4 Annotation and Variant Calling . . . 198
10.5 Data Curation . . . 199

11 Searching Viral Sequence Data . . . 201
11.1 Requirements Analysis . . . 202
11.1.1 Lessons Learnt . . . 203
11.2 Web Interface . . . 204
11.3 Example Queries . . . 208
11.4 Discussion . . . 210
11.5 Related Works . . . 212

12 Future Directions of the Viral Sequence Repository . . . 213
12.1 Research Agenda . . . 214
12.2 ViruSurf Extensions . . . 214
12.3 Visualization Support: VirusViz . . . 216
12.4 Active Monitoring of SARS-CoV-2 Variations . . . 219
12.5 Integrating Host-Pathogen Information . . . 221
12.5.1 The Virus Genotype – Host Phenotype Connection . . . 221
12.5.2 The Host Genotype – Host Phenotype Connection . . . 222

Part III Epilogue

13 Conclusions and Vision 229
13.1 Summary of Thesis Contributions 230
13.2 Achievements within GeCo Project 233
13.3 Outlook 234

A META-BASE tool configuration 237
A.1 User Manual 237
A.2 Process Configuration 238
A.3 Mapper Configuration 240

B Experimental setup GEO metadata extraction 243

C Mappings of viral sources attributes into ViruSurf 245
References 251

Acronyms

API Application Programming Interface
BAM Binary Alignment Map
BCR Biospecimen Core Repository
BED Browser Extensible Data
BPMN Business Process Model and Notation
BRCA BReast invasive CArcinoma
CARD COVID-19 Analysis Research Database
CCLE Cancer Cell Line Encyclopedia
CDE Common Data Elements
CDS CoDing Sequence
CHLA Children's Hospital, Los Angeles
CL Cell Ontology
CM Conceptual Modeling
CNGBdb China National GeneBank DataBase
COG-UK COronavirus disease 2019 Genomics UK Consortium
COSMIC Catalogue Of Somatic Mutations In Cancer
COVID-19 HGI COVID-19 Host Genetics Initiative
CPM Center for Personalized Medicine
CSV Comma-Separated Values
dbGaP database of Genotypes and Phenotypes
DCC Data Coordination Center
DDBJ DNA Data Bank of Japan
DFA Deterministic Finite State Automata
DNA DeoxyriboNucleic Acid
EBI European Bioinformatics Institute
EDACC Epigenomics Data Analysis and Coordination Center
EFO Experimental Factor Ontology
EGA European Genome-phenome Archive
EMBL European Molecular Biology Laboratory
ENCODE Encyclopedia of DNA Elements
ExAC Exome Aggregation Consortium

FAIR Findable, Accessible, Interoperable, Reusable
FPKM Fragments Per Kilobase of transcript per Million mapped reads
FPKM-UQ Fragments Per Kilobase of transcript per Million mapped reads - Upper Quartile
GA4GH Global Alliance for Genomics and Health
GCM Genomic Conceptual Model
GDC Genomic Data Commons
GDM Genomic Data Model
GeCo data-driven Genomic Computing
GEO Gene Expression Omnibus
GISAID Global Initiative on Sharing All Influenza Data
GKG Genomic Knowledge Graph
GMQL GenoMetric Query Language
gnomAD Genome Aggregation Database
GO Gene Ontology
GTEx Genotype-Tissue Expression
GTF Gene Transfer Format
GWAS Genome-Wide Association Studies
HeTOP Health Terminology/Ontology Portal
HGNC HUGO Gene Nomenclature Committee
HUGO HUman Genome Organisation
ICGC International Cancer Genome Consortium
IEDB Immune Epitope Database and Analysis Resource
IGSR International Genome Sample Resource
IHEC International Human Epigenome Consortium
INSDC International Nucleotide Sequence Database Collaboration
ISB Institute for Systems Biology
JSON JavaScript Object Notation
LDACC Laboratory, Data Analysis, and Coordinating Center
LKB Local Knowledge Base
MERS-CoV Middle East Respiratory Syndrome-related CoronaVirus
NCBI National Center for Biotechnology Information
NCBITaxon NCBI Taxonomy Database
NCBO National Center for Biomedical Ontology
NCI National Cancer Institute
NCIT National Cancer Institute Thesaurus
NFA Non-deterministic Finite Automata
NGS Next-Generation Sequencing
NHGRI National Human Genome Research Institute
NIH National Institutes of Health
NMDC National Microbiology Data Center
NW Needleman-Wunsch algorithm
OBI Ontology for Biomedical Investigations
OGG Ontology of Genes and Genomes
OLS Ontology Lookup Service

PANGOLIN Phylogenetic Assignment of Named Global Outbreak LINeages
QC Quality Control
RDF Resource Description Framework
REP Roadmap Epigenomic Project
RNA RiboNucleic Acid
SARS-CoV Severe Acute Respiratory Syndrome-related CoronaVirus
SARS-CoV-2 Severe Acute Respiratory Syndrome CoronaVirus 2
SO Sequence Ontology
SRA Sequence Read Archive
TADs Topologically Associating Domains
TARGET Therapeutically Applicable Research to Generate Effective Treatments
TCGA The Cancer Genome Atlas
UBERON Uber Anatomy Ontology
UCSC University of California, Santa Cruz
UMLS Unified Medical Language System
URI Uniform Resource Identifier
URL Uniform Resource Locator
UTR UnTranslated Region
VCF Variant Call Format
VCM Viral Conceptual Model
XML Extensible Markup Language
XSD XML Schema Definition

Chapter 1
Introduction

"Genomics is at risk of becoming the sport of kings, where only the wealthy can play."

— Anthony Philippakis - Broad Institute

Genomics was born in relatively recent times. The first publication, concerning the double helix model of DNA by the Nobel prizes James Watson and Francis Crick, was in 1953 in Nature; the first draft of the human genome (resulting from the Human Genome Project[1]) was completed and published in April 2003 [313].

In the last two decades, the technology for DNA sequencing has made incredible steps; Figure 1.1 shows the trend of the sequencing cost in the years 2021–2020; it has followed a pattern similar to the one of computing hardware, approximately halving every two years (see Moore's Law indication in the graph). That trend has changed since the beginning of 2008: we observe a big drop corresponding to the introduction of Next-Generation Sequencing (NGS), a high-throughput technology based on massively parallel image capturing [361], bringing increasing amounts of genomic data of multiple types, together with microarray and single-cell technologies. With such technological enhancements, sequencing costs perceived a significant reduction, less than 20 years after the Human Genome Project; producing a complete human sequence costs less than 1000 US$ in 2020 and is expected to become even more affordable for individuals in the next few years, as many companies are promising [1, 320]. In terms of processing time, the first attempt to sequence a whole human genome took over a decade; to date, the same operation can be accomplished in one single day [60].

A sequencing session is able to produce a huge mass of raw data (i.e., "short reads" of genome strings), reaching a typical size of 200 Gigabytes per single human genome, once it is stored. According to [376], genomics is going to generate the biggest "big data" problem for mankind: between 100 million and 2 billion human genomes are expected to be sequenced within 2025.

[1] The Human Genome Project, funded by the National Institutes of Health (NIH), was the result of a collective effort involving twenty universities and research centers in the USA, UK, Japan, France, Germany, Canada, and China.

A. Bernasconi: *Model, Integrate, Search... Repeat*, LNBIP 496, pp. 1–15, 2023.
https://doi.org/10.1007/978-3-031-44907-9_1

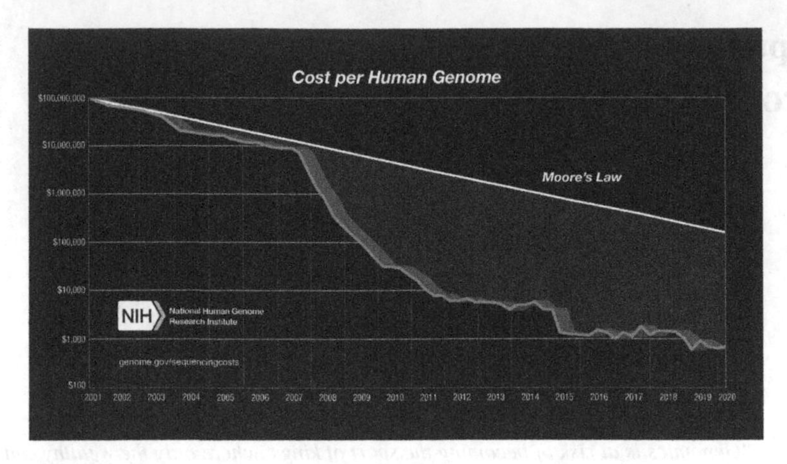

Fig. 1.1 Cost of DNA sequencing per genome. Courtesy: National Human Genome Research Institute (`https://www.genome.gov/`) [333].

From being a long string of nitrogenous bases, encoding adenine (A), uracil (U), cytosine (C), guanine (G), and thymine (T) representing the raw data, our concept of genomic data has by now evolved, starting to include also the *signals* produced by the living system represented in the genome; signals can be integrated and interpreted, to better understand carried information. A variety of data types, collected within numerous files that are heterogeneous both in formats and semantics, are available, mainly expressing DNA or RNA features:

1. *Mutations*, specific positions of the genome where the code of an individual differs from the code of the "reference" human being. Mutations are associated with genetic diseases, which are inherited in specific positions of the chromosomes, and other diseases such as cancer, produced during human life and related to external factors such as nutrition and pollution.
2. *Gene expression*, indicating in which conditions genes are active (i.e., they transcribe a protein) or inactive; the same gene may have intense activity in given conditions and no activity in others.
3. *Peaks of expression*, specifying the positions of the genome with higher read density due to a specific treatment of DNA; these in turn indicate specific biological events (e.g., binding of a protein to the DNA).
4. *Annotations*, a peculiar kind of "signal", representing known information, such as positions of genes, transcripts, or exons;
5. *Structural properties* of the DNA (e.g., breakpoints, where the DNA is damaged, or junctions, where the DNA creates loops).

All signals are aligned to a reference genome, a standard sequence characterizing human beings that is constantly improved and updated by the scientific community. Signals on a genome browser may appear, for example, as in Figure 1.2, where tracks of different colors describe gene annotations, gene expression, peaks of expressions, and mutations. The browser is open on a window of a given number of bases (from

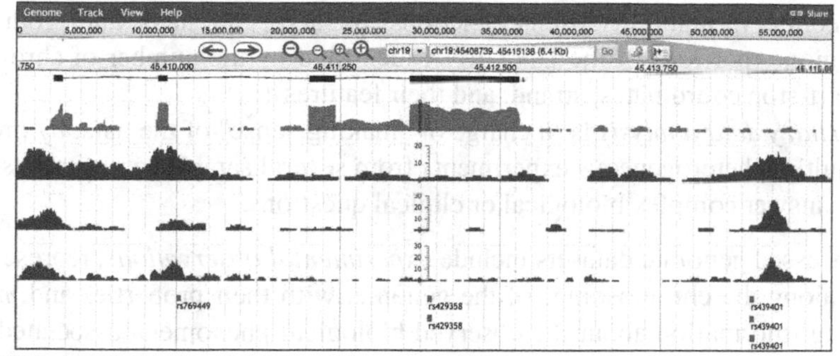

Fig. 1.2 Signals on a genome browser, corresponding to genes (*annotation* track where four genes are shown in black), gene expression (one track with red signal), peaks (three tracks with blue signals) and mutations (three tracks with red segments pointing to mutated positions of the genome).

a few bases to millions of them), and signals are presented as separated tracks in the window; each track displays the signal by showing its position and possibly its intensity.

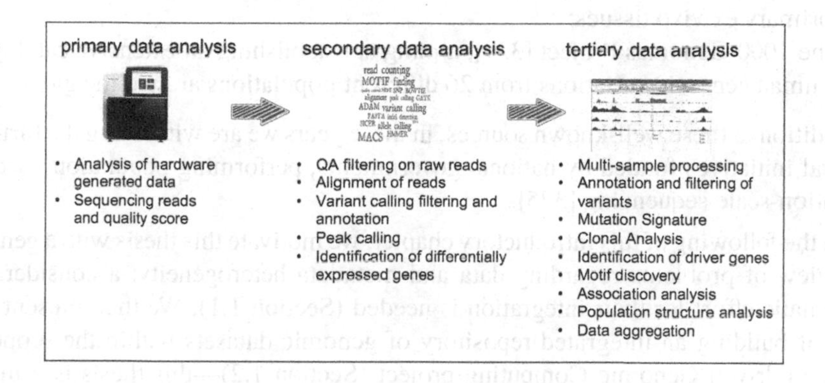

Fig. 1.3 Primary, secondary, and tertiary data analysis for genomics.

Signals can be loaded on the browser and analyzed to extract statistical and novel biological knowledge only after being produced by a long and complex bioinformatics pipeline. In particular, the analysis of NGS data is composed of three phases (see Figure 1.3):

1. sequencing machines perform *primary data analysis*, producing raw datasets of sequences of DNA/RNA bases;
2. *secondary data analysis* includes alignment of reads to a reference genome (typical for a biological species), variant/peak calling (i.e., the process of determining the differences between a sample and the reference genome, such as single nu-

cleotide variants, insertions, deletions, or larger variants), and production of regions (portions of the DNA/RNA identified by the number of chromosome, start-stop coordinates, strand, and their features);
3. *tertiary data analysis* is in charge of "making sense" of the data, by processing multiple heterogeneous experiments from several samples and patients in order to answer complex biological or clinical questions.

Processed genomic datasets include *experimental observations*, representing regions along the chromosomes of the genome, with their properties and *metadata*, carrying information about the observed biological phenomena associated clinical elements, and technological/organizational aspects. Thousands of these processed datasets are becoming available every day, typically produced within the scope of large cooperative efforts, open for public use and made available for secondary research use, such as

1. the Encyclopedia of DNA Elements (ENCODE) [330], the most general worldwide repository for basic biology research comprising hundreds of epigenetic experiments of processed data in humans and mice;
2. The Cancer Genome Atlas (TCGA) [397], a full-scale effort to explore the entire spectrum of genomic changes involved in human cancer;
3. the Roadmap Epigenomics Project [233], a repository of "normal" (not involved in diseases) human epigenomic data from NGS processing of stem cells and primary ex vivo tissues;
4. the 1000 Genomes Project [329], aiming at establishing an extensive catalog of human genomic variations from 26 different populations around the globe.

In addition to these well-known sources, in these years we are witnessing the birth of several initiatives funded by national governments, performing population-specific or nation-scale sequencing [375].

In the following of this introductory chapter, we motivate this thesis with a general overview of problems regarding data and metadata heterogeneity; a considerable systematic effort for their integration is needed (Section 1.1). We then present our goal of building an integrated repository of genomic datasets within the scope of the data-driven Genomic Computing project (Section 1.2)—this thesis is a major reference point of the data integration achievements of the project. In Section 1.3 we discuss how our presented approach to data integration in the human genomic domain has proven extremely effective when, nine months ago – in the middle of the COVID-19 pandemic – we applied the experience gained in the previous work, on the new domain of viral sequences. In a critical period for scientific research, we provided a well-defined contribution to support virologists' everyday activities (searching data, understanding sequence variation, and building preliminary hypotheses). Finally, we outline the thesis organization, by overviewing each Chapter's focus and related publications (Section 1.4).

1.1 Genomic Data Integration

As genome sequencing data production is by now considered a routine activity, research interest is rapidly moving toward its analysis and interpretation, only recently becoming possible thanks to the generous amount of available data. Although the potential collective amount of available information is huge, the effective combination of genomic datasets from disparate public sources is hindered by the inherent heterogeneity of datasets, their lack of interconnectedness and considerable heterogeneity, spanning from download protocols and formats, to notations, attribute names, and values.

The lack of organization and systematization is partly due to the metadata authoring process (i.e., the preliminary compilation of information describing datasets); research practitioners often prioritize publishing their research rather than leaving a legacy of reusable data (note that funding agencies do not always explicitly require data sharing and documentation). Moreover, systematic data curation – considered cumbersome, tedious, and time-consuming – is not considered as rewarding as other downstream activities. As a result, researchers have to face substantial obstacles when searching for data that suits their analysis.

Therefore, prior to data analysis and biological knowledge discovery, *data and metadata integration* is gaining irrefutable priority, with pressing demands for enhanced methodologies of data extraction, normalization, matching, and enrichment, to allow building multiple perspectives over the genome, possibly leading to the identification of meaningful relationships, not perceivable otherwise [343]. Genomic data integration for processed data includes technological efforts on the data representing signals and (mainly) conceptual efforts on the metadata.

Many approaches are already showing the benefits of data integration, at times performed with *ad-hoc* techniques [398, 82, 344], others with automatic methods [58, 202, 420]. However, several issues are still to be addressed, including:

1. the need for always updated data, to guarantee a higher quality of results [240],
2. the lack of normalization/harmonization between the processing pipelines [177],
3. the limited structured metadata information and agreement among models [318], and
4. the unsystematic use of controlled terminology to allow interoperability [171].

In general, shared and systematic solutions are missing and – in the majority of cases – the necessary integration efforts are solved by the authors themselves.

There have been broader attempts at integration, but mostly inside the single institutes' walls (Broad Institute's Terra,[2] EMBL-EBI,[3] NIH-NCBI,[4] Seven Bridges[5]). All these consortia provide portals for data access, but integrated access is not provided. When data are handled within the scope of the consortia that coordinate

[2] `https://terra.bio/`

[3] `https://www.ebi.ac.uk/`

[4] `https://www.ncbi.nlm.nih.gov/`

[5] `https://www.sevenbridges.com/`

their production, their downstream impact will be much less powerful and relevant with respect to scenarios where data can be managed all together, addressing cross-source differences—a single laboratory may produce data of excellent quality for one specific technology, but not for many.

In view of scientific advancement, important results are most likely achieved when combining multiple signals; for this, input data is needed also from other technologies and their integration is an essential ingredient, where considerable timely efforts should be concentrated on current research.

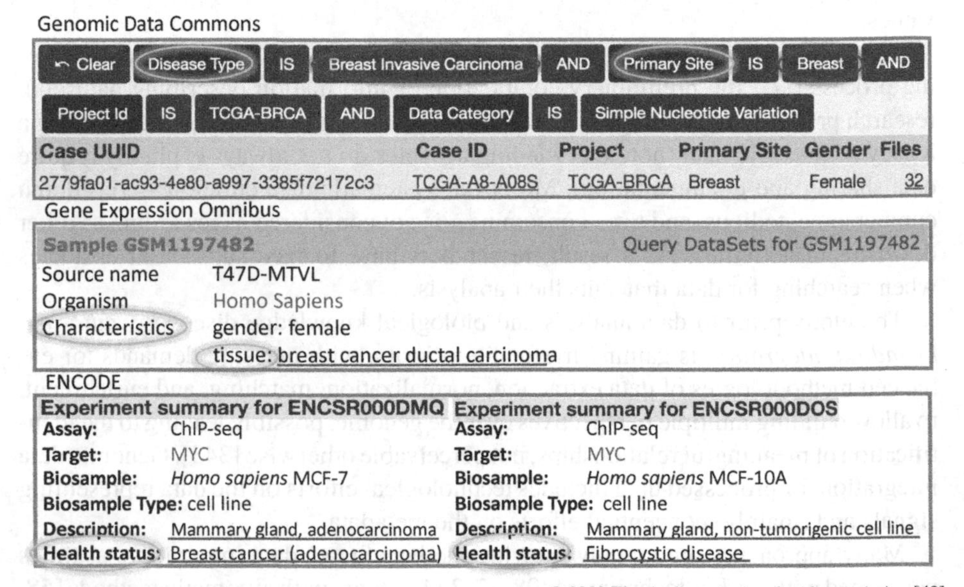

Fig. 1.4 Example of Web interfaces of data sources: GDC, GEO, and ENCODE. Yellow circles highlighted how information regarding the category *disease* observed in an experiment is named in different sources. The red circles and lines show how similar disease values are used in the sources.

Motivating example. Let us consider a researcher who is looking for data to perform a comparison study between a human non-healthy breast tissue, affected by carcinoma, and a healthy sample coming from the same tissue type. Exploiting her previous experience, the researcher locates three portals having interesting data for this analysis (see Figure 1.4). For the diseased data, describing gene expression, the chosen source is the portal of Genomic Data Commons (GDC) [177], an important repository of human cancer mutation data. As it can be seen on the top of Figure 1.4, one or more cases (i.e., datasets) can be retrieved by composing a query that allows to locate variation data on "Breast Invasive Carcinoma" from the "Breast" primary site. To compare such data with references, the researcher chooses additional datasets coming from cell lines, a standard benchmark for investigations. A tumor cell line data is found on the Gene Expression Omnibus (GEO) [31] web interface (middle rectangle of Figure 1.4) where, by browsing thousands of human samples, the researcher locates one with analyzed cell type "T47D-MTVL" and observed disease "breast cancer ductal carcinoma". On ENCODE, the researcher chooses both a tumor cell line (bottom left of Figure 1.4) and a normal cell line (bottom right of Figure 1.4), to make a control comparison. "MCF-7" is a cell line from a diseased tissue affected by "Breast cancer (adenocarcinoma)", while "MCF-10A" is its widely considered non-tumorigenic counterpart.

From the point of view of attributes (i.e., how each piece of information is identified), when searching for disease-related information, we find many possibilities: "Disease type" in GDC, "Characteristics–tissue" in GEO, and "Health status" in ENCODE (see yellow circles in Figure 1.4). From the point of view of values, instead, when searching for breast cancer-related information, we find multiple similar expressions, possibly pointing to comparable samples: "Breast Invasive Carcinoma" (GDC), "breast cancer ductal carcinoma" (GEO), "Breast cancer (adenocarcinoma)" (ENCODE) (see red lines in Figure 1.4). It can be noted that "Breast Invasive Carcinoma" and "breast cancer (adenocarcinoma)" are related sub-types of "breast carcinoma" (as observed in specialized ontologies); this allows the researcher to compare GDC's data with the dataset from ENCODE. For what concerns the *cell line*, researchers typically query specific databases or dedicated forums to discover tumor/normal matched cell line pairs. This kind of information is not encoded in a unique way over data sources and is often missing. Considerable external knowledge is necessary in order to find the above-mentioned connections, which cannot be obtained on the mentioned portals. Besides disease, tissue, and cell line-related aspects, many other metadata aspects are involved.

1.2 Thesis Contribution within the GeCo Project

My Ph.D. research has evolved within the scope of the Genomic Computing (GeCo) project[6], led by Stefano Ceri at Politecnico di Milano, who received an ERC Advanced Grant project (No. 693174), starting in 2016 and ending in September 2021. The project is set in the context of tertiary data analysis for genomics and aims to provide a new focus on data extraction, querying, and analysis by raising the level of abstraction of models, languages, and tools.

In Figure 1.5 we summarize short/mid/long-term goals of the GeCo project divided into six Working Packages. Initially, a new core model for genomic processed data [276] has been devised (**WP1**); based on this, a data engine for genomic region-based data and metadata was developed to support the novel GenoMetric Query Language (GMQL) [277] (**WP2**), also available as a Python library (pyGMQL, [295]) and an R-Bioconductor package (RGMQL[7]); the associated cloud-based query sys-

[6] `https://cordis.europa.eu/project/id/693174`

[7] `https://bioconductor.org/packages/release/bioc/html/RGMQL.html`

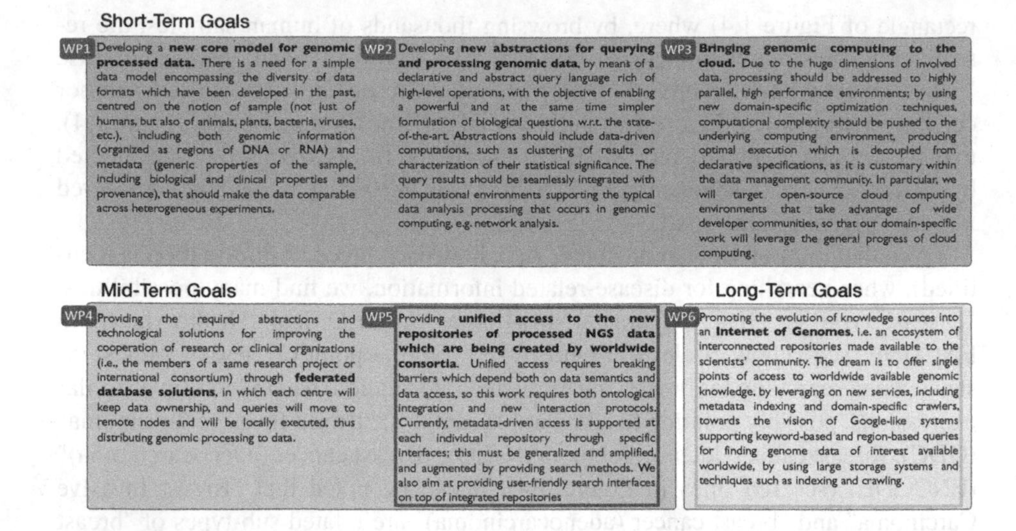

Fig. 1.5 Description of working packages of the ERC AdG project on data-driven Genomic Computing 2016-2021. WP5 is highlighted as it is the main focus of this thesis; here we also set the basis to achieve WP6.

tem, using Apache Spark[8] on arbitrary servers and clouds was proposed in [275] (**WP3**), followed by many works on optimization [216, 180, 200]; a solution to share and process genomic datasets with a federated version of the GMQL system was published in [78], allowing its adoption in clinical organizations and in cooperating scenarios between distinct research institutes (**WP4**).

My thesis focuses on GeCo project's **WP5**, while also setting the basis to achieve **WP6**; it is concerned with building a repository of genomic datasets to provide unified access to a large number of well-known processed NGS data created and continuously updated by worldwide consortia. To this end, we analyzed thoroughly the issues related to the integration of heterogeneous data coming from multiple open data sources, and we followed a systematic approach:

- *Model*: first we analyze the domain thoroughly, by means of a complete state-of-the-art analysis, comprehensive of online resources scouting (as many systems do not have connected research papers), reading of their documentation, and testing of functionalities (focusing on the methods they expose to retrieve their data). The data is studied and understood, with the objective of proposing a conceptual model that captures the main characteristics shared by relevant data sources in the field; we target completeness but favor simplicity, as the goal is to produce easy-to-use systems for the community of biologists and genomic experts.
- *Integrate and build*: we select a number of interesting open data sources for the domain, and we build solid pipelines to download data from them while

[8] `https://spark.apache.org/`

transforming it into a standard interoperable format. We include descriptions into a relational database which allows for interoperability of schemata and of instances, after a process of data normalization and semantic annotation using specialized ontologies available for biomedical terms. The result is a repository of homogenized data, that can be used seamlessly from a unique endpoint, allowing for integrated and complex biological querying.

- *Search*: we then target the end-users of the repository, i.e., experts of the domain who browse the repository in search of the right datasets to prove or disprove their research hypotheses, or just explore data types to find interesting data for their analysis. Interfaces need to take into account the specific needs of the particular users: very knowledgeable in biology, limited knowledge of programming languages, and limited time to understand complex features as opposed to more intuitive functionalities.

All in all, this thesis addresses the need for a general and systematized approach to data and metadata integration for genomic datasets, targeted to an improved interoperable search and analysis; it advocates that many efforts thus far proposed are not as general as necessary, only solving the problem in specific sub-areas of genomics (e.g., epigenomics, cancer genomics, or annotation data) or are not published at all. In this sense, we have fully achieved the indications of WP5 and started a long-term study towards the so-called "Internet of Genomes", which bridges many systems and queries genomic regions along with their metadata.

1.3 The Recent COVID-19 Pandemic

During the first phase of the COVID-19 epidemic, in March and April 2020 we responded proactively to the call to arms addressed to the broad scientific community. We first conducted an extensive requirement analysis by engaging in interdisciplinary conversations with a variety of scientists, including virologists, geneticists, biologists, and clinicians, including – in alphabetical order – Ilaria Capua (University of Florida, US), Matteo Chiara (University of Milano-Statale, IT), Ana Conesa (University of Florida, US), Luca Ferretti (University of Oxford, UK), Alice Fusaro (Istituto Zooprofilattico Sperimentale delle Venezie, IT), Susanna Lamers (BioInfo-Experts, US), Stefania Leopardi (Istituto Zooprofilattico Sperimentale delle Venezie, IT), Alessio Lorusso (Istituto Zooprofilattico Sperimentale Abruzzo e Molise, IT), Francesca Mari (University of Siena, IT), Carla Mavian (University of Florida, US), Graziano Pesole (University of Bari, IT), Alessandra Renieri (University di Siena), Anna Sandionigi (University of Milano-Bicocca, IT), Stephen Tsui (Hong-Kong Chinese University, HK), Limsoon Wong (National University of Singapore, SGP), Federico Zambelli (University of Milano-Statale, IT).

This preliminary activity convinced us of the need for a structured proposal for viral data modeling and management. We reapplied the previously proposed methodology of modeling a data domain, integrating many sources to build a global repository, and finally making its content searchable to enable further analysis. This experience suggests that the approach is general enough to be applied to any domain of life sciences and encourages broader adoption.

1.4 Thesis Structure

This thesis is organized into two parts:

- Part I '**Human Genomic Data Integration**' is dedicated to the description of the data integration problem in the field of genomic tertiary analysis, where data are complex and heterogeneous. We proposed a model for their description, an integration pipeline, a continuously updated repository, and two search systems upon its content, tested by domain experts and members of our research community. The chapters of this part are dedicated to "data" assuming the general meaning of the word (including data and their descriptions); at times we focus specifically only on metadata—when this is the case, it is stated in the discussion.
- Part II '**Viral Sequence Data Integration**' is dedicated to understanding the world of viral sequences and their descriptions in terms of collected samples, characteristics of host organisms, variants, and their impact on the related disease. We describe a data model for metadata and integrate several pieces of information into a unique schema, exploited by a powerful search system over viral sequences, that will be potentially expanded to include also genotype and phenotype information regarding the host organism (with particular attention to humans, as this is critical for future epidemics events).

In Part III, the thesis concludes with a vision of how the two debated parts can indeed be included in one single system, where different organisms' data can be interconnected to drive more powerful biological discovery. In the following, we outline the content of each chapter of the thesis, specifying when they are based on contributions that have been recently published.

Chapter 2 '**Genomic Data Players and Processes**' provides an analysis of the most important data contributors and distributors currently active in the landscape of genomic research, describing the data integration practices implemented in their pipelines. We give a brief explanation of the technological pipeline through which genomic data are produced and manipulated before being exposed and used via interfaces. Then we propose a taxonomy to characterize genomic data actors and describe about thirty important data sources according to this model. This analysis has been conducted on a wide landscape of genomic data sources and has been instrumental to a more informed development of the following parts of this thesis, which are our original contributions. This chapter is based on the review journal article:

[48] **A. Bernasconi**, A. Canakoglu, M. Masseroli, and S. Ceri. The road towards data integration in human genomics: players, steps and interactions. *Briefings in Bioinformatics*, 22(1):30–44, 2021. https://doi.org/10.1093/bib/bbaa080

In Chapter 3 '**Modeling Genomic Data**' we describe a general modeling option for genomic data (which has been introduced in previous works of the GeCo research group) and we propose a novel and concise modeling solution for metadata, by means of a compendious conceptual model for the representation of the most meaningful attributes documenting data samples in many genomic data sources; this model may be used to motivate and drive schema integration of such documenting information. The work presented in this chapter is based on the conference article:

[54] **A. Bernasconi**, S. Ceri, A. Campi, and M. Masseroli. Conceptual Modeling for Genomics: Building an Integrated Repository of Open Data. In *Proceedings of the International Conference on Conceptual Modeling*. ER, 2017. `https://doi.org/10.1007/978-3-319-69904-2_26`

Chapter 4 '**Integrating Genomic Data**' provides a broad explanation of how our modeling principles have been employed to drive an ambitious and extendable integration pipeline, which is able to solve, at the same time, heterogeneity at multiple levels. The described architecture includes all the integration steps for data and metadata. For what concerns data collection, we follow a partition-driven approach to synchronize our local instances with the origin ones. We do not re-process data, but perform many transformation tasks, homogenizing formats, and cleaning metadata. We consolidate data into a relational database (realized after our unique conceptual representation), resolving schema integration issues. Then, we achieve value interoperability among collected instances, by annotating them with specialized ontologies' terms and their hierarchies, in order to instrument a semantically enriched search of datasets linked to such metadata. The work presented in this chapter is based on two workshop papers and one journal publication, respectively describing the following aspects.

The metadata integration vision.

[39] **A. Bernasconi**. Using Metadata for Locating Genomic Datasets on a Global Scale. In *International Workshop on Data and Text Mining in Biomedical Informatics*, 2018. `http://ceur-ws.org/Vol-2482/paper5.pdf`

The ontological annotation of metadata.

[45] **A. Bernasconi**, A. Canakoglu, A. Colombo, and S. Ceri. Ontology-Driven Metadata Enrichment for Genomic Datasets. In *International Conference on Semantic Web Applications and Tools for Life Sciences*, 2018. `http://ceur-ws.org/Vol-2275/paper6.pdf`

The comprehensive formal description of the metadata integration architecture.

[47] **A. Bernasconi**, A. Canakoglu, M. Masseroli, and S. Ceri. META-BASE: a novel architecture for largescale genomic metadata integration. *IEEE/ACM Transactions on Computational Biology and Bioinformatics*, 19(1):543–557, 2022. `https://doi.org/10.1109/TCBB.2020.2998954`

Chapter 5 '**Snapshot of the Data Repository**' shows the result of the application of the META-BASE architecture: we built an integrated repository of genomic data that contains many datasets from relevant data sources used in the domain and that is periodically updated. Here, we describe each included source more in-depth and give a quantitative view of the repository composition. With respect to other compared integrative repositories, our approach is the only one that joins together such a broad range of genomic data, which spans from epigenomics to all data types typical of cancer genomics (e.g., mutation, variation, expression, etc.), until annotations. For two data sources, we provide additional details on the methods used for extracting metadata and including them in the general META-BASE pipeline, as they required extra effort with respect to the formalization proposed in the preceding chapter. Part of the work presented in this chapter is based on one journal paper and one conference paper:

Integration of Genomic Data Commons.

[81] E. Cappelli, F. Cumbo, **A. Bernasconi**, A. Canakoglu, S. Ceri, M. Masseroli, E. Weitschek. OpenGDC: Unifying, modeling, integrating cancer genomic data and clinical metadata. *Applied Sciences*, 10(18):6367, 2020. https://doi.org/10.3390/app10186367

Extraction of structured metadata from Gene Expression Omnibus.

[79] G. Cannizzaro, M. Leone, **A. Bernasconi**, A. Canakoglu, and M. J. Carman. Automated Integration of Genomic Metadata with Sequence-to-Sequence Models. In *Proceedings of the 24th European Conference on Machine Learning and Principles and Practice of Knowledge Discovery in Databases (ECML PKDD 2020)*. https://doi.org/10.1007/978-3-030-67670-4_12

An important side effect of providing a global and integrated view of data sources is the ability to build user-friendly query interfaces for selecting items from multiple data sources. In Chapter 6 '**Searching Genomic Data**' we overview the critical aspects that search interfaces over integrative genomic systems need to address. We describe the implemented search methods employed on top of the integrated repository, to make the results of this effort available to the bioinformatics and genomics community.

We propose two different search systems, both for users that are searching for datasets upon which performing data analysis and knowledge discovery. Different personas are targeted: 1) a graph-based representation of the repository targets expert users with knowledge of semantic interoperability concepts, with exploration objectives; 2) a fast metadata-based search engine targets users with basic knowledge of the underlying technologies with an interface designed to allow user-friendly interaction – positively evaluated by an audience of bioinformaticians and genomics practitioners. The work presented in this chapter is based on three publications.

The genomic knowledge graph for metadata.

[44] **A. Bernasconi**, A. Canakoglu, and S. Ceri. From a Conceptual Model to a Knowledge Graph for Genomic Datasets. In *Proceedings of the International Conference on Conceptual Modeling*. ER, 2019. https://doi.org/10.1007/978-3-030-33223-5_29

The Web metadata-based search engine.

[76] A. Canakoglu, **A. Bernasconi**, A. Colombo, M. Masseroli, and S. Ceri. GenoSurf: metadata driven semantic search system for integrated genomic datasets. *Database: The Journal of Biological Databases and Curation*, 2019. https://doi.org/10.1093/database/baz132

The validation of the web search engine.

[43] **A. Bernasconi**, A. Canakoglu, and S. Ceri. Exploiting Conceptual Modeling for Searching Genomic Metadata: A Quantitative and Qualitative Empirical Study. In *2nd International Workshop on Empirical Methods in Conceptual Modeling (EmpER)*, 2019. https://doi.org/10.1007/978-3-030-34146-6_8

Chapter 7 '**Future Directions of the Data Repository**' includes prospective directions of the work described in the preceding chapters. We aim to feed the repository with new data sources in a continuous fashion in the next years, in addition to running periodic updates of the currently included datasets. A novel perspective that I am currently investigating targets data quality dimensions during the process of collecting and joining different datasets of genomic features together. Moreover, we are embedding the technologies developed for the repository within a broader project that aims to provide biologists and clinicians with a complete data extraction/analysis environment, guided by a conversational interface, breaking down the technological barriers that are currently hindering the practical adoption of some of our systems. We also plan to add semantic interoperability support and conceive a marketplace service to suggest best practice queries for data extraction and analysis. Part of the work presented in this chapter has been described in two journal papers.

[40] **A. Bernasconi**. Data quality-aware genomic data integration. Computer Methods and Programs in Biomedicine Update, 1, 100009, 2021. https://doi.org/10.1016/j.cmpbup.2021.100009

[114] P. Crovari, S. Pidò, P. Pinoli, **A. Bernasconi**, A. Canakoglu, F. Garzotto, and S. Ceri. GeCoAgent: a conversational agent for empowering genomic data extraction and analysis. ACM Transactions on Computing for Healthcare (HEALTH), 3(1), pp.1-29, 2021. https://doi.org/10.1145/3464383

Chapter 8 '**Viral Sequences Data Management Resources**' overviews the latest scenario on integrative resources dedicated to SARS-CoV-2, the virus responsible for COVID-19, in addition to other viruses. We compare characteristics of data, interfaces, and openness policies, discussing how these impact the success of data integration efforts. The work presented in this chapter is based on the review journal article:

[49] **A. Bernasconi**, A. Canakoglu, M. Masseroli, P. Pinoli, and S. Ceri. A review on viral data sources and search systems for perspective mitigation of COVID-19. *Briefings in Bioinformatics*, 22(2):664–675, 2021. https://doi.org/10.1093/bib/bbaa359

In Chapter 9 '**Modeling Viral Sequence Data**' we propose a novel conceptual model for describing viral sequences, with a particular focus on aspects that are relevant for SARS-CoV-2 and other RNA single-stranded viruses: their genes, mutations, and metadata. The work presented in this chapter is based on a conference publication:

[50] **A. Bernasconi**, A. Canakoglu, P. Pinoli, and S. Ceri. Empowering virus sequence research through conceptual modeling. In *Proceedings of the International Conference on Conceptual Modeling*. ER, 2020. https://doi.org/10.1007/978-3-030-62522-1_29

Chapter 10 '**Integrating Viral Sequence Data**' and Chapter 11 '**Searching Viral Sequence Data**' report how we collected requirements from experts and built the pipeline that imports viral sequence data from many sources, computes variants with respect to a reference sequence, and feeds a web engine with complex query capabilities. The work presented in these two chapters is based on a journal publication:

[77] A. Canakoglu, P. Pinoli, **A. Bernasconi**, T. Alfonsi, D. P. Melidis, and S. Ceri. Virusurf: an integrated database to investigate viral sequences. *Nucleic Acids Research*, 49(D1):D817–D824, 2021. https://doi.org/10.1093/nar/gkaa846

Chapter 12 '**Future Directions of the Viral Sequence Repository**' concludes Part II by describing the contribution of this thesis to the viral sequence integration problem. In particular, we plan to systematize our requirements elicitation process (tailored at emergency times), we extend the perspective to future schema and content additions, we preview our visualization support proposal, we sketch our assets to establish active monitoring of viral sequences, and we present the idea of linking our database with data regarding the host organism characteristics, including both genomic features and phenotype elements (e.g., clinical picture). This kind of integration is currently hindered by a lack of connections in the input data. Part of the future developments presented in this chapter has been described in two papers.

The research preview on extreme requirements elicitation.

[41] **A. Bernasconi**. Extreme Requirements Elicitation for the COVID-19 Case Study. In *International Working Conference on Requirement Engineering: Foundation for Software Quality*, 2021. https://ceur-ws.org/Vol-2857/pt2.pdf

The visualization support.

[57] **A. Bernasconi**, A. Gulino, T. Alfonsi, A. Canakoglu, P. Pinoli, A. Sandionigi, S. Ceri. VirusViz: Comparative analysis and effective visualization of viral nucleotide and amino acid variants. *Nucleic Acids Research*, 49(15):e90, 2021. https://doi.org/10.1093/nar/gkab478

Chapter 13 '**Conclusions and Vision**' ends this dissertation by highlighting the realized final result: a big integrated repository of data and metadata consisting of an important asset for the GeCo project and all its users in the bioinformatics community. This can be used to locate interesting datasets for analysis both on the original sources and on our system. Processed datasets available in several sources, are provided with compatible metadata; Processed region datasets of many other sources in the future can be imported into the repository with minimal effort. Search methods will automatically apply to all imported data. Analysis can be further carried out using other GeCo's modules or any analysis tool that operates with standard format data files.

The same protocol has been reapplied to another kind of data, which is very relevant at this critical time of the COVID-19 pandemic. We built similar pipelines and a system to support virological research and encourage interoperability with patients' data, which – when put in place – would lead to a very powerful analysis setting, able to uncover genetic determinants of the disease, either connected to particular mutations of the virus sequence or specific genetic characteristics of the human individual hosting the virus.

Part I
Human Genomic Data Integration

Chapter 2
Genomic Data Players and Processes

"In the life sciences, we do not have a big data problem. We have the 'lots of small data' problem."
— Vijay Bulusu, Head, Data & Digital Innovation - Pfizer

My journey across different genomic data types and actors (including consortia, integrators, and analysis initiatives) started in 2017 at the beginning of my Ph.D. First, we studied the most used data sources within the GeCo group (e.g., ENCODE, TCGA, Roadmap Epigenomics); then we moved our attention towards the understanding of alternative projects (such as 1000 Genomes, Gene Expression Omnibus, GWAS, and genomic annotations of RefSeq and GENCODE), finally, we reviewed approaches more similar to ours, i.e., studying mechanisms to combine data to enable powerful analysis workflows (e.g., DeepBlue, Cistrome, Broad Institute's Terra). After three full years of data scouting and field experience, in 2020 we were able to understand more pragmatically the role of different initiatives and finally extract an organized perspective of the whole genomic players landscape. The content of this chapter should be perceived as such an organized compendium, considering the *bottom-up* strategy that has guided the acquisition of the necessary knowledge and understanding of our position in the genomic world. GeCo (together with its main products GMQL and GenoSurf) is finally a valuable provider of integration systems that can help the genomic research community. We constructed this result step by step.

In the following, we provide readers with a complete background representation of the context where the contribution of Part I of the thesis should be set.

Chapter organization. Section 2.1 is dedicated to the *technological pipeline of genomic data*, from data production to final use of genomic datasets. We here describe the steps required for data and metadata production and integration. Each data source and platform may perform only some of these steps. We also discuss services and characteristics of the data access options provided to end users. Section 2.2 presents a *taxonomy of the players involved in data production and integration*; it describes the five main roles of such players and the relationships that exist between them. Specifically, we identified the following types: *contributors* are laboratories that produce the wet-lab data and associated information; *repository hosts* are organizations handling primary and secondary data archives—such as the well-known GEO [31]; *consortia* are international organizations who have agreed on broad data collection

A. Bernasconi: *Model, Integrate, Search... Repeat*, LNBIP 496, pp. 19–38, 2023.
https://doi.org/10.1007/978-3-031-44907-9_2

actions (the ENCODE [330] is a notable example); *integrators* are initiatives whose main objective is to combine data collections from other players and provision high-quality access to integrated resources; finally *consumers* represent the actual users of the exposed data platforms and pipelines. We also discuss the interactions among different players. In Section 2.3 we describe the *main players* in the three central categories (including 4 repository hosts, 12 consortia, and 13 integrators), specifying which parts of the technological pipeline – discussed in Section 2.1 – they address. In particular, we anticipate a description of the integrative strategy operated by our group within the GeCo project, which has dedicated huge efforts to the whole genomic data integration problem, mainly described in this Ph.D. thesis.

2.1 Technological Pipeline of Genomic Data

Data and their corresponding descriptions, i.e., metadata, are first produced, then integrated. In this section, we give an overview of the relevant technological phases towards final use, distinguishing between data, metadata, and also services and access interfaces built on top of them. Relevant steps are highlighted in the following in bold and comprehensively depicted in Figure 2.1 (data steps are in grey, metadata ones in purple, service/access ones in green), along with supporting objects (in orange) that guide the definition of each step.

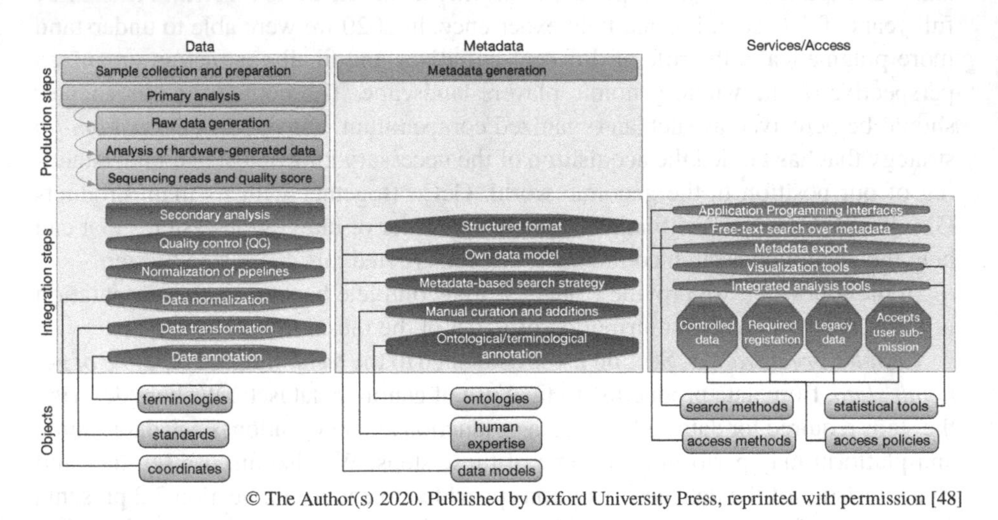

Fig. 2.1 Diagram of production and integration steps for data and metadata in genomics.

2.1.1 Production

Every genomic research study starts with nucleic acid **sample collection and preparation**; ensuring high-quality samples is important to maximize research efforts and validity of data analysis. This phase deals with privacy issues, for example, related to the use of clinical samples in research; it is impossible to create fully anonymized samples and this leads to issues of identifiable population data.

Methods that determine the nucleotide sequence of DNA and RNA molecules are called "sequencing". Next Generation Sequencing (NGS) is a high-throughput sequencing technology that enables the reading of billions of nucleotides in parallel. Sequencing (also referred to as "**primary analysis**") includes: (i) raw data generation; (ii) analysis of hardware-generated raw data; (iii) generation of sequencing reads and their quality score, i.e., billions of short sequencing reads that are stored in text files in FASTQ[1] format.

Typically, production is not driven by any imposed wet-lab standard, unless laboratories are guided by a consortium or other organization (e.g., ChIP-seq can have antibody standards, RNA-seq and DNase-seq[2] can have specific protocols and replicate numbers). The **metadata generation** is usually performed by the laboratories that generate the raw data; they document it in a rich way, yet approximate in the structure and possibly imprecise in the content. Basic information, about the performed assay, the used sequencing platform, and the analyzed biological material, is collected.

Researchers can then submit data through one of the several data brokers that act as links between production laboratories and ingestion APIs provided by collecting platforms—at times these include web interfaces or web services. Upon submission, the ingestion services sometimes perform basic quality assurance and checking of format consistency and then deposit the data into their data stores.

2.1.2 Integration

We describe as part of "data integration" all the steps that follow data production and their preliminary publication. Along the way, a number of issues may be encountered. Thus we hint at existing methodological solutions adopted by players addressing the mentioned aspects.

During data processing, also referred to as "**secondary analysis**", genomic sequences are reconstructed in a computational way by exploiting overlaps between short sequencing reads. After quality assurance filtering on raw reads, the data processing workflow typically includes alignment of reads to a reference genome, which

[1] A text file that stores both a biological nucleotide sequence and its corresponding quality scores.

[2] ChIP-sequencing is a method to analyze protein interactions with DNA; RNA sequencing is a technique used to reveal the presence and quantity of RNA in a biological sample at a given moment; DNase I hypersensitive sites sequencing is a method to identify the location of regulatory regions.

produces BAM files.[3] The differences between the sequenced genome and the reference one can be identified, for example, by performing variant calling and filtering, which produce VCF files. Other secondary analysis workflows output different file formats,[4] e.g., BED files from peak calling or GTF files from the identification of differentially expressed genes.

The following steps are at times performed together with the previous ones, other times delegated to other data players that follow up with other data manipulation practices.

Quality Control (QC) is vital for NGS technology experiments. It can be performed during three phases: on the initial extracted nucleic acids (in case they are degraded), after the sequencing library preparation (to verify that the insert size is as expected and that there are no contaminating substances), and after sequencing (most common tools are Sequence Analysis Viewer[5] and FastQC[6]). The more time and effort are spent on QC, the better quality results will be. Many players report some kind of QC check in one of these phases; sometimes even just producing quality studies and reports is referred to as QC.

Some players may decide to reprocess portions of the data collected elsewhere. Reasons for taking this approach may be several and of different nature, mainly including the need for normalized pipelines, as a means to obtain more homogeneous data ready for analysis. The **normalization of the pipelines** deals with the problem of converting raw data to numerical data such that any expression differences between samples are due solely to biological variation, and not to technical variation introduced experimentally; for example, in microarrays-based experiments, technical bias can be introduced during sample preparation, array manufacture and array processing. The selected data types require some processing to achieve compliance with standards (e.g., alignment to a reference sequence, uniform peak calling, thresholding of signal peaks, consistent signal normalization, consistency check between replicates, ...).

Among post-processing activities tailored at enhancing interoperability among different datasets, we mention **data normalization** procedures (such as format conversions like normalization of coordinates or re-formatting into narrowPeak or broadPeak standard formats in ENCODE), **data transformation** (e.g., matrix-based data formatted as BED data), and **data annotation**. Examples of the latter include: (i) providing positional information (i.e., genomic coordinates) and associated known genomic regions (e.g. genes) in a standardized framework; (ii) allowing joined use of different data types (e.g. gene expression and methylation) based on common gene and sequence identifiers, such as gene IDs from HUGO Gene Nomenclature Committee (HGNC) [405], Entrez Gene[265] or Ensembl [413] terminologies; (iii)

---

[3] The reconstruction of a genome is facilitated by using a reference genome to which the sequencing reads are systematically aligned; the use of reference genomes is possible since representatives of a species are genetically highly similar.

[4] https://genome.ucsc.edu/FAQ/FAQformat.html

[5] https://support.illumina.com/sequencing/sequencing_software/sequencing_analysis_viewer_sav.html

[6] https://www.bioinformatics.babraham.ac.uk/projects/fastqc/

merging together in the same files multiple expression measures obtained through different calculations, such as FPKM [384],[7] FPKM-UQ,[8] and counts in gene expression data.

For what concerns metadata, information associated with produced data, may be organized in a **structured format**. In some cases, integrators apply tailored integration pipelines to extract the needed information to fill their **own** agreed **data models**, thus performing schema-level integration. Generally, the idea is to redistribute metadata over a few essential entities, e.g., *Project-Sample-File* or *Investigation-Study-Assay* [349], as proposed in the general-purpose ISA-Tab format for structuring metadata [348] and now adopted by the FAIRsharing resource [347]. A number of questions are usually answered during this process: *Is this set of entities minimal? Is it enough to hold all information? Is something lost at this granularity?* Note that, depending on the specific sources, metadata elements have been linked to different entities, and a different base entity has been selected at times. To mention some notable choices:

- ENCODE [330] has centered everything on the *Experiment*, which includes a number of *Biosamples*, from which many *Replicates* are produced, to which *Items* belong (sometimes with a many-to-many cardinality as files may be combined from multiple replicas).[9]
- GDC [177] is centered on the *Patient* concept, from which multiple *Samples* are derived. From another perspective, data are divided by *Project*, associated with a *Tumor Type*, for which many *Data Types* are available.
- GEO [31] is organized into *Series* that include *Samples* (whereas these latter ones can be employed in multiple *Series*), sequenced with a *Platform*. A higher-order classification organizes *Series* and *Samples* in *Datasets* and *Profiles*.
- Sequence Read Archive (SRA) [227] is organized into *Studies* that include Samples, used for many *Experiments*, which derive from multiple *Runs*.
- The International Cancer Genome Consortium (ICGC) [415] includes entities for *Donor*, *Exposure*, *Family*, *File*, *Project*, *Sample*, *Specimen*, *Surgery*, and *Therapy*.

Most platforms offer a **metadata-based search strategy**, exploiting the querying possibilities over the new metadata schema. However, sometimes such a search functionality is available even when no new data schema has been applied, using a simple string matching search.

Some players, especially the ones working in connection with a Data Coordination Center (DCC), perform **manual curation and additions** to metadata. Within such activities, we mention in particular two: *assigning labels to replicas* and *cleaning metadata names*. The first activity is used for managing technical and biological replication, which is a common and recommended practice in genomic experiments [237, 404, 360]. In a data model where information is organized based on

[7] FPKM = Fragments Per Kilobase of transcript per Million mapped reads

[8] FPKM-UQ = Fragments Per Kilobase of transcript per Million mapped reads - Upper Quartile

[9] A complete list of entities is available at `https://www.encodeproject.org/profiles/`.

a hierarchy (e.g., *Experiment/Replicate/ File*), it is very likely that metadata will be replicated inside each element. Metadata formats such as JSON or XML have a hierarchical structure and can easily represent such data models, by encapsulating each element inside its parent. When a de-normalization of such structures is produced (e.g., to associate with a materialized file also information about the ancestor *Replicate* or *Experiment*), integrators face the problem of assigning labels to metadata in such a way that the one-to-many or many-to-many relationship, which is implicit in the JSON or XML syntax, can be explicit in the data. To overcome this problem while flattening hierarchical formats, it is customary to assign labels to metadata that belong to ancestors, in such a way that they can be recognized also in the de-normalized version. The second activity, cleaning metadata names, is needed because after label assignment attribute names may become too long. It is highly preferable to obtain minimal names so that they still express their information without losing their meaning, even if the semantics of nesting is removed; in this way, they are more easily usable within a metadata-based search system and in the connected analysis platforms. As an example, a rather complicated attribute such as `replicates.biosample.donor.organism.scientific_name`, derived from flattening five hierarchical levels in a JSON document, may be simplified into `donor.organism` to facilitate understanding. Redundant information, including duplicated attributes deriving from a comprehensive download approach from the source, may also be removed using similar rule-based mechanisms.

Other widely adopted processes to enhance metadata interoperability include **ontological/terminological annotation** on top of the original or curated metadata. Annotation is a means to achieve metadata normalization, essential for comparing metadata terms. Genomics, like many other fields in Bioinformatics, is greatly helped by specialized ontologies, which mediate among terms and enable interoperability. A considerable number of key ontologies are used by many genomic actors: Uber Anatomy Ontology (UBERON) [291] for tissues, Cell Ontology (CL) [280] and Experimental Factor Ontology (EFO) [268] for primary cells and cell lines, Ontology for Biomedical Investigations (OBI) [25] for assays, Gene Ontology (GO) [331] for biological processes, molecular functions, and cellular components. All these are employed by ENCODE, which has dedicated great efforts to the systematization of official term names for the description of its data. EFO is used by the Genome-Wide Association Studies (GWAS) Catalog [74] that curates all trait descriptions by mapping them to terms of this ontology. Moreover, GDC enforces a standardization using the National Institutes of Health (NIH)[10] Common Data Elements (CDE)[11] rules. Many attributes present codes referencing terms from the CDE Repository controlled vocabularies. Other relevant resources include: the National Cancer Institute (NCI)[12] Thesaurus [121] for clinical care, translational/basic research, and administrative/public information, and the National Center for Biotechnology Infor-

[10] `https://www.nih.gov/`

[11] `https://cde.nlm.nih.gov/`

[12] `https://www.cancer.gov/`

mation (NCBI)[13] Taxonomy [144], providing curated nomenclature for all of the organisms in the public sequence databases.

Several search services, which integrate a high number of ontologies, are employed in the landscape of genomic data integration. Examples include: BioPortal [399] and Ontology Lookup Service (OLS) [215], two repositories of biomedical ontologies and terminologies that provide services to annotate search keywords with ontological terms; Ontology Recommender [270], a BioPortal service that annotates free text with a minimal set of ontologies containing terms relevant to the text; Zooma[14], an OLS service providing mappings between textual input and a manually curated repository of text-to-ontology-term mappings. Annotation can also include adding external identifiers pointing to different databases that contain same real-world entities. While redundancy is not accepted within a single source, in the genomics domain it is common across sources, provided that resources are well interlinked and representations are coherent between each other (i.e., metadata values have the same level of detail).

2.1.3 Services and Access

Organizations operating in the integration field also, provide interfaces to access the result of their service. To this end, they must address the issues related to the synchronization of their local database with the original data one. As data size is significant, when updating the interface content, downloading everything from scratch from the original sources should be avoided. Instead, it is necessary to precisely define metrics to compare contents: *What is new, what has been updated, and what is not present anymore?* One possible strategy is defining a partitioning schema. In many cases, this is not the simplest possible one (i.e., file by file) since information is typically structured in a complex and hierarchical way. For example, when considering the source ENCODE, metadata can be used to partition the source data repository. API requests can be composed in order to extract always the same partition of data, specifying parameters such as "`type = Experiment`", "`organism = Homo Sapiens`" and "`file.status = released`". Consequently, a list of corresponding files is downloaded; the identifying characteristics of the files (typically including size, last update date, and checksum) can be compared with the ones saved in the local database at a previous download session of the same partition. Making a distinction between genomic region data and metadata, the latter are typically smaller in size; in case comparing versions becomes too complicated, metadata may be downloaded each time, as often there are no such things as a data release version or pre-computed checksum values to be checked.

Besides offering updated content, genomic players that host data and make it available through any kind of interface, usually offer also other supporting services.

[13] `https://www.ncbi.nlm.nih.gov/`

[14] `https://www.ebi.ac.uk/spot/zooma/`

Typically these include: **application programming interfaces** to directly download and extract specific portions of data, or perform rich and structured queries; **free-text search over metadata**, sometimes only on selected kinds of metadata, such as gene names or functional annotations; direct **metadata export**, at times, included within the API options, other times as bulk download; **visualization tools** or ready-to-use connections to common visualization browsers (e.g., UCSC and Ensembl genome browsers [222, 203]); embedded **integrated analysis tools** to further process and analyze the results retrieved in the interface (diverse use cases include clustering [369], spatial reconstruction [353], visualization [256], and graph-based analysis [27]); possibility to perform **computations on cloud** in a dedicated space, with reserved computational resources.

In addition to openly available datasets, some sources also feature **controlled data**, whose access is only given upon authorization; some just **require a registration step** to access the download functionality. Many players make available **legacy versions of data**, rarely of metadata.

A few players **accept user-data submissions**, either to be included as future content of the platform or to be processed together with publicly available datasets in further computations and analysis.

2.2 Taxonomy of Players Involved in Data Production and Integration

The landscape of institutions, private actors, and organizations within the scope of genomics is broad and quite blurred. The authors of [31] had previously proposed a tentative classification of sources: *primary resources* publish in-house data; *secondary resources* publish both in-house data and collaborator data; *tertiary resources* accept data to be published from third, unrelated parties. More recently, the Global Alliance for Genomics and Health (GA4GH) [381], an international, nonprofit alliance formed in 2013, built the Catalogue of Genomic Data Initiatives,[15] where they include the following types, not mutually exclusive: *Biobank/Repository*, *Consortium/Collaborative Network*, *Database*, *GA4GH Driver Project*, *Industry*, *National Initiative*, *Ontology or Nomenclature Tool*, *Research Network/Project*, *Standards*, and *Tool*.

We expand the taxonomy of [31], whereas we compact the one proposed by GA4GH – which in any case only includes initiatives under the alliance's umbrella – by identifying five categories to classify every entity that plays a role in this field, named *genomic data player*. In general terms, data are produced at laboratories (corresponding to the player: *contributor*), deposited at data archives (player: *repository host*), harmonized within programs (player: *consortium*), integrated by systems or platforms that aggregate data from different sources and add value to it (player: *integrator*), and employed by end users, mainly biologists and bioinformaticians (player:

[15] `https://www.ga4gh.org/community/catalogue/`

consumer). These categories are not intended to completely represent the whole possibilities, nor to be exclusive with respect to each other. In the following, we detail each category's characteristics and the interactions among categories, carrying genomic data from production to its integrative use.

Contributor. A contributor generates raw data with any high-throughput platform, using next-generation sequencing or alternative technology; it takes care of annotating wet-lab experiment data with a set of descriptive metadata, as well as encrypting and uploading data to archives. A contributor can be a laboratory or hospital, which reports directly to a Principal Investigator holding an independent grant and leading the grant project. In other cases, laboratories are part of a bigger program, led by a consortium or national institution. In both cases, it is standard for laboratories to send their data to other players, who carry on the publication and integration process.

Repository host. We call "repository hosts" the organizations standing behind primary data archives (also referred to as "data storage"), recently grown exponentially in size. They host data not only from independent laboratories and companies that gain visibility in this way but also from consortia that wish to make their data available from such general archives. Moreover, it is customary for authors of biological publications to deposit their raw and processed datasets on these repositories—some journals even require it upon submission [132]. Primary data archives currently face a number of challenges:

- Their primary goal is to pool disparate data into a single location, giving priority to quantity and typically not demanding a structure. However, without any homogenization effort, data is barely useful, impeding analysis and cross-comparison that build an added value with respect to individual experiments [29].
- They archive raw sequencing data, which is usually not immediately usable by the scientific community. The majority of these archives do not provide access to pre-processed published data, leaving this cumbersome task to individual scientists who need to analyze them.
- Usually metadata deriving from contributors' submissions is not sufficient to ensure that each dataset/experiment is reproducible and that the data can be re-analyzed. As new technologies, protocols, and corresponding annotation vocabularies are constantly emerging, new metadata fields are required and need curation to accurately reflect the data.

Consortium. Consortia provide evolved forms of primary repositories. They usually include many participants and projects, which have to abide by certain policies (see, for example, policies of GDC[16]) and operational conventions for participation (see, for example, experiment guidelines of ENCODE[17]). These policies have to ensure agreement among the parties about sensitive matters such as data access, data submission, and privacy. Guidelines guarantee compatibility among datasets,

[16] https://gdc.cancer.gov/about-gdc/gdc-policies/

[17] https://www.encodeproject.org/about/experiment-guidelines/

in order to establish an infrastructure that enables data integration, analysis, and sharing.

Many consortia refer to a DCC in charge of data and metadata normalization and cleaning, and of all the activities that stand between production and publication. Most well-known DCCs include the ones of ENCODE [199], BLUEPRINT [4], ICGC, and 1000 Genomes; Roadmap Epigenomics Consortium has an Epigenomics Data Analysis and Coordination Center (EDACC)[18] and Genotype-Tissue Expression (GTEx) [255] has a Laboratory, Data Analysis, and Coordinating Center (LDACC). Along with repository hosts, consortia are required by their own policies to submit raw sequencing reads and other primary data to controlled access public repositories. The main ones that serve this purpose are the European Nucleotide Archive [15], the European Genome-phenome Archive (EGA) [239], and the database of Genotypes and Phenotypes (dbGaP) [385].

Integrator. An integrator may be a platform, an initiative, or a project whose objective is to overcome the constant need of users to learn how to navigate new query interfaces and transform data from different sources to be integrated with the analysis. As a secondary purpose, an integrator usually aims at providing visualization and integrative analysis tools for the research community. Integrators do not point to raw data; they instead always reference the sources of their data (either with links to the source portals or by reporting original identifiers for each data unit).

Consumer. Genomic data and metadata are finally used by biologists, bioinformaticians, and data scientists, who download them from sources' platforms and FTP servers to feed a wide variety of tertiary analysis pipelines, including applications in pharmacology, biotechnology, and cancer research.

Interactions among genomic data players are described in Figure 2.2. Experimental genomic data and metadata are first produced – occasionally also preliminarily processed – by contributors, then published on repositories or directly on consortia's platforms. Within consortia themselves, re-processing may happen, as their pipeline for raw data processing uses community-agreed or consortium guideline-based algorithms. Intermediate, derived results are generated to be later published. In some cases, data curated by consortia are re-published also on general archives, such as ENCODE and Roadmap Epigenomics on GEO.[19] Data are finally collected by integrators that expose them to tertiary interfaces, tailored to enhancing interoperability and use. Simpler interfaces are provided to consumers also by many repository hosts and consortia. In rare cases, not depicted in Figure 2.2 as they are exceptional, integrators may consider it important to re-process data of some sources with normalized pipelines to enhance the possibilities of integration. Two long arrows show the taxonomy from different points of view: from a *data perspective*, contributors deal with raw data, repository hosts and consortia with processed data, while integrators make data interoperable and fit for use of consumers; from a *process perspective*, data

[18] `http://www.roadmapepigenomics.org/overview/edaac`

[19] ENCODE: `https://www.ncbi.nlm.nih.gov/geo/info/ENCODE.html`; Roadmap Epigenomics: `https://www.ncbi.nlm.nih.gov/geo/roadmap/epigenomics/`.

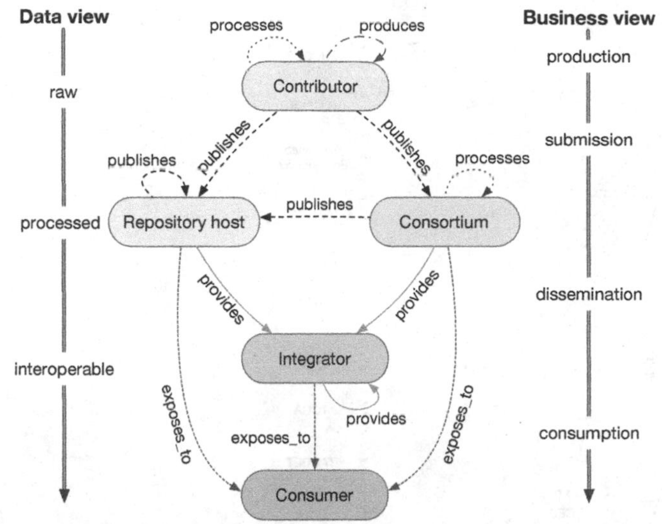

Fig. 2.2 Diagram of interactions among genomic data players. Nodes are players; arrows, with different colors and textures represent their interactions. Contributors publish either on repositories or consortia platforms; data are then integrated. Consumers retrieve data from repositories, consortia, or integrators. The *data view* represents the perspective of data stages, while the *business view* shows the process applied to genomic datasets.

is produced by contributors, submitted to aggregating platforms, that take care of dissemination to tertiary players, who make it available for its consumption.

We instantiate the taxonomy described by the diagram in Figure 2.2 with a number of relevant genomic players, which will be described thoroughly in the following. Figure 2.3 thus shows interactions between example players, starting from the laboratories where data are primarily generated, throughout repositories where data are deposited, consortia where they are curated, and finally integrator interfaces where they are used and explored. The used notation and colors reflect the ones adopted in Figure 2.2.

We drew the relationships between these players according to their specifications in the documentation and relevant publications, to the best of our knowledge at the time of writing. Some consortia, for instance TCGA [397] and GDC, accept submissions both from laboratories gathered under the same organization and from individual submitters that observe the submission guidelines. There are labs that contribute to more projects. Raw experimental data are usually deposited to SRA [227], while GEO (the most used by researchers) and ArrayExpress [21] are employed for publication of data at later stages of processing; complete studies are uploaded to BioStudies [351]. Note that, in the Figure 2.3 diagram, even primary archives reference each other.

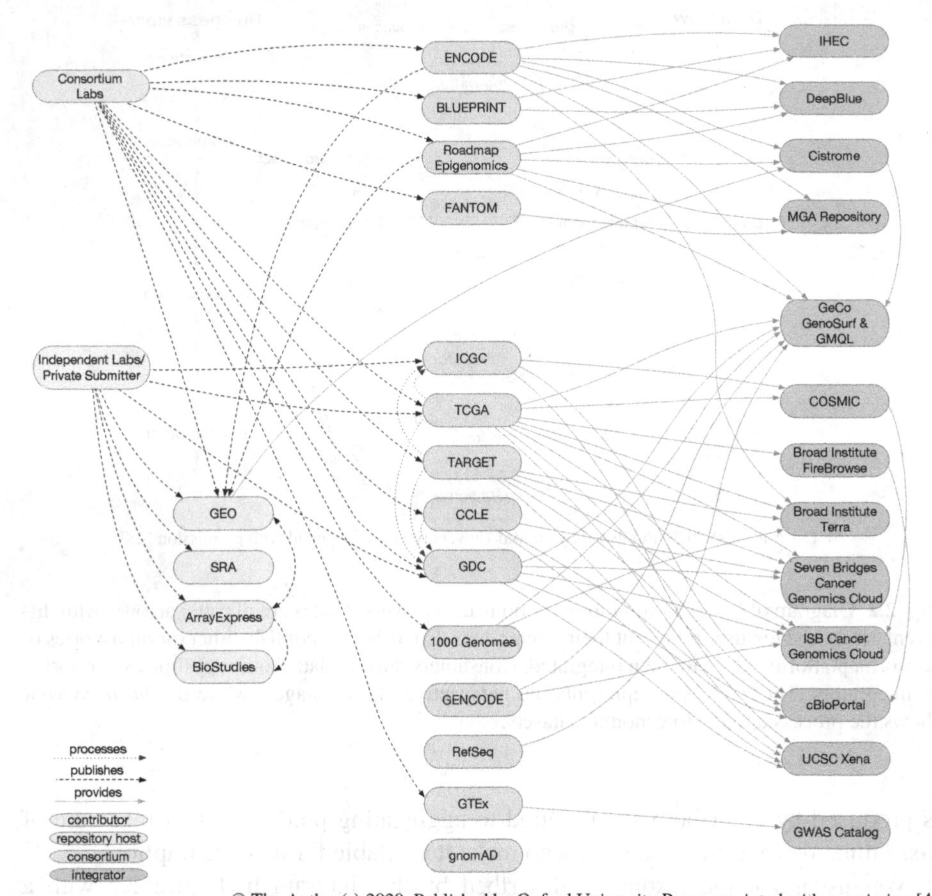

Fig. 2.3 Diagram of example players and their most important interactions.

2.3 Main Genomic Data Players

We propose a systematic overview of a number of genomic data players, guided by Table 2.1. The first column of the table contains a list of data sources that contribute to producing, integrating, and promoting the use of genomic data for research, grouped by the categories of the taxonomy introduced in the previous section. The list is in no way meant to be comprehensive but should be received as a starting reference. For each mentioned player we show which steps/functionalities are provided. The following columns in Table 2.1 represent steps described in Section 2.1 in bold font and outlined in Figure 2.1.

Inside Table 2.1 cells, the notation × indicates a step included by the player; an empty cell stands for a step not provided by the player; ~ is used when an answer is only partially positive (e.g., some parts of the step are performed while others

		Production			Integration																			
					Data					Metadata					Services						Access			
		sample collection and preparation	primary analysis	metadata generation	secondary analysis	quality control	pipeline normalization	data normalization/transformation	data annotation	structured format	own data model	metadata-based search strategy	manual curation/additions	ontology/terminology annotation	application program interfaces	meatadata-free-text search	metadata export	visualization tools	integrated analysis tools	computations on cloud	include controlled data	required registration	legacy versions	accept data submissions
Contrib.	Laboratories	×	×	×	~	~																		
Reposit.	ArrayExpress [21]					×			×	×			×		×	×	?	×	×		×		×	×
	BioStudies [351]											×		×		×	?				×	~	~	×
	GEO [31]					×				~	~	~	×				×	×	×				×	×
	SRA [227]					~				×	×	×			×		×						?	×
Consortium	1000 Genomes [329]				×	×	×	×	×	~		~			~		×	~					×	
	BLUEPRINT [4, 147]				?	?	?	×	?	×	×	×		×	×		×	×	×				?	
	CCLE [164]				×	×	×	×	×			×				×	×	×	×			×	×	
	ENCODE [120]				×	×	×	×	×	×	×	×	×	×	×	×	×	?	?		×		×	
	FANTOM [253]				×	×	×	?	×									×	×				×	
	GENCODE [153]					×	?	×	×						~			~					×	
	gnomAD [218]				×	×	×											×					×	
	GDC [177]				×	×	×	×		×	×	×	×	×	×		×	×			×		×	×
	GTEx [255]				×	×	×	×	×						×			×			×		×	
	ICGC [415]					×	×	×	×	×	×	×		×	×		×		×		×	~	×	×
	RefSeq [301]					×	×	×	×				×										~	
	Roadmap Ep. [233]				×	×	×	?	×	×		~					×	×						
Integrator	Firehose/FireBrowse				?	?	?	?	?			×			×	~		×	×				×	×
	Terra									×		×		~		~	×	×	×	×	×	×		×
	cBioPortal [88]				×		×	×	×			×			×	~	×	×	×				×	~
	Cistrome [418]				×	×	×		×	×	×	×	×	×		×	×	×	×			×		
	COSMIC [380]								×	×	×	×	×	×		×	×	×	×			×	×	
	DeepBlue [9]								×	×	×	×	×	×	×	×	×		×					×
	GenoSurf/GMQL [76]							×	×	×	×	×	×	×	×	×	×	×	×	×			×	×
	GWAS Catalog [74]					×		×	×	×	×	×	×	×	×	×	×	×						×
	IHEC [72]					×	×	×		×	×	×	×	×	×		×	×	×		×	~	×	×
	ISB CGC [336]									×		×			×			×	×	×	×	×		×
	MGA [130]				×		×	×	×	×			×		×	×		×	×					
	Seven Bridges CGC [241]									×		×	?	?	×	~	×	×	×	×	×	×	×	×
	UCSC Xena [168]									×	?	×	?		×	×		×	×	×			×	×

Table 2.1 Overview of the steps towards data integration included by genomic data players. Rows represent players (with reference to their flagship publication, when available) and are grouped by player type. Columns represent the steps for genomic data integration described in Section 2.1 and are grouped according to their progression in a typical pipeline. Used notation: × indicates that a certain step is included/performed by the player; ~ indicates an uncertain answer (i.e., in some cases the service/step is provided just in a few studies or only for some data types); ? indicates that the player's documentation and publications did not allow us to determine a sharp answer; empty cell indicates that the service/step is not provided. Note that, with respect to nodes referring to consortia in Figure 2.3, two rows are omitted here: TCGA and TARGET. Indeed, currently, their data is only made available through other platforms (the most important being GDC and ICGC); the old TCGA portal was dismissed and TARGET does not have its own.

are not, or the step is only performed under certain conditions); ? is used for an unknown answer, when the documentation and publications describing the player and its services did not allow us to determine an answer. All information contained in the table is filled up to the best of our knowledge at the time of writing (March

2020), being retrieved from the player's main publications or from the linked online documentation. Our discussed overview is divided into subsections, one for each player type.

2.3.1 Contributors

As it can be observed from Table 2.1, contributors – including independent labs, private submitters, and consortium labs (depicted in Figure 2.3 – are the only players in charge of sample collection, preparation, and primary analysis, followed by generation of metadata; only in some cases they also apply quality control measures before or during secondary analysis activities.

2.3.2 Repository Hosts

Gene Expression Omnibus (GEO) [31] is the most general and widely used among repositories. It started in 2002 as a versatile, international public repository for gene expression data [131]; it consequently adopted a more flexible and open design to allow submission, storage, and retrieval of a variety of genomic data types, such as from next-generation sequencing or other high-throughput technologies. To include also non-expression data, in 2008 GEO created a new division called "Omix", standing for a mixture of "omic data" [30].

Data can also be deposited into **Sequence Read Archive (SRA)** [227] as supporting evidence for a wide range of study types: primarily raw sequence reads and alignments generated by high-throughput nucleic acid sequencers (BAM file format), now expanded to other data including sequence variations (VCF file format) and capillary sequencing reads. As a part of the International Nucleotide Sequence Database Collaboration (INSDC), the SRA is materialized in three instances, one at the EBI,[20] one at the NCBI [355], and one at the DNA Data Bank of Japan (DDBJ) [226].

ArrayExpress [21] was first established in 2002 only for microarray data. It is now an archive of functional genomics data ranging from gene expression and methylation profiling to chromatin immunoprecipitation assays. Recently, it also increased the number of stored experiments investigating single cells, rather than bulk samples (i.e., single-cell RNA-seq).

The EBI **BioStudies** [351] database holds high-level metadata descriptions of biological studies, with links to the underlying data databases hosted at the EBI or elsewhere, including general-purpose repositories. Also those that have not been already deposited elsewhere can be hosted at BioStudies.

[20] `https://www.ebi.ac.uk/`

Discussion. By observing the repository-related rows of Table 2.1, we conclude that repositories are quite diverse with respect to the data integration steps included in their practice. Generally, they do not perform specific steps on data, however, they often require submitters to ensure quality control checks, as GEO and ArrayExpress do, while SRA mentions it as future work. Metadata is not treated uniformly; some organization is enforced, but much information is left also in an unstructured format. While services offered by their interfaces are various, they all allow submissions from any user; this is a characterizing feature of the repositories.

Since repositories are growing [337] in diversity, complexity and (unexpressed) interoperability, the need for organization and annotation of the available data has become urgent. In response to this, NCBI and EBI have implemented additional, complementary, initiatives on top of repositories. NCBI BioProject and BioSample databases [28] and EBI Biosamples [110] were initiated to help address these needs by facilitating the capturing and managing structured metadata and data for diverse biological research projects and samples represented in their archival databases.

2.3.3 Consortia

Consortia are usually focused on particular aspects of what we generically call "genomics". The following four work on matters related to epigenomics.

The **Encyclopedia of DNA Elements (ENCODE)** Consortium [330] is an ongoing international collaboration of research groups funded by the National Human Genome Research Institute (NHGRI). The primary goal of the project is to characterize functional features in DNA and RNA expression in a wide number of cell lines. The project's updated portal is presented in [120].

BLUEPRINT [4] is an EU-funded consortium under the umbrella of the International Human Epigenome Consortium (IHEC.) It was set up to develop new high-throughput technologies to perform epigenome mapping, and to analyze diverse epigenomic maps comprehensively, making them available to the scientific community as an integrated resource. Besides being available through IHEC resources, the BLUEPRINT built its own Data Analysis Portal [147], as the first platform based on EPICO, an open access reference set of libraries to be used to develop data portals for comparative epigenomics.

The **Roadmap Epigenomics Consortium** [233] was born in 2015 from the NIH with the aims of: (i) understanding the biological functions of epigenetic marks and evaluating how epigenomes change; (ii) designing and improving technologies, i.e., standardized platforms, procedures, and reagents, that allow researchers to perform epigenomic analysis and to study epigenetic marks efficiently; (iii) creating a public resource of disease-relevant human epigenomic data to accelerate the application of epigenomics approaches.

FANTOM [253] is an international research consortium created to perform functional annotations of the mammalian genomes, including – but not limited to – the *Homo Sapiens* one. The objective of the project has recently moved from understanding the transcripts to understanding the whole transcriptional regulatory network.

The following three consortia are instead focused on problems related to cancer genomics.

Genomic Data Commons (GDC) [177] is an information system for storing, analyzing, and sharing genomic and clinical data from cancer patients. It aims to give democratic access to such data, improve sharing and promote approaches of precision medicine that can diagnose and treat cancer. Ultimately, the goal is to become the one-stop cancer genomics knowledge base; however, consolidation and harmonization of genomic and clinical data are ongoing and they will require a long process. GDC was created mainly to help individual investigators and small programs to meet NIH and the NCI genomic data sharing requirements, and thus to store their data in a permanent home. In addition, GDC now includes data from big cancer programs, such as The Cancer Genome Atlas (TCGA) [397] and the Therapeutically Applicable Research to Generate Effective Treatments (TARGET), also shown as consortia nodes in Figure 2.3. While GDC is technically a cancer knowledge network, we classify it as a *consortium* as it has a very broad mission: it accepts user submissions, it performs quality control, it provides storage and it also redistributes the data. What distinguishes it particularly from a simple *repository host* or an *integrator* is the considerable effort dedicated to harmonizing data (standardizing metadata, re-aligning data and re-generating tertiary analysis data using new pipelines [158]) deriving from incoming submissions and from the included cancer programs. TCGA, instead, is a terminated program; it no longer accepts samples for characterization. It used to expose the data by means of its own portal, while now it relies on the GDC infrastructure. As its concluding project, in 2018 the TCGA program produced the Pan-Cancer Atlas [197], a collection of analyses performed cross-cancer type. In addition to including many single-cancer-type projects, the last datasets that GDC platform makes available are the ones produced within the context of the Pan-Cancer project.

The **International Cancer Genome Consortium (ICGC)**[415] was established in 2011 to launch and coordinate a large number of research projects de-centralized in many countries of the world, sharing the common goal of explaining the genomic changes present in many forms of cancer. Its data portal hosts data from other large-scale projects focused on cancer research, such as TCGA and TARGET, as shown by the two dotted incoming arrows in Figure 2.3.

The **Cancer Cell Line Encyclopedia (CCLE)** [164], for almost 1,500 human cancer cell lines, collects gene expression, chromosomal copy number, and massively parallel sequencing data.

The last five consortia we mention are instead focused on various matters, such as variation across populations, transcriptomics, exome sequencing, and annotation.

Launched to become one of the largest distributed data collection and analysis projects in genomics, the goal of the **1000 Genomes Project** [329] was to find most genetic variants with frequencies of at least 1% in the studied populations. In 2015 the International Genome Sample Resource (IGSR) [105] was established to expand and improve the legacy inherited from the 1000 Genomes Project.

The **Genotype-Tissue Expression (GTEx)** Consortium [255], supported by the NIH Common Fund, aims at establishing a resource database and associated tissue bank to study the relationship between genetic variation and gene expression and other molecular phenotypes in multiple reference tissues. The results of this transcriptomics-focused project help the interpretation of findings from Genome-Wide Association Studies (GWAS) by providing data and resources on the expression of quantitative trait loci in many tissues and diseases.

The **GENCODE** project [153] produces high-quality reference gene annotations and experimental validation for human and mouse genomes. It aims at building an encyclopedia of genes and gene variants, by identifying all gene features in the human and mouse genomes, using a combination of computational analysis and manual annotation.

The NCBI **RefSeq** project [301] provides a comprehensive manually annotated set of reference sequences of genomic DNA, transcripts, and proteins—including, for example, genes, exons, promoters, enhancers, etc... Exploiting the data from the INSDC, it provides a stable reference for genome annotation, analysis of mutations and studies on gene expression.

The Exome Aggregation Consortium (ExAC) [244] unites a group of investigators who are aggregating and harmonizing exome sequencing data from other large-scale projects. According to the most updated news (end of 2016, [219]), the ExAC provided sequences from almost 61,000 individuals belonging to studies about different diseases and populations. Recently, the ExAC browser has been dismissed in favor of the Broad Institute **Genome Aggregation Database (gnomAD)** [218], which more than doubles the previous sample size.

Discussion. From Table 2.1 we observe that consortia are generally concerned with the coordination of data transformation, from secondary analysis activities to quality control filtering and normalization/annotation of data. Almost all the analyzed consortia definitely include a pipeline normalization step in their activities, as this is the characterizing step of this type of player (only BLUEPRINT and GENCODE did not mention information about uniform workflows in their documentation). Instead, the approach towards metadata curation and tertiary analysis tools is diversified and does not show a unique trend. As to the "metadata-based search strategy" column in Table 2.1, we specify that some consortia just provide a very limited functionality of this kind (e.g., 1000 Genomes can only filter by population, technique and data collection, and Roadmap Epigenomics by tissue and data type only). While ENCODE and GDC present sophisticated search interfaces, other ones are quite basic. As to providing APIs and visualization tools, GENCODE and 1000 Genomes were assigned the ~ symbol since they do not offer such services natively, but exploit the ones of Ensembl.

2.3.4 Integrators

In a similar way as consortia, also integrators tend to cluster based on the specific sub-branch of genomics they cover. This happens mostly because rules and common practices are better shared within the same area, following similar purposes, while cross-branch projects are rare. We first mention four integration organizations that collect data from epigenomics-focused consortia.

The **International Human Epigenome Consortium (IHEC)** [72] coordinates large-scale international efforts towards the production of reference epigenome maps. For a wide range of tissues and cell types, the regulome, methylome, and transcriptome are characterized. As a second phase, the consortium is expanding its focus from data generation to the application of integrative analyses and interpretation on these datasets, with the goal of providing a standardized framework for the clinical translation of epigenetic knowledge. Counter-intuitively, we classified the IHEC as an *integrator* rather than a *consortium* as the normalization work is mainly carried on by the members' institutions (or consortia themselves) that are part of it (for instance ENCODE, BLUEPRINT, Roadmap Epigenomics). The main outcome of the IHEC is instead its Data Portal, which can be used to view, search, download, and analyze the data already released by the different associated projects.

DeepBlue [9] is a data server that was developed to mitigate the lack of mechanisms for searching, filtering and processing epigenomic data, within the scope of the IHEC. DeepBlue made a precise work of data integration by homogenizing many epigenomic sources, including data from ENCODE, BLUEPRINT, Roadmap Epigenomics among others. It uses a clear distinction between region data and metadata, manages both experiment and annotation-related datasets, defines a set of mandatory metadata attributes – while storing additional ones as key-value pairs – and uses metadata to locate region data.

In the **Cistrome** Data Portal [281] users can find data relevant to transcription factor and chromatin regulator binding sites, histone modifications, and chromatin accessibility. Such data is useful to a number of studies, including differentiation, oncogenesis, and cellular response to environmental changes. As to the last available publication [418], its database contains about 100,000 samples, both for human and mouse organisms. It includes data of ChIP-seq and chromatin accessibility from ENCODE, Roadmap Epigenomics, and GEO, which has been carefully curated and homogeneously re-processed with a new streamlined analysis pipeline.[21] Comparison between Cistrome enriched region signal peaks and the ones in ENCODE, which they are derived from, showed that they are significantly different.

The **MGA repository** [130] is a database of both NGS-derived and other genome annotation data, which are completely standardized and equipped with metadata. It does not store raw sequence files, but instead lists of base positions in the genome corresponding to reads from experiments, e.g., ChIP-seq. Ten model organisms are represented.

[21] The analysis pipeline is detailed at `http://cistrome.org/db/#/about/`

The following seven integrators are mainly working in the field of cancer genomics. Notice that, in Figure 2.3, the cluster formed by consortia and integrators working in the cancer domain is the most connected—integrators retrieve datasets from the most important consortia portals.

The **Catalogue Of Somatic Mutations In Cancer (COSMIC)** [380] catalog is the most comprehensive global resource for information on somatic mutations in human cancer. It contains 6 million coding mutations across 1.4 million tumor samples, which have been (primarily) manually curated from over 26,000 publications.

Broad Institute maintains both **Firehose/FireBrowse**[22] and **Terra**[23] platforms as aggregators of genomic data. The first one mainly imports TCGA data and offers a number of visualization options over it. The second one is a new large-scale project that also includes cloud computational environments.

Along with the Broad Institute, the **Seven Bridges Cancer Genomics Cloud** [241] and the **Institute for Systems Biology (ISB) Cancer Genomics Cloud** [336] are the other two systems funded by the NCI to store massive public datasets (first of all TCGA ones) and together provide secure scalable computational resources for analysis.

The **cBio Cancer Genomics Portal** (cBioPortal, [88]) was designed to address the data integration problems that are specific to large-scale cancer genomics projects, such as TCGA—including the Pan-Cancer Atlas datasets, TARGET, and ICGC. In addition, it also makes the raw data generated by large projects more easily and directly available to cancer researchers.

UCSC Xena [168] is a high-performance visualization and analysis tool that handles both large public repositories (e.g., CCLE, GDC's Pan-Cancer, TARGET and TCGA datasets) and private datasets. Its characterizing aspects are strong performances and a privacy-aware architecture, working across multiple hubs simultaneously. Target users are cancer researchers with and without computational expertise.

The NHGRI-EBI **GWAS Catalog** [74] is a collection of all published genome-wide association studies that enable investigations to identify causal variants, understand disease mechanisms, and establish targets for novel therapies. It adds manually curated metadata for publication, study design, sample, and trait information. Much information from GTEx is also integrated.

As a last player, we consider our own project, **data-driven Genomic Computing (GeCo)** [89], we analyzed thoroughly the issues related to the integration pipeline described in Section 2.1, and we proposed a unique integration approach that is thoroughly described in Chapter 4. As indicated in Table 2.1, we include all the integration steps for data and metadata that follow the secondary analysis. Among the analyzed integrators, the GeCo approach is the only one that joins together a broad range of genomic data, which spans from epigenomics to all data types typical of cancer genomics (e.g., mutation, variation, expression, etc.), until annotations.

Discussion. Integrators, as reported in Table 2.1, are in general concentrated on metadata and services; however, some of them do re-process also data (e.g., Cistrome),

[22] Firehose: `https://gdac.broadinstitute.org/`; FireBrowse: `http://firebrowse.org/`
[23] `https://terra.bio/`

and many of them transform and augment it in various ways. Table 2.1 also proves that, for genomic data players, performing data integration corresponds to applying the pipeline described in Section 2.1.

We should note that, in the table, information about the Broad Institute's Fire-Browse and Terra is not as accurate as for other players, since currently there is no publication and insufficient documentation that specify official information for these two integrators. In the specific case of metadata ontological annotation, we highlight that in FireCloud, which has now been embedded into Terra, the connection to the Disease Ontology [37] has been implemented.

Some integrators have been assigned the ~ symbol to the "metadata free-text search" column since this only works on one attribute at a time. IHEC Data Portal does not exactly require registration for downloading. Indeed, only the raw data files for most of the IHEC datasets are located in controlled access repositories (e.g., EGA or dbGaP), while processed data are openly visible. cBioPortal does not allow direct user submission but indeed encourages suggestions of inclusions of potential interest.

Chapter 3
Modeling Genomic Data

"The eternal mystery of the world is its comprehensibility... The fact that it is comprehensible is a miracle."

— Albert Einstein

Advancements in database research should be conveyed by paradigm shifts in the modeling of inputs, transformations, and outputs. Following the consolidated expertise of our group in data modeling (even in domains that are very distant from genomics and life sciences [84, 90]), our approach to genomics has started precisely from understanding the true nature of the data at hand, by ensuring simple, powerful and orthogonal abstractions, and by applying well-known techniques such as Entity-Relationship modeling, also referred to as Conceptual Modeling [98, 32].

Chapter organization. Even before my Ph.D. research started, the GeCo group had dedicated large efforts to the problem of modeling genomic-related data. The first important contribution has been the Genomic Data Model (GDM), presented in [276], achieving the novel results of modeling processed data, their supporting metadata (while other models used in the past only focused on data), and providing seamless interoperability among heterogeneous datasets. We briefly discuss this model in Section 3.1. During my Ph.D., we consolidated the mentioned work, by continuing the implementation of several modules for integrating more datasets to be used conjunctively in the repository. More importantly, my doctoral research has been concerned with giving a more central role to metadata, i.e., the descriptions of genomic datasets. We understood that not only was it important to support the joint use of data with their metadata, but that metadata deserved to become first-class citizens in the GeCo system, needing their own model, semantics, and a plan to achieve interoperability among sources. My efforts to model metadata are described in Section 3.2. Section 3.3 concludes the chapter with an overview of literature modeling approaches in genomics and comparable fields.

A. Bernasconi: *Model, Integrate, Search... Repeat*, LNBIP 496, pp. 39–51, 2023.
https://doi.org/10.1007/978-3-031-44907-9_3

3.1 The Genomic Data Model

The Genomic Data Model (GDM) [276] is based on the notions of *datasets* and *samples*; datasets are collections of samples. Samples are the basic unit of information, containing experimental data that corresponds to a given individual and preparation (e.g., cell line and antibody used) that first undergoes NGS sequencing (producing "raw data"), then alignment and calling processes (producing "processed data"). Each sample includes DNA segments or regions (possibly the whole genome) – called *region data* in the following – and it is associated with information about the performed experiment, i.e., *metadata* describing the general properties of the sample.

Genomic region and feature data describe many molecular aspects, which are measured individually; the resulting variety of formats hampers their integration and comprehensive assessment. GDM provides a schema to the genomic features of DNA/RNA regions, making heterogeneous data self-describing and interoperable.

Original files are imported into the GDM format, which has a fixed part – representing the genomic coordinates – that guarantees the comparability of regions produced by different kinds of processing, and a variable part, i.e., data-type-specific attributes, describing region properties, reflecting the process of feature calling that produced the regions with their features specific of the particular processing experiment.

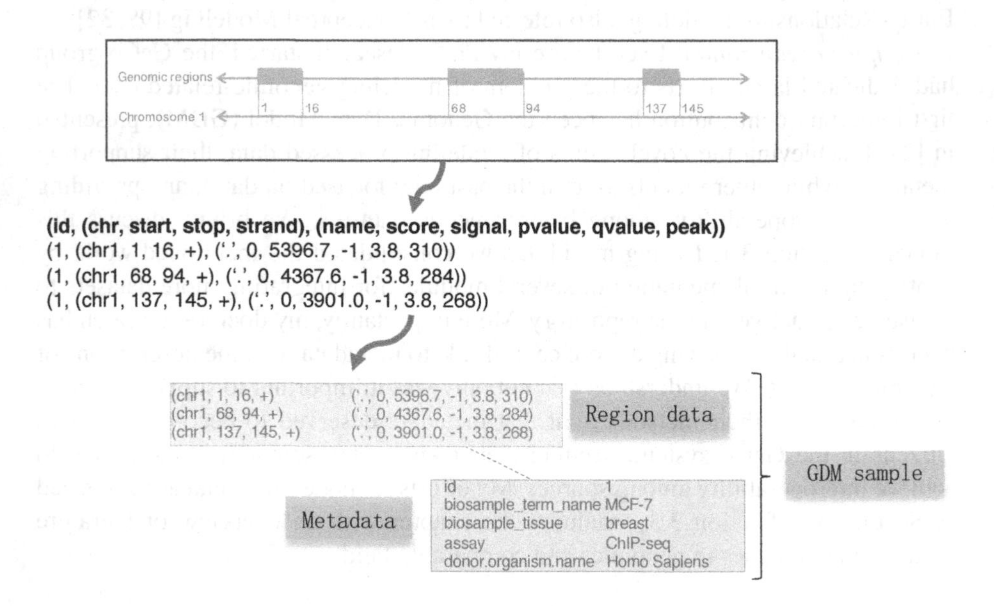

Fig. 3.1 Genomic Data Model.

Moreover, GDM copes with the lack of widely agreed standards for metadata by representing them using a free arbitrary semi-structured attribute-value pairs structure: attributes can have multiple values (e.g., the Disease attribute can present both "Cancer" and "Diabetes" values).[1]

Figure 3.1 shows how a processed data genome track is modeled into one GDM sample. Each blue rectangle becomes a region following the schema indicated in bold, where `id` is unique for each sample, `chr/start/stop/strand` are the fixed part, and `name/score/signal/pvalue/qvalue/peak` are the variable one. Each region data file is tightly linked to its metadata file (with the same identifier). The metadata part includes free attribute-value pairs; in GDM, attributes are freely associated with samples, without specific constraints. A typical GDC dataset contains thousands of samples like the represented one.

GDM is tightly liked to GenoMetric Query Language (GMQL), a closed algebra over datasets with the ability to compute distance-related queries along the genome, seen as a sequence of positions. GMQL is part of previous work of the GeCo group [277] and has been consolidated in later extensions to which I have also contributed [275, 364]. GMQL is capable of expressing high-level queries for these genomic computations and evaluate them on big datasets over a cloud computing system.

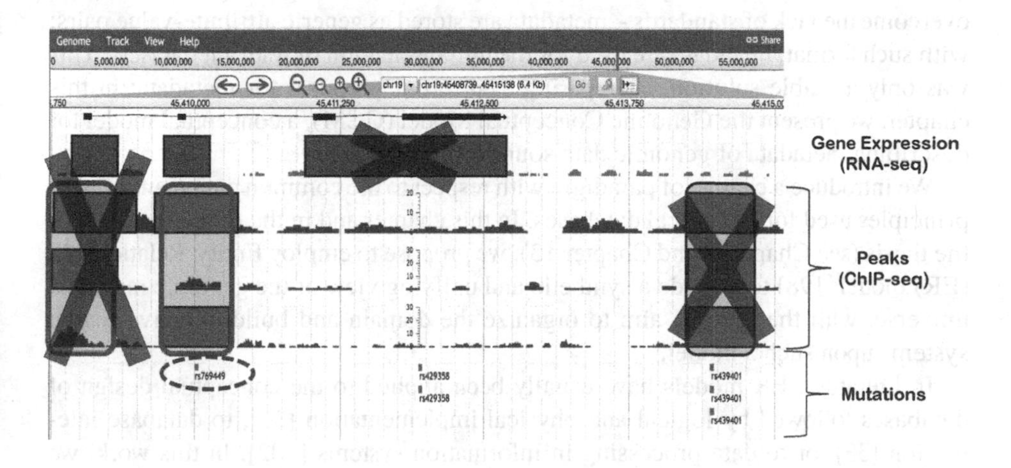

Fig. 3.2 Example of GMQL query to retrieve MYC-bounded promoters with somatic mutations in recurrent breast cancer patients.

Figure 3.2 depicts an example of genomic computation that can be performed by GMQL over heterogeneous genomic signals. Three types of signals have been selected on a genome browser; the distinct tracks respectively represent one RNA-seq experiment, extracting gene expression levels (in red, in correspondence of black annotated genes on top), three ChIP-seq experiments, extracting peaks of expression (locations where a protein interacts with DNA), and mutations. A typical question

[1] The reader should note that this simplified metadata representation will be outdone along the course of the next chapters. However, it will be maintained in parallel, to provide the possibility to biologists to operate on purely file-based datasets, with their usual tools.

looks for mutations (one is highlighted in green) upon expressed genes that overlap with "confirmed peaks" (in at least two experiments).

3.2 The Genomic Conceptual Model

While a lot of efforts are made for the production of genomic datasets, much less emphasis is given to the structured description of their content, collectively regarded as *metadata*. There is no standard for metadata, thus each source or consortium enforces some rules autonomously; a conceptual design for metadata is either missing or, when present, overly complex (see the ENCODE model[2]) or useless (see the GDC and GEO models[3]). Genomic metadata are lacking a conceptual model for understanding which sources and datasets are most suitable for answering a genomic question.

Since its beginning, the GeCo project has pursued the goal of developing an integrated repository of open processed data, supporting both structured and search queries; the first repository has been based on the GDM paradigm, where – to overcome the lack of standards – metadata are stored as generic attribute-value pairs; with such format, metadata are used for the initial selection of relevant datasets. This was only a viable solution, a preparing step to our proposal for metadata: in this chapter, we present the Genomic Conceptual Model (GCM), a conceptual model for describing metadata of genomic data sources.

We introduce a change of paradigm with respect to the common data management principles used for biological databases. In this chapter and in the following parts of the thesis (see Chapter 9 and Chapter 13), we propose to employ Entity-Relationship (ER) models [98] to provide a synthetic and unifying view of the genomic metadata universe, with the specific aim to organize the domain and build effective search systems upon such a model.

In literature, ER models have usually been applied to the conceptual design of databases followed by logical and physical implementation [32], to database integration [33], or to data processing in information systems [302]. In this work, we use conceptual models as *tools to model reality and to build systems upon them.* As our purpose is to simplify and organize the domain, our conceptual models result in essential use of entities, orthogonal use of attributes, poor generalization/specialization relationships and in relationships' labels (since they are trivial "has"), as

[2] At `https://www.encodeproject.org/profiles/graph.svg` see the conceptual model of ENCODE, an ER schema with tens of entities and hundreds of relationships, which is neither readable nor supported by metadata for most concepts.

[3] At `https://gdc.cancer.gov/developers/gdc-data-model/` see the conceptual model of GDC, a graph data model without relationships cardinality or attributes that is far from practical for use. At `https://www.ncbi.nlm.nih.gov/books/NBK159736/figure/GEO.F1/` see the elementary conceptual model of GEO, containing only a hint to relationship cardinality and no attributes.

they describe typical data warehouse structures. We will later enforce semantics by linking relevant attributes to existing biomedical ontologies (see Section 4.7.2).

In this Section we first explain our analysis of attributes in sources and the characteristics they may hold in the conceptual schema (Section 3.2.1), then we detail the entities and attributes of the model (Section 3.2.2) and show its validation against ER schemas of important sources (Source 3.2.3).

3.2.1 Analysis of Metadata Attributes

As a first step in developing GCM, we defined a taxonomy of the main properties of metadata attributes; we then systematically applied the taxonomy to each considered source, so as to better characterize its content. According to our taxonomy, attributes can be:

- **Contextual (C)** when they are present (or absent) only within specific contexts, typically because another attribute takes a specific value. In such cases, there is an *existence dependency* between the two attributes.
- **Dependent (D)** when the domain of their possible values is restricted, typically because another attribute takes a specific value. In such cases, there is a *value dependency* between the two attributes.
- **Restricted (R)** when their value must be chosen from a controlled vocabulary.
- **Single-valued (S)** when they assume at most one value for each specific experiment.
- **Mandatory (M)** when they must have a value, either for all experiments or within a specific context.

The resulting taxonomy is shown in Table 3.1; it includes orthogonal features and we targeted both completeness and minimality. By default (and in most cases), attributes do not have any of the above properties. Very few attributes are mandatory and unfortunately, sources do not always agree on them; in many cases, they are named and typed somehow differently.

We use the first five categories to describe the attributes that are included in the conceptual model, as explained in the next section; we label the attributes with a feature vector, e.g. $\textit{BiosampleType}^{[RSM]}$ denotes *BiosampleType* as an attribute which is restricted, single-valued and mandatory, while $\textit{Feature}^{[D(\textit{Technique})RSM]}$ denotes *Feature* as a restricted single-valued mandatory attribute with a value dependency from the attribute *Technique*.

We examined several sources among the ones previously described in Section 2.3. Three of them were chosen as the most representative for the formal preliminary analysis:

- TCGA (from its GDC distribution, as explained in Section 2.3) reports many experiment pipeline-specific metadata attributes; out of them, we selected 22 attributes, common to all pipelines, which are the most interesting from a biological point of view (Table 3.2).

Level	Symbol	Feature	Default
Source	C	Contextual	Non-contextual
	D	Dependent	Independent
	R	Restricted	Free
	S	Single-valued	Multi-valued
	M	Mandatory	Optional

Table 3.1 Taxonomy of features for metadata attributes.

- ENCODE includes both a succinct and an expanded list of metadata attributes; while the expanded list has over 2000 attributes, the succinct list has 49 attributes for experiments, 44 attributes for biosamples, and 28 attributes for file descriptions.
- GEO is the hardest to describe as its metadata attributes are expressed through semi-structured fields; their values are in many cases free texts.

Overall, TCGA and ENCODE have proven to be the most metadata-rich sources. Later, the model has been validated on many other sources, which were integrated into the repository one by one, as Chapters 4-5 will clarify. Differently, the GEO source is at the same time a very rich public repository of genomic data (as most research publications include links to experimental data uploaded to GEO), but is also a very poor source of metadata, which is not well structured and often lacks information; hence our mapping effort is harder and less precise for GEO than for the more organized TCGA and ENCODE sources.[4]

3.2.2 Model Design

The driving design principle of GCM is to recognize a limited set of concepts that are supported by most data sources, although with very different names and formats. GCM is centered on the notion of the *experiment item*, typically a file containing genomic regions and their properties. Four sub-schemata (or views) depart from the central entity, recalling a classic star-schema organization that is typical of data warehouses [67] (but here we allow for many-to-many relationships and hierarchies of dimensions):

- The *technology* used in the experiment, including information about item containers and their formats.
- The *biological* process observed in the experiment, in particular, the sample being sequenced (derived from a tissue or a cell culture) and its preparation, including its donor.

[4] Textual analysis to extract semantic information from the GEO repository is reported in [79] and summarized in Section 5.3; we plan to capitalize on this work and import into the repository relevant subsets of this important source.

C	D	R	S	M	Dependency	Attribute
			×	×		clinical.demographic.id
			×			clinical.demographic.year_of_birth
		×	×	×		clinical.demographic.gender
		×	×	×		clinical.demographic.ethnicity
		×	×	×		clinical.demographic.race
			×	×		biospecimen.sample.id
		×	×	×		biospecimen.sample.sample_type
	×	×	×		sample_type	biospecimen.sample.tissue_type
			×	×		generated_data_files.data_file.⟨type⟩.id
		×	×	×		generated_data_files.data_file.⟨type⟩.data_type
	×	×	×	×	data_type	generated_data_files.data_file.⟨type⟩.data_format
			×	×		generated_data_files.data_file.⟨type⟩.file_size
	×	×	×	×	data_type	generated_data_files.data_file.⟨type⟩.experimental_strategy
×		×	×		data_type	generated_data_files.data_file.⟨type⟩.platform
	×	×	×	×	data_type	analysis.⟨workflow⟩.workflow_type
			×	×		analysis.⟨workflow⟩.workflow_link
			×	×		case.case.id
			×			case.case.primary_site
	×		×		primary_site	case.case.disease_type
			×	×		administrative.program.name
			×	×		administrative.project.name
			×			administrative.tissue_source_site.name

Table 3.2 TCGA metadata attributes analysis.

- The *management* of the experiment, describing the organizations/projects which are behind the production of each experiment.
- The *extraction* parameters are used for the internal selection and organization of items. It describes the containers available in the repository for storing items that are homogeneous for data analysis.

These views are recognized in most sources and enable powerful query processing for extracting relevant datasets from each source. The conceptual schema is designed top-down, based on a systematic analysis of metadata attributes and of their properties in many genomic sources, building the entity-relationship schema represented in Figure 3.3.[5]

3.2.2.1 Central Entity: Item

We next describe the attributes of the ITEM entity and associate each of them with their feature vector. The *SourceId*$^{[D(Source)SM]}$ denotes the item identifier within

[5] The current version of the ER diagram, which can be appreciated in Figure 3.3, is slightly different from the one originally proposed in [54], due to the practical experience we gained in the field.

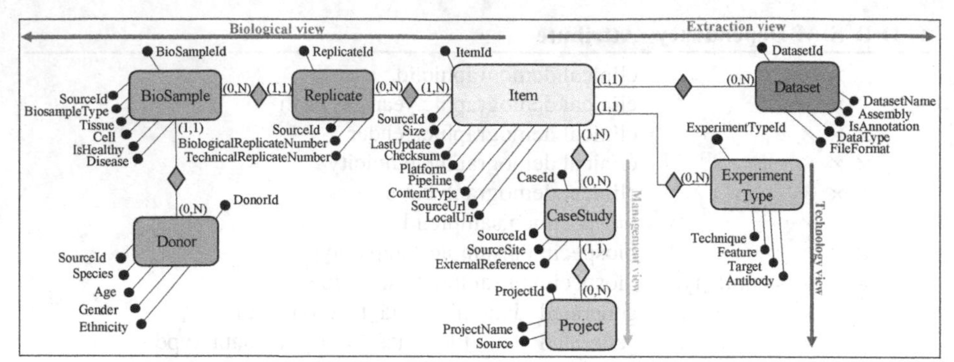

Fig. 3.3 Genomic Conceptual Model.

the source (from which it adopts the specific naming convention) and must always be included, along with its *Size*$^{[SM]}$. When available, also the *LastUpdate*$^{[S]}$ and the *Checksum*$^{[S]}$ should be included to help distinguish copies of the item downloaded at different times. *Platform*$^{[C(DataType)RSM]}$ is used to illustrate the NGS platform used for sequencing and depends on the *DataType* of the item. *Pipeline*$^{[D(Technique)S]}$ is a descriptor of the specific parameters adopted in the methods used for producing the processed data; it is a single-valued attribute interpreted in the general context of the *Technique* used for producing several items of the same kind. The single-valued attribute *ContentType*$^{[C(DataType)D(IsAnnotation)RS]}$ accepts a restricted number of values, e.g., "peaks"/"hotspots"/"exon quantifications" when the contained regions are experimental, or e.g., "gene"/"transcript"/"promoter" when they are annotations. For this, it is dependent on the *isAnnotation* attribute (in addition to being contextual w.r.t. the *DataType*). The Item is physically available for download at the *LocalURI*$^{[SM]}$ and, in its original form, at the *SourceURI*$^{[M]}$, where possibly multiple endpoints are stored. Providing parameters and references to the original data is relevant in the case of processed data, as sometimes biologists need to resort to original raw data for reprocessing.

3.2.2.2 Biological View

This view consists of a chain of entities: Item-Replicate-BioSample-Donor describing the biological process leading to the production of the Item. An Item is associated with one or more Replicates (N:N relation), each originated by a BioSample (functional relation), each derived from a Donor (functional relation).

Donor has the attribute *SourceId*$^{[D(Source)SM]}$ (donor identifier related to a source) and it represents an individual – characterized by the optional attributes *Age*$^{[S]}$, *Gender*$^{[RS]}$ and *Ethnicity*$^{[RS]}$ – or a strain of a specific organism (*Species*$^{[RSM]}$) from which the biological material was derived or the cell line was established.

BioSample describes the material sample taken from the biological entity represented by the Donor and used for the experiment; it is identified within the original source by the mandatory *SourceId*$^{[D(Source)SM]}$. *BiosampleType*$^{[RSM]}$ is restricted to values such as "cell line", "tissue", or "primary cell", depending on the kind of material sample used for the experiment. Based on the value of this attribute, either *Tissue*$^{[C(BiosampleType)D(Gender)SM]}$ or *CellLine*$^{[C(BiosampleType)SM]}$ becomes mandatory. *Cell* includes information on (single) cells in their natural state, immortalized cell lines, or cells differentiated from specific cell types. *Tissue* includes information regarding a multicellular component in its natural state or the provenance tissue of the *Cell*(s) of a biosample. In animal samples, some tissues are gender-specific. *IsHealthy*$^{[RS]}$ is Boolean and denotes a healthy (normal/control) or non-healthy (e.g., tumoral) sample; *Disease*$^{[C(IsHealhty)D(Tissue,Gender)]}$ stores information about the disease investigated with the sample and its values depend on *Tissue* and *Gender* because given diseases can only be related to given tissues and gender-related organs. It is contextual w.r.t. *IsHealthy* because it is mandatory when the sample is marked as diseased.

Replicate, identified by *SourceId*$^{[D(Source)SM]}$ is useful to model cases where an assay is performed multiple times on similar biological material. If repeated on separate biological samples, the generated items are biological replicas of the same experiment; if repeated on two portions from the same biological sample (treated for example with the same growth, excision, and knockdown), the items are technical replicates. This occurs only in some epigenomic data sources (such as ENCODE and Roadmap Epigenomics) that perform assay replication. Different replicas of the same experiment are associated with a distinct item and progressive numbers (indicated as *BiologicalReplicateNumber*$^{[S]}$ and *TechnicalReplicateNumber*$^{[S]}$).

3.2.2.3 Technology View

This view describes the technology used to produce the data Item and contains the ExperimentType entity. An Item is associated by means of a one-to-many relationship with a given ExperimentType, which refers to the specific methods used for producing each item. It includes the mandatory attribute *Technique*$^{[RSM]}$, which describes the assay, i.e., the investigative procedure conducted to produce the items (e.g., ["ChIP-seq", "DNase-seq", "RRBS", . . .]). *Feature*$^{[D(Technique)RSM]}$ is a mandatory manually curated attribute that we add to denote the specific aspect studied within the experiment (e.g., "gene expression", "Copy Number Variation", "Histone Modification", "Transcription Factor"). Epigenomic experiments (i.e., *Technique* = "ChIP-seq"), usually require two additional attributes to be fully characterized. These experiments typically analyze a protein, which we call *Target*$^{[C(Technique)RSM]}$. Instead, the *Antibody*$^{[C(Technique)D(Target)RSM]}$ is the protein employed against such target. The *Target* value is usually aligned to the Antibody Registry,[6] when available. The *Antibody* value depends on the *Target* since it is specific against that antigen.

[6] `http://www.antibodyregistry.org/`

3.2.2.4 Management View

This view consists of a chain of entities: ITEM-CASESTUDY-PROJECT describing the organizational process for the production of each item and the way in which items are grouped together to form a case. An ITEM is associated with one or more CASESTUDIES (N:N relation), each originated by a PROJECT (functional relation).

CASESTUDY represents a set of items that are gathered together because they participate in the same research objective (or study). Providing a precise definition for this entity is complex because it represents a broad concept, usually named in different ways in distinct sources;[7] the grouping of items into cases can be very useful in many applications. *SourceId*$^{[D(Source)SM]}$ and *ExternalReference* contain identifiers respectively taken from the main original source and other sources that contain the same data. While the first is unique and mandatory the second may optionally contain multiple values. The *SourceSite*$^{[S]}$ represents the single physical site where the material is analyzed and experiments are physically produced (e.g., universities, biobanks, hospitals, research centers, or just laboratory contact references when a broader characterization is not available).

PROJECT represents the initiative responsible for the production of the item. It provides a single point of reference to find diverse data types generated in the same research context. *Source*$^{[SM]}$ describes the programs or consortia responsible for the production of genomic items (e.g., TCGA, ENCODE, Roadmap Epigenomics, RefSeq, GENCODE...). Within a source, items may be produced within a specific initiative, specified in the *ProjectName*$^{[SM]}$, which uniquely references the project; it is particularly relevant in the context of TCGA data, where items are organized based on the type of tumor analyzed in the specific project (e.g., TCGA-BRCA identifies a set of items regarding the Breast Invasive Carcinoma study), or of annotation projects (such as the RefSeq reference genome annotation). This entity is inspired by the BioProject concept introduced by NCBI.[8]

3.2.2.5 Extraction View

This view includes the entity DATASET, used to describe common properties of homogeneous items. It gathers groups of items stored within a folder named *DatasetName*$^{[D(Source)SM]}$, which embeds the name of the *Source* as a substring. Dataset items are homogeneous as they share a specific data type, format, and assembly. The *DataType*$^{[RSM]}$ has values such as "peaks", "copy number segments", "gene expression quantification", "methylation levels". The *FileFormat*$^{[D(DataType)RSM]}$ denotes the standard format of the items dictating the genomic region data schema, including the number and semantics of attributes (e.g., "bed", or more specific ones such as "narrowPeak" and "broadPeak"); it depends on *DataType* (e.g. "nar-

[7] More in detail, with CASESTUDY we capture the concepts of "experiment" from ENCODE, of "case/patient" from TCGA/GDC, of "epigenome" from Roadmap Epigenomics, and of "annotation/data file" for GENCODE, RefSeq, and ICGC.

[8] https://www.ncbi.nlm.nih.gov/bioproject/

rowPeak" format is compatible with "peak" and not compatible with "mutations"). The ($Assembly^{[D(Species)RSM]}$) is restricted to a smaller vocabulary according to the *Species*. The Boolean attribute $IsAnnotation^{[RSM]}$ distinguishes between datasets containing experimental data (describing arbitrary genomic regions) and datasets storing genomic annotations (describing known genomic regions), currently defined in the ITEM's *ContentType* field.

3.2.3 Validation: Source-Specific Views of GCM

Arbitrary queries on GCM can be propagated to sources, using a *global-as-view* approach [245]. Mapping rules are used to describe how data are loaded from specific sources (with local schemata) into the global schema of the integrated repository. We will describe this operation in detail in Section 4.6.

Our conceptual schema was verified bottom-up, on TCGA, ENCODE, and GEO; we show that ER schemas describing these sources can be constructed as subsets of GCM. We verify that the global-as-view approach really captures the three considered data sources, by showing them as *subsets of GCM* in Figure 3.4; we use the following notation:

- We place the attributes of each source in the same position as in GCM, but we use for them the name that we found in the documentation of each source; missing attributes correspond to white circles.
- We cluster the conceptual entities corresponding to a single concept in the original source by encircling them within grey shapes. The entity names corresponding to the original source are reported with a bold bigger font on the clustered shape (e.g. Series in GEO) or directly on the new entity (e.g., Case in TCGA) when this corresponds to an entity present in our GCM.
- We indicate specific relationship cardinalities where GCM differs from the source, using a bold font (e.g., see (1,1) from ITEM to CASESTUDY in ENCODE).
- We enclose fixed human-curated values in inverted commas and use the functions notation *tr*, *comb*, and *curated* to describe a transformation of a source field, a combination of multiple source fields, and curated fields, respectively.

3.3 Related Works

The use of conceptual modeling to describe genomics databases dates back to the late nineties, including a functional model for DNA databases named "associative information structure" [300], a model representing genomic sequences [279], and a set of data models for describing transcription/translation processes [312]. In the subsequent literature, many works [97, 220, 392, 210, 136, 327] – and many more listed by Idrees and Khan [206] – employ conceptual models' expressive power for explaining biological entities and their interactions in terms of conceptual data

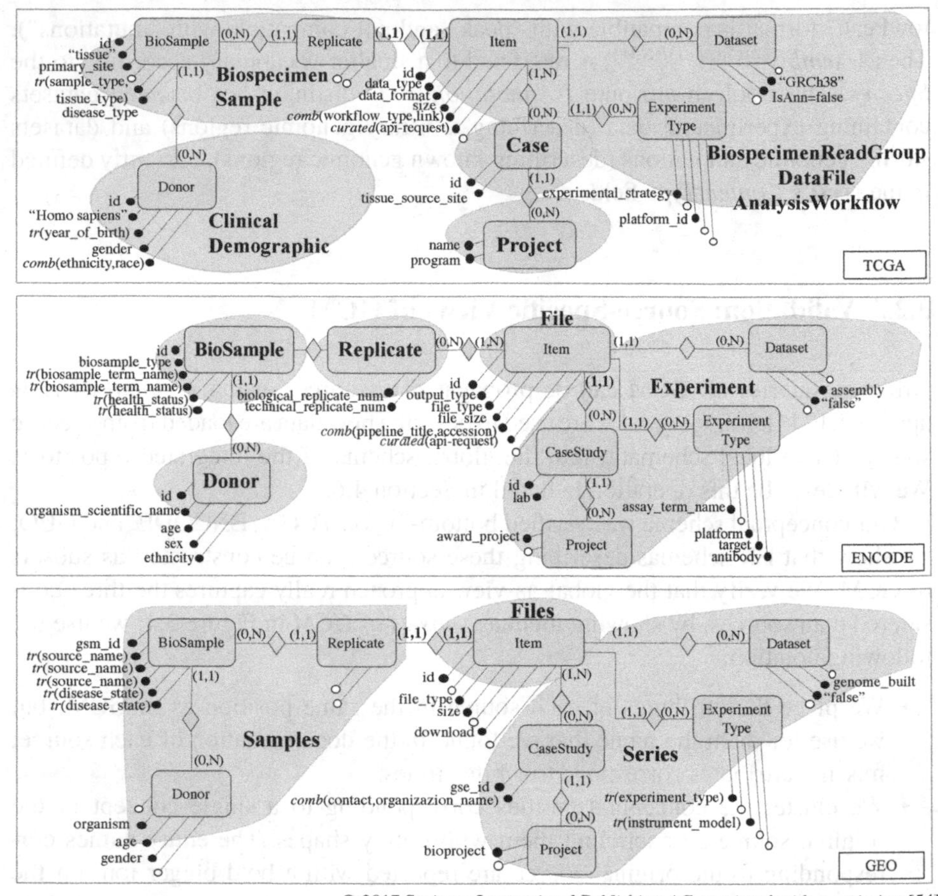

Fig. 3.4 Source-specific views of GCM for TCGA, ENCODE, and GEO.

structures. For example, in [69] Bornberg-Bauer and Paton use ER modeling (and UML class diagrams) to describe protein structures and genomic sequences, with rather complex concepts aiming at completely representing the underlying biology.

More recently there has been a solid stream of works dedicated to data quality-oriented conceptual modeling: [340] introduces the Human Genome Conceptual Model; [307] applies it to uncover relevant information hidden in genomics data lakes; [311] highlights the need for reliability-driven management of genomic data and implements an applicable solution using conceptual modeling.

While the works cited until here can be considered ER schemas of what we call region data in this thesis, others have targeted descriptions of experiments (i.e., metadata) with a different purpose: providing integrated access to the underlying genomic region data, deriving from heterogeneous sources. Surveys of preliminary works are [195, 257], demanding more cross-influences between genomics and data integration efforts.

A common approach in integrated data management is data warehousing, consisting of *offline* integration and reconciliation of data extracted from multiple sources, such as in EnsMart/BioMart [187, 372]. Along this direction, [274] describes a warehouse for integrating genomic and proteomic information using generalization hierarchies and a modular, multilevel global schema to overcome differences among data sources. Older conceptual modeling-based data warehouses include: the GEDAW UML Conceptual schema [179] (for a gene-centric data warehouse), the Genome Information Management System [109] (a genome-centric data warehouse), the GeneMapper Warehouse [127] (integrating expression data from genomic sources), and BioStar [392]—a biomedical data warehouse supporting a data model capturing the semantics of biomedical data and providing some extensibility to cope with the evolution of biological research methodologies. Buneman *et al.* [73] described the problem of querying and transforming scientific data residing in structured files of different formats. Along that work, BioKleisli [119] and K2 [118] describe early systems supporting queries across multiple sources. BioKleisli was a federated database offering an object-oriented model; its main limitation was the lack of a global schema, imposing users to know the structure of underlying sources. To improve this aspect, K2 included GUS (Genomics Unified Schema [23]), an extensive relational database schema supporting a wide range of functional genomics data types.

Finally, conceptual models have also been used to characterize the processes and objects during related analysis workflows in [326].

From a different perspective, the genomics community has always made great use of specialized ontologies, that can serve different purposes than ER models. The collective OBO Foundry [373] includes many fundamental ontologies that support everyday biologists' research (such as the Gene Ontology [331] or the Sequence Ontology [135]). These are commonly used for annotating data and experiments and making them interoperable (i.e., finding common ground between the descriptions and terminologies used in different sources); this application will be discussed in Section 4.7, as part of our own approach.

A classic work [189] proposed a general Genomics Ontology, while a more recent one [148] promotes the use of foundational ontologies to avoid errors while creating and curating genomic domain models for personalized medicine.

Overall, conceptual modeling has been mainly concerned with aspects of the human genome, even when more general approaches were adopted; in Section 3.2 we presented GCM, describing metadata associated with genomic experimental datasets available for model organisms. With our approach, we use conceptual modeling for driving the continuous process of metadata integration and for offering high-level query interfaces on metadata for locating relevant datasets, under the assumption that users will then manage these datasets for solving biological or clinical questions.

Chapter 4
Integrating Genomic Data

"Problems worthy
of attack
prove their worth
by hitting back."
— 'Problems', p. 2 in GROOKS, by Piet Hein

After proposing in Chapter 3 a modeling approach for genomic data, which mediates the most important and complex data sources, we then address the more practical problem of integrating real-world datasets into the proposed global schema. This chapter is focused on the process required to generate the GCM content and on the resulting repository.

This process encounters many challenges, including solving heterogeneity aspects of data models among various data sources at the schema and instance level, dealing with semantically *equivalent* concepts that are described differently across sources or semantically *different* concepts (possibly generalizations/specializations) that are defined by using the same description.

We propose META-BASE, a novel architecture for the integration of genomic datasets; it is deployed as a generic pipeline of six progressive steps for data integration, applicable to arbitrary genomic data sources providing semi-structured metadata descriptions. Four steps are driven by source-specific modules, the others are source-independent. Two steps are assisted by tools that help the designer in the progressive creation and adaptation of data management rules, with the general objective of minimizing the cognitive effort required from integration designers. The result generated by META-BASE is a very large integrated repository of tertiary genomic datasets. Along with the theoretical presentation of the steps, we use examples focused on sources that we integrated into the repository, starting from the ones that feature more complex metadata scenarios. Some of the challenges that are specific to each source are discussed in Chapter 5. It is our intention to make the META-BASE repository grow continuously in the next years, responding to current and upcoming biological and clinical needs.

Every step of the META-BASE pipeline produces a data ingestion program that can be applied to data sources after an initial design; these programs need to be

A. Bernasconi: *Model, Integrate, Search... Repeat*, LNBIP 496, pp. 53–97, 2023.
https://doi.org/10.1007/978-3-031-44907-9_4

adapted only in case of structural changes in the data sources. The process is extensible, as the designer who wants to add a new source has just to add new definitions and rules to the data integration framework.

Within the data enrichment step of the META-BASE pipeline, we also use selected ontological sources for improving value matching, which is extended from exact match to semantic match inclusive of the use of synonyms, hyponyms, and hyperonyms; they enable simple value conversion strategies, which capture some value mismatches that may occur in different repositories.

Queries upon the META-BASE repository are supported by GenoSurf [76], a user-friendly, attribute-based, and keyword-based search interface (described in Chapter 6), producing, as a result, the URIs of the relevant data in the source repositories; scientists can build over them an arbitrary genomic computation, using any bioinformatics system and resource. In this way, the META-BASE repository provides a conceptual entry point to the supported genomic data sources. In addition, the META-BASE pipeline and repository feed an architecture for genomic data processing, defined in [275], providing portable and scalable genomic data management on powerful servers and clusters;[1] in such distinct environment, metadata can be queried together with their respective datasets using the GMQL high-level domain-specific query language [277].

The most innovative aspects of this integration work are: from a computer science perspective, the design and engineering of an end-to-end pipeline whose steps make novel use of rewrite rules for data cleaning, mapping, normalization, enrichment, and constraint verification; from a biological perspective, the partitioning schemes for each data source, the selection of a common genomic base unit into which different formats can be transformed, and the selection of the ontologies providing enrichment for specific GCM attributes; from an application perspective, the possibility of the broad community of computational biology and bioinformatics feeding new interfaces/systems with solid data architecture.

Chapter organization. Chapter 4.1 relates the theoretical framework within which our integration effort should be set: design choices in this chapter are properly introduced and justified. Section 4.2 describes our generalized approach to metadata management and integration. Sections 4.3-4.8 describe the pipeline to extract metadata from original sources and to prepare it for integration, including it into a relational database, performing ontological enrichment and integrity constraints checks. Section 4.9 explains the realized architecture. Section 4.10 validates the approach and discusses its effectiveness. Section 4.11 provides an overview of other literature works on genomic data integration.

[1] Based on Apache Spark `http://spark.apache.org/`

4.1 Theoretical Rationale

The integration pipeline described in this chapter is the central contribution of this Ph.D. thesis. As such, we provide readers with an introduction to the theoretical bases upon which our choices have been performed.

Data Structure. The GCM schema proposed in Chapter 3.2 is an ambitious and functional attempt of systematizing the various and disorganized world of annotations describing genomic experiments. It is ambitious because data sources are many and they all propose their in-house methods, it is functional because it is designed with attention to the practical implementation that immediately follows its conception.

We chose to use a data warehouse-like representation; as stated in [67, 225], a data warehouse structures data by means of a star schema, with a central fact table (files of genomic regions) and a set of smaller dimension tables set as a radial pattern around the fact (specifying notable characteristics of the experiment and sample from which the genomic file was derived). Typically, the fact table establishes a one-to-many relationship from each dimension row to the (many) fact tuples sharing the same value as the key of the dimension. Through a classical normalization process, each dimension may be transformed into a hierarchy of tables, yielding to a more complex schema, i.e., snowflake schema.

Our GCM design has led the foundations to address the more practical data warehouse schema design problem: determining a suitable schema, which started from a star, evolved towards a snowflake, then admitted two many-to-many relationships (as files may be organized in many case studies and may be derived from several biological replicates), with the final goal of enabling easy search and aggregation of information on genomic datasets, while – at the same time – being efficiently supported by the schema of the underlying operational database. Evidently, this basic schema would need further modifications if it were to accommodate future advanced applications such as data mining analysis.

Integration Paradigm. When evaluating different options for integrating database systems, we observed that, differently from other data integration settings, our priority was not the timeliness of available information. Instead, our first focus should be the interoperability of offered datasets. The strength of the proposed system is the idea of offering a self-contained database, with an integrated search system and bio-computational engine, that can be employed one after the other in continuous iterations. Under this assumption, we did not consider the implementation of virtual non-materialized databases, as data at the sources may change too rapidly, whereas we aim to provide users with a consistent environment of processed genomic data to perform their analysis. Instead, we chose to perform an Extract-Transform-Load (ETL) process that merges data in a new materialized database, which is suitable to be maintained at our site and offers the possibility of physical optimizations and manual curation for low-quality metadata.

During this process of data integration of genomic experiments we solve several problems connected to schema reconciliation and value-level integration via semantic enrichment, whereas we do not address data fusion [297], as we intend to keep the specificity of the terminology used in the original data. Indeed, researchers are used to accessing data sources directly, thus it is important that they can recognize the original forms. In addition, we support them with a rich interoperability infrastructure that allows matching existing ontological terms in well-established and validated resources (with synonym/hierarchical mechanisms). Semantic enrichment is performed by transforming bio-ontology portions into relational structures, rather than employing an interoperable format such as Resource Description Framework (RDF), for consistency with the design choices that defined the GCM and its relational implementation; this will be detailed in Section 4.7.2.

To devise the mapping function from data sources into our global schema we employed an approach similar to Global-as-view (GAV), as opposed to Local-as-view (LAV) or to hybrid solutions such as GLAV [245]. GAV is the typical choice when data sources are stable, and no extensions to new data sources are foreseen, as otherwise, the global schema needs to be reconsidered every time; this does not correspond to our scenario. However, our priority was given to achieving easier mapping and query-answering solutions (without including reasoning), which are guaranteed by GAV, as opposed to LAV/GLAV. This makes our GCM not so flexible: if we change the model all the mappings for data sources need to be revised; in partial mitigation of this drawback, GCM can accommodate new information easily in the form of unstructured $\langle key, value \rangle$ pairs.

Granularity of the Model. When modeling data, integrating databases that observe phenomena from (even slightly) different perspectives, tailored to their precise goals, one problem that comes to attention is that of *granularity*. In our scenario, we have two different problems of granularity: i) on the genomic region data (described in Section 3.1); ii) on the genomic metadata describing datasets of regions (see Section 3.2). We worked towards understanding what our *urelement*[2] is in both i) and ii) levels.

In the first case, our genomic data is organized in datasets, which contain samples, each containing a set of genomic regions. To the benefit of integration between different sources, we chose the *GDM sample* as our "genomic basic data unit". This choice is discussed in more detail in Section 4.4, explaining its implications on the data transformation phase.

In the second case, the metadata Genomic Conceptual Model, the approach is different. We refer to Keet [221], who articulates granularity based on categories of non-scale dependency (NSD), relationships between levels, set theory, and mereology. NSD granularity provides levels ordered through primitive relations such as the *part_of* (structural information), the *contained_in* (spatial information), and the *is_a* (subsumption information). Starting from the $\langle key, value \rangle$ metadata pairs deriving from a first import of data sources, we defined the *urelement* as the *GCM item* (which is the same as GDM sample) and further elaborated this level by adding synonyms and ontological hierarchies, providing what Keet would define as four levels of gran-

---

[2] In mathematics set theory, a urelement is an object that contains no elements, belongs to some set, and is not identical with the empty set [288].

ularity, allowing users to see metadata from different (from more detailed to more general) levels.

Notably, our urelement in i), i.e., the GDM sample, and our urelement in ii), i.e., the GCM item, coincide. They both represent conceptually a file of genomic regions, our basic data unit for integration, search, and analysis activities.

Governance of integration process. In the following of this chapter, we focus on the formalization and implementation of a rich architecture for importing genomic datasets from sources, transforming them into a common format, and enriching their descriptions semantically. In the discussion we do not detail the evolution of the system; however, the governance of the process is tracked by registering all the performed steps in a database that traces the timings and the observed partitions of data. Such a persistence layer allows us to memorize the system at specific time points (see Section 4.9.1). Future work may exploit the content of this database as the backbone of an "integration process governance dashboard", focusing on the temporal information of the process and defining meta-queries to understand the evolution of the system; this could inform the development of a value-based strategy for further integration efforts.

4.2 Approach Overview

The high-level description of the proposed approach to data and metadata integration is shown in Figure 4.1, articulated in the six phases of the META-BASE framework. Through downloading, data and metadata are imported at the repository site in their original formats (Section 4.3). During transformation, they are translated into the GDM format, i.e., data are regions with coordinates and other properties, and metadata are raw attribute-value pairs (Section 4.4). Metadata are then cleaned, thereby producing a collection of clean pairs for each source, to improve raw attribute names and to filter irrelevant metadata (Section 4.5). During data mapping, a syntactic transformation is applied on cleaned metadata, which is mapped into the global relational schema implementing the Genomic Conceptual Model (GCM) (Section 4.6). GCM values are then normalized – resorting to generic term-ids that may take specific sets of values – and enriched of term labels, references, hyponyms, hyperonyms and synonyms, by means of external ontologies (Section 4.7). The constraints checker provides methods for reporting integrity constraints' violations, based on specifications of legal values in the repository (Section 4.8).

For exemplifying the META-BASE framework, we consider important and complex data sources among the ones mentioned in Chapter 2. These represent consortia/players that publish open data and enforce their own data models and data management principles, which are strict and not easily interoperable with different sources. Their maintained repositories are subject to rapid changes, as each source is a continuously evolving system. Luckily, most changes are additive and use already existing metadata in their descriptions. For this reason, we approach each source with

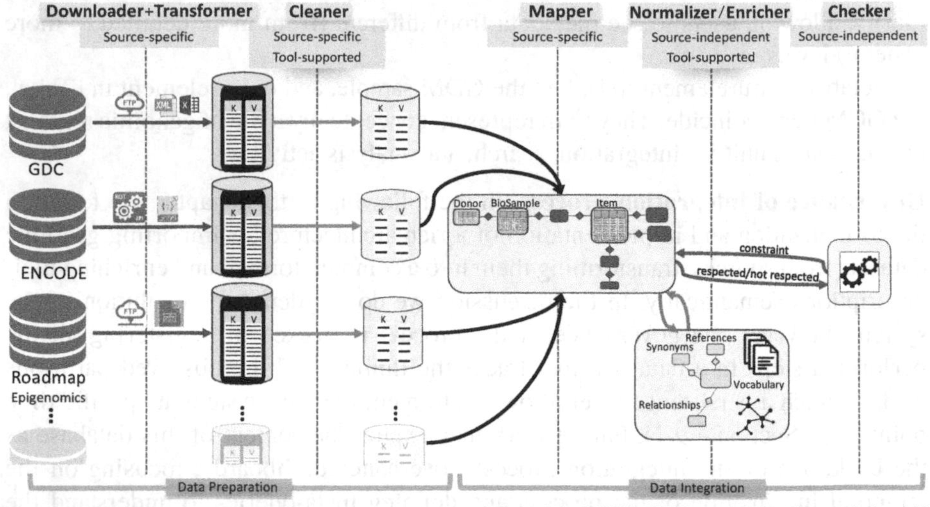

Fig. 4.1 The overall data preparation and integration process.

an *initial activity* for the production of a source-specific set of data and metadata manipulation rules, followed by periodic *data integration sessions*, where new items are discovered and their metadata are modeled.

4.3 Data Download

The *Downloader* module produces files – both for the genomic data and its metadata – in original source-specific format, at the processing site hosting our repository. We build a collection of protocol-specific modules with few parameters to adapt them to new sources; tunings for each specific source may be necessary.

All open sources generally allow for direct download from their web page, while they offer heterogeneous protocols for programmatic access (GTEx and GWAS Catalog are an exception): ENCODE exposes HTTPS GET requests to retrieve a list of files corresponding to the set filters;[3] GDC uses both GET and POST requests to retrieve the list of files;[4] ICGC provides many API endpoints for requesting filtered sets of files; Roadmap Epigenomics, GENCODE, RefSeq, and 1000 Genomes store all their files on FTP websites that can be navigated programmatically; GEO provides a variety of methods both through its own portal and from alternative interfaces.

The *Downloader* addresses the issues related to the synchronization of our local repository with the original ones. As data size is significant, downloading everything

[3] Other API requests can be used to retrieve files one by one.

[4] Consequently, each of these files can be downloaded using an HTTPS GET/POST request.

from scratch from the original sources should be avoided. Instead, it is necessary to precisely define metrics to compare contents: *What is new, what has been updated, and what is not present anymore?* More in general: *how can changes on genomic data sources be taken into account to be reflected on our regularly updated repository?* When targeting integrated systems up to date, the main difficulty is to identify a specific data partitioning scheme at each source; in this way, each partition can be repeatedly accessed and source files that are added to or modified within the partition can be selectively recognized, avoiding the download for those source files that are unchanged.

ENCODE	List of `file_id`	Protocol HTTP GET: https://www.encodeproject.org/metadata/?type=Experiment&⟨`params`⟩/metadata.tsv
	Example `params`	assembly=hg19 & file.status=released & project=ENCODE & ... & files.file_type=bed+narrowPeak
	Download file	https://www.encodeproject.org/files/⟨`file_id`⟩/@@download/⟨`file_id`⟩.bed.gz
GDC	List of `file_id`	Protocol HTTP POST: https://api.gdc.cancer.gov/files with ⟨`params`⟩ in Payload
	Example `params`	field:cases.project.project_id-value:["TCGA-ACC"], field:files.data_type-value:["Copy Number Segment"], ...
	Download file	https://gdc-api.nci.nih.gov/data/⟨`file_id`⟩
Roadmap Epigenomics	`dir` paths	Protocol FTP: http://egg2.wustl.edu/roadmap/data/byFileType/peaks/consolidated/`dir`
	Example `dir`	broadPeak
	Download file	http://egg2.wustl.edu/roadmap/data/byFileType/peaks/consolidated/⟨`dir`⟩/⟨`file_name`⟩.⟨`dir`⟩.gz

Table 4.1 Endpoints for data download from sources and example invocations.

Suppose we are interested in downloading a certain updated ENCODE portion (e.g., narrowPeak samples on human tissue, aligned to reference genome hg19). We then produce an API request to the endpoint `https://www.encodeproject.org/matrix/`, specifying the parameters `type` = `Experiment`, `assembly` = `hg19`, `file.status` = `released`, `project` = `ENCODE`, `replicates.library.biosample.donor.organism.scientific_name` = `Homo+sapiens`, and `files.file_type` = `bed+narrowPeak`. Table 4.1 illustrates the endpoints for data download used for three major sources, with their protocol, request format, and example parameters for invocation. The example partitioning scheme for the ENCODE data source is illustrated in Figure 4.2, with its specific set of parameters used during download, corresponding to a partition.

Formalization. For a given source i, a *Downloader* is a method $\mathcal{D}_i = \langle \mathrm{D}_i, P_i \rangle$ for importing genomic data and metadata from the specific partition P_i. At each invocation of the method, a new set D_i of files (each representing region data, metadata, or both) is retrieved at the repository site, and associated with a signature $\langle dataset\_name, source, endpoint, parameters \rangle$; parameters include the timestamp t_h of the download operation. In this way, future invocations of $\mathcal{D}_i$ at time $t_k > t_h$ will be used to download information from P_i and then start a data integration session by tracking the changes that occurred to P_i at the data source between time t_h and t_k. The set of files D_i may be allocated to a single GDM dataset $D_{i,j}$ such that $\mathrm{D}_i = D_{i,j}$ or instead be partitioned into multiple GDM datasets, i.e. $D_{i,1}, ..., D_{i,j}, ..., D_{i,N} \in \mathrm{D}_i$ such that $D_{i,1} \cap D_{i,j} \cap D_{i,N} = \emptyset$. This decision is made by the integration designer, based on the fact that genomic files contained in a GDM dataset must adhere to

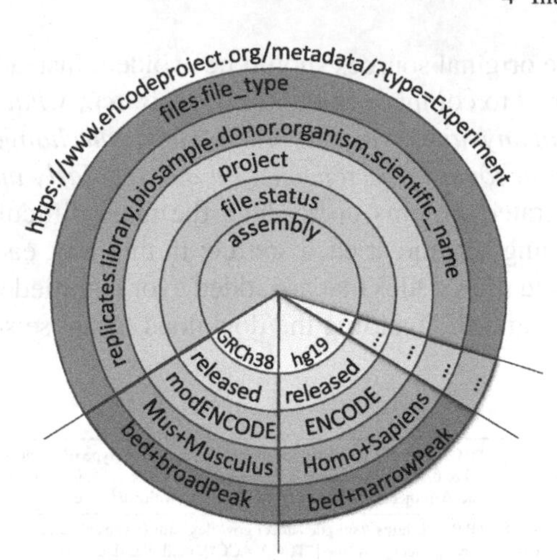

Fig. 4.2 Selection of portions from ENCODE. In the upper area, we specify parameter names, in the two small bottom slices we specify example values, defining a partition of the source.

a unique schema: if all the downloaded files can be expressed by using the same schema, then the simplest possible solution is employed, i.e. we just produce one downloaded dataset $D_{i,j}$.

Method. Each download module first connects to the data source servers and retrieves the list of the identifiers of the files that belong to the partition to be downloaded. Many sources provide (semi-)programmatic methods to translate a query composed on their portal visual interface into an API request or a downloadable list of files corresponding to the search; otherwise, this step has to be programmed *ad-hoc*.

For each file, the *Downloader* typically retrieves its *Size, LastUpdate* and *Checksum*, denoting identifying properties; these are provided by most sources.[5] We match these values with data that is stored in our local database – corresponding to the previous download session of the same partition – using the file unique identifier. The matching allows us to pinpoint:

- *New files*: they are stored as genomic data files and are processed, together with their metadata, by invoking the pipeline discussed in this section.
- *Matching files*: when they have the same *Size*, *LastUpdate* and *Checksum* values as their local values stored in the database, we reprocess just the metadata by invoking the pipeline discussed in this section – we avoid the download of region data, which is typically much bigger in size; when any one of the *Size*, *LastUpdate* and *Checksum* values is different, we re-download also the genomic data files.

[5] If some of them are unavailable, we either compute them at the source or accept a less precise matching by using fewer parameters.

- *Missing files*, i.e., files whose identifier was present at the previous invocation but is no longer present: these files are deprecated, the genomic data and metadata are copied to an archive, which can only be inspected by archive lookups (but they are no longer retrieved by standard queries).

Eventually, we collect into the set of files D_i all data and metadata files relative to new or changed items, divided into the datasets $D_{i,j}$; these downloaded files are then used in the next steps of the META-BASE pipeline.

Example 1. The 1000 Genomes Project stores its files in an FTP repository.[6] They represent big sets of variants discovered in control population samples (i.e., healthy individuals' genomes compared against the reference one). Variants are stored in 25 enormous files in VCF format, one for each human chromosome from 1 to 22, plus chromosomes X, Y, and MT (i.e., mitochondrial). Each file contains many variants (represented by rows) that may or may not be present in each sample/patient (columns). Metadata are contained in 4 single files (scouted around their FTP complex structure[7]); we save such files together with data files inside the same dataset folder.


```
{"accession": "ENCSR635OSG",
 "assembly": ["hg19"],
 "award": {
        "pi": {
          "lab": {"name": "michael-snyder",...},
          ...},
 ...},
 "dbxrefs": [],
 "files":[
     {"accession": "ENCFF134AVY",
       "biological_replicates": [1],...},
     {"accession": "ENCFF429VMY",
       "biological_replicates": [1,2],
       "file_type": "bed narrowPeak",...},
       ...
 ],
 "replicates": [
     {
        "@id": "/replicates/4874c170-7124-4822-a058-4bb/",
        "biological_replicate_number": 1,
        "library": {
            "biosample": {
               "donor": {"age": "6",...},
               "health_status": "healthy",
            ...},
```


[6] `http://ftp.1000genomes.ebi.ac.uk/vol1/ftp/`

[7] Metadata about the indexes of the alignment files are at `http://ftp.1000genomes.ebi.ac.uk/vol1/ftp/data_collections/1000_genomes_project/1000genomes.sequence.index`; information on the 26 population participating to the project are at `http://ftp.1000genomes.ebi.ac.uk/vol1/ftp/phase3/20131219.populations.tsv`; characteristics of the individuals donating the samples can be found at `http://ftp.1000genomes.ebi.ac.uk/vol1/ftp/release/20130502/integrated_call_samples_v2.20130502.ALL.ped`; sequencing strategies are detailed at `http://ftp.1000genomes.ebi.ac.uk/vol1/ftp/technical/working/20130606_sample_info/20130606_sample_info.txt`.

```
        ...},
        "antibody": {"lot_id": "940739",...},
        ...
    },
    {
        "@id": "/replicates/d42ff80d-67fd-45ee-9159-25a/",
        "biological_replicate_number": 2,
        "library": {
            "biosample": {
                "donor": {"age": "32",...},
                "health_status": "healthy with non-
                    obstructive coronary artery disease",
            ...},
        ...},
        "antibody": {"lot_id": "940739",...},
        ...
    }
],
...}
```


Listing 4.1 Excerpt from example JSON file retrieved for *ENCODE* experiment ENCSR635OSG.

Example 2. When the ENCODE *Downloader* is invoked, it calls the ENCODE data/metadata endpoint (see Table 4.1 again), extracting a list of files, each belonging to one experiment. We then download: i) one metadata file for each experiment (which contains information about multiple region data files); ii) all region data files described in the metadata such that they satisfy the filters set in the API call (note that experiments may also contain other kinds of files as well). After retrieval, the identifiers of the files belonging to the considered partition are recorded together with their size, last update date, and checksum.

In Listing 4.1, we observe the metadata of a genomic experimental study, with accession ENCSR635OSG: a hierarchically structured JSON file, including several embedded elements: information about the whole experimental study, arrays of "files" elements (a list of items included in the experimental study) and of "replicates" elements, along with other information. Here multiple files are included, namely ENCFF134AVY and ENCFF429VMY, where the former one has one replica and the latter one has two. Correspondingly, region data files ENCFF134AVY.bed and ENCFF428VMY.bed have been downloaded.

As GDM samples correspond to files (not experiments), the unique metadata file will be accordingly partitioned into single files, each reflecting only one data file. We discuss this in the next step, i.e., transformation.

4.4 Data Transformation

The *Transformer* deals with the lack of agreement towards a standard data unit for tertiary analysis. *Can genomic data be expressed using a unique model that is general enough to represent all analyzed formats and that also allows ease of operation?*

We argue that such *basic genomic data unit* is missing in the current practice: a single – self-contained – piece of information that contains genomic regions with their properties and is identifiable with an entity that is interesting for downstream analysis (e.g., a patient, a biological sample, a reference epigenome...). For this purpose, we propose to use the "sample" of the Genomic Data Model (GDM) [276], in contrast with other complex/hierarchical solutions. The *Transformer* module takes, as its input, the data and metadata files resulting from the download phase and transforms them into a GDM-compliant format. When represented in this form, data and metadata from different sources can be queried integratively, regardless of their format in the source of origin. More specifically, this module resolves two kinds of heterogeneity of genomic files: 1) the different data units; 2) the different data schema within each unit (e.g., the schema of the GDM sample). We wrote transformers for the most used formats in origin (meta)data. Additional ones can be easily added.

Formalization. For a given source i, a *Transformer* is a source-specific method $\mathcal{T}_i = \langle \mathrm{D}_i, \mathrm{T}_i \rangle$. For each dataset $D_{i,j} \in \mathrm{D}_i$, it produces a corresponding dataset $T_{i,j} \in \mathrm{T}_i$, including transformed *sample-pairs* $< t_{data}, t_{meta} >$ such that:

- each t_{data} is a *data file* that adheres to the GDM schema (i.e., one row per region, each with 4 genomic coordinates and other variable columns, depending on the schema), which is set for dataset $T_{i,j}$.
- each t_{meta} is a *metadata file* that contains a list of $\langle key, value \rangle$ pairs, compatible with the GDM format.

We split the process into two steps: 1) the transformation from each $D_{i,j}$ dataset into its related $T^0_{i,j}$ dataset applies *data unit transformation* operations; 2) the transformation from each $T^0_{i,j}$ dataset into its related the $T_{i,j}$ dataset applies *data format transformation* operations. Depending on the transformation relation cardinality (i.e., 1:1, 1:N, N:1, N:M) of both data and metadata (potentially different from one another), the adopted algorithm is different.

Method: data unit transformation. Some sources provide a data file for each experimental event, others include more complex formats, such as MAF, VCF, and gene expression matrices. As to the associated metadata information, in some cases, they follow the same scheme as region data files (i.e., each data file has a corresponding metadata file). In other cases, a single metadata file describes a collection of experimental files (see Listing 4.1 shown before).

Under this evidence, we claim that any set of downloaded files – with their input format – should be convertible through a *transformation relation* into a set of genomic basic data units. We define as *transformation relation cardinality* the pair $X : Y$, where X is the cardinality of the set of files from the input source and Y is the cardinality of the output set of *basic units* into which the input is transformed; $X : Y$ is a fraction in the lowest terms. Our Y output set is a set of GDM samples.[8]

[8] Threat to validity: there is not only one correct way to transform data from the source into GDM format. Most of the time the most convenient choice corresponds to preparing one GDM data file for each patient/donor/individual/cell line belonging to the source. Other times such a clear classification is not present, therefore other heuristics are considered, such as the reasonable size

The cardinality of the transformation depends both on the data type and on particular choices made by each data source. For example, typically gene and miRNA expression data are provided in TSV data matrices where rows represent genes (or transcripts, exons...) and columns represent patients or sample IDs. When the Variant Call Format (VCF) format is used, typically we find collections of files, one for each chromosome: rows are a long list of known variations, while columns are the sample IDs. Some sources provide a data file for each experimental event, for example, ENCODE. In this case, the transformation has a 1:1 cardinality, i.e., to each ENCODE-produced file, it corresponds one GDM sample. Other sources include everything in one big file; the transformation phase takes care of compiling one single data file for each patient or univocally identified sample in the origin data. The transformation cardinality is thus 1:N, N being the number of patients or biological samples. An example instance is ICGC mutation data, where data is provided in big files where each row records a mutation in the sequence of a given patient. A patient may have more associated rows (i.e., mutations). From such a file, we build a number of GDM samples corresponding to the number of cancer patients.

Metadata also features diverse data units in the analyzed sources; transformations apply with the same criteria as for region data. A metadata file is transformed according to the following cases:

i) hierarchical formats (JSON, XML, or equally expressive) require applying a flattening procedure to create key-value pairs—the key results from the concatenation of all JSON/XML elements from the root to the element corresponding to a value;
ii) tab-delimited formats (TSV, CSV, or Excel/Google Spreadsheets) strictly depend on the semantics of rows and columns (e.g., 1 row = 1 epigenome, 1 row = 1 biological sample)—they often require pivoting tab-delimited columns into rows (which corresponds to creating key-value pairs);
iii) two-columns tab-delimited formats (such as GEO's SOFT files) are translated into GDM straightforwardly;
iv) completely unstructured metadata formats, collected from Web pages or other documentation provided by sources needs case-specific manual processing.

Table 4.2 shows transformation relation cardinalities regarding both data and metadata input formats, targeting the GDM output format. We analyzed different data types in a number of important data sources, that possibly include files with different formats.[9] In all these scenarios, our *Transformer* module should produce multiple data files, one for each experimental event (or annotated entity, such as genes, transcripts, exons), each paired with its own metadata file.

---

of GDM data files, partition by meaningful metadata (e.g., for annotation data, *annotation type*), source-specific concepts (e.g., epigenomes for Roadmap Epigenomics). As a future work, we would like to provide a more precise – semantically sound – definition of the GDM sample and of general files transformation into this format.

[9] Note that, while for descriptive purposes we indicate physical formats (e.g., TSV, TXT, JSON), the indication of cardinalities also embeds semantic information: how many data units are represented in one file.

Source	Data type	Data format, cardinality[1]	Metadata format, cardinality[1]
ENCODE	peaks transcription transcription	BED, 1:1 TSV, 1:1 GTF, 1:1	JSON, 1(experiment):#samples JSON, 1(experiment):#samples JSON, 1(experiment):#samples
GDC	mutations gene expression methylation, cnv, quantifications	MAF, 4:1 TXT, 3:1 TXT, 1:1	JSON, 4:1 JSON, 3:1 JSON, 1:1
ICGC	mutation, methylation, miRNA/gene expression	TSV, 1:#donors	TSV, 1:#donors
Roadmap Ep.	peaks transcription	BED, 1:1 TSV, 2:#epigenomes	Spreadsheet, 1:(#samples×#epigenomes)[2] Spreadsheet, 1:(#samples×#epigenomes)[2]
GENCODE	annotations	GTF, 1:#annotation_types	region data file + webpage, X:1[3]
RefSeq	annotations	GFF, 1:#annotation_types	region data file + webpage, X:1[3]
1000 Genomes	variation	VCF, 23:#individuals	TSV, 4:#individuals
GEO	expression	BED, 1:1	HTML/SOFT, 1:#files_from_sample
GTEx	expression	GCT, 1:#donors	TXT, 1:#donors
GWAS Cat.	associations	TSV, 1:1	-
CISTROME	peaks	BED, 1:1	TSV, 1:1
CCLE	various	GCT/TXT, 1:#cell_lines	TXT, 1:#cell_lines
COSMIC	various	TSV, 1:#individuals	TSV, X:1[3]

[1] Expressed as $X : Y$; this ratio represents the number X of data (resp. metadata) units used in the origin source to compose Y data (resp. metadata) file(s) in GDM format.
[2] Each reference epigenome is used for many data types, thus many GDM samples. The same epigenome-related metadata is replicated into many samples. [3] In these cases it is difficult to build a numerical relation—many meta are retrieved from the data files themselves, in addition to manually curated information.

Table 4.2 Census of 13 important data sources reporting for each: the processed data types that can be downloaded (along with metadata), their physical formats, and the semantic cardinality of the transformation relation with respect to the GDM output format [276].

Method: data format transformation. The *Transformer* also solves the heterogeneity of data formats and prepares the GDM *datasets* as sets of GDM *samples* that are uniform and adhere to a predefined schema. This phase applies procedures such as format conversions, normalization of coordinates, re-formatting into schema standards (e.g., narrow/broad peak), as well as data annotation, including:

1. providing positional information (i.e., genomic coordinates) and associated known genomic regions (e.g., genes) in a standardized framework;
2. transforming 0-based coordinates into 1-based ones or vice versa;
3. allowing joined use of different data types (e.g., gene expression and methylation) adding common gene identifiers, such as HUGO or HGNC gene symbol, Entrez or Ensembl gene ID;
4. merging together in the same files multiple expression measures obtained through different calculations, such as RPKM, FPKM, FPKM-UQ, and counts in gene expression data;

5. completing missing information, in case it is mandatory (e.g., the missing strand is filled with the wildcard character "*").

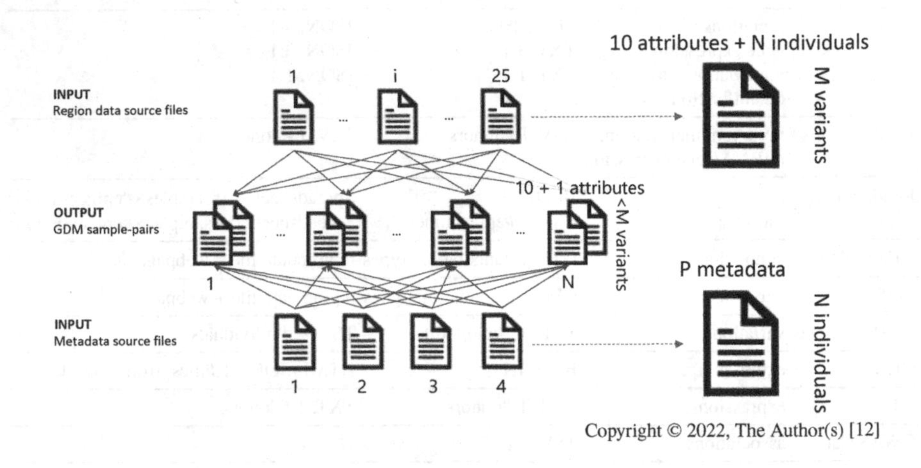

Fig. 4.3 Description of transformation process for 1000 Genomes files.

Example 1. The process of data unit transformation applied to 1000 Genomes files can be appreciated in Figure 4.3. We use 25 region data files corresponding to chromosomes, each representing a matrix of variants (in rows), which are either present or absent in the individuals represented in columns; variants also have about 10 other characterizing attributes. In addition, we consider four files containing metadata in matrices where rows are individuals (identified by the same sample id used in region data file columns) and each column represents one type of information (regarding the population of the individual, parent/child relationships, instrument platforms, etc.). The *Transformer* builds N GDM pairs, such that N is the number of individuals represented in the input files. The number of rows of region files will be smaller than M (the number of total variants represented in the input files), as obviously, not every individual presents every variant. The schema of region files, defining their columns, corresponds to the attributes characterizing variants (around 10 attributes) including the chromosome, position, origin and alternative nucleotide, etc. As GDM schema requires both start and stop coordinates and variants are one base long, we add the stop position, simply as start + 1.

Example 2. The output of metadata transformation for ENCODE is shown in Listing 4.2; it is obtained by considering as input the portion of the JSON file from Listing 4.1 and describes the transformed content of the specific item with accession ENCFF429VMY (with two replicates) of experiment ENCSR635OSG. First-level elements are translated directly to $\langle key, value \rangle$ pairs (e.g., ⟨accession, ENCSR635OSG⟩); nested elements are flattened (e.g., "`name`" inside "`lab`", inside "`pi`", inside "`award`" becomes `award__pi__lab__name`, where double underscore `__` is used to separate levels of nesting); arrays are translated in one ⟨*key*, *value*⟩ pair for each value in the array (e.g., see `file__biological_replicates`); empty arrays are not translated (e.g., `"dbxrefs"`).

Note that several replicates can be associated with each file; in such a case, a progressive naming scheme tracks the replicate to which each ⟨*key*, *value*⟩ pair relates. In the specific example, the file has two biological replicates, each with five associated key-value pairs (in Listing 4.2 other pairs are omitted for brevity). All elements in the replicate element with id `4874c170-7124-4822-a058-4bb` are transformed into keys that start with `"replicate__1__"`. Vice versa, elements in replicate `d42ff80d-67fd-45ee-9159-25a` are transformed into keys that start with `"replicate__2__"`.

The corresponding region data file, ENCFF429VMY.bed is instead copied as is, already being GDM compliant.

```
accession ENCSR635OSG
assembly hg19
award__pi__lab__name michael-snyder
file__accession ENCFF429VMY
file__biological_replicates 1
file__biological_replicates 2
file__file_type bed narrowPeak
replicates__1__@id /replicates/4874c170-7124-4822-a058-4bb/
replicates__1__biological_replicate_number 1
replicates__1__library__biosample__donor__age 6
replicates__1__library__biosample__health_status healthy
replicates__1__antibody__lot_id 940739
replicates__2__@id /replicates/d42ff80d-67fd-45ee-9159-25a/
replicates__2__biological_replicate_number 2
replicates__2__library__biosample__donor__age 32
replicates__2__library__biosample__health_status healthy with non-
    obstructive coronary artery disease
replicates__2__antibody__lot_id 940739
```

Listing 4.2 Excerpt from example transformed file corresponding to *ENCODE* file accession ENCFF429VMY.

4.5 Data Cleaning

After the transformation step, we dedicate our attention only to metadata, while data files are kept in their form as obtained in T_i. A typical metadata key is a long string, e.g., `replicates__1__library__biosample__donor__age`. As this information applies to a single data file, a simpler attribute name can be derived, e.g., `donor__1__age`. Such a name is later used to map values in the conceptual schema and is a much simpler key.

The *Cleaner* module applies transformation rules to complex attribute names, so as to simplify them. For illustration purposes, rules are indicated with the notation *antecedent* ⇒ *consequent*. The antecedent of rules uses the formalism of *regular expressions*, matching a set of keys: it recognizes the strings that compose a complex attribute. The consequent, which is an action encoded in the form of *pattern matching replacement strategy*, builds a simpler output string. The use of regular expressions brings a simple formalization of cleaning algorithms through language containment and language-recognizing automata. Rules are source-specific, as they depend on the specific way in which attribute names are encoded at each source; after an initial design, they are applied to each transformed file. Rules may require adjustments when the attribute names change or new attributes are created. We provide a tool for rule design and ordering, which assists designers in rule creation and maintenance.[10]

The rule's antecedent contains parentheses, which group parts of regular expressions in order to either apply a quantifier or to restrict alternation to the entire group, and positionally identify the rule's parameters, used in the rule's consequent as numbered capturing groups.[11] Some parameters are typed, e.g., "`[0-9]`" denotes a sequence of digits; some keys may be equivalently used, e.g., "`(age|sex)`" denotes an alternative. The consequent can contain strings of characters or special dollar ($) symbols, which positionally refer to the content of the antecedent's variables. The consequent can be empty, in which case no cleaned key is generated for the transformed key, and the corresponding pair is removed.

Let us now consider an example cleaning rule, to understand its mechanism: `replicates(__[0-9]__)library__biosample__(donor)__(age|sex)(.*)` ⇒ `$2$1$3$4`. When `replicates__1__library__biosample__donor__age` is considered as the input key, `$2$1$3$4` stands for a concatenation of the content of the second variable "`donor`", with the first one "`__1__`", with the third one "`age`" and finally with the fourth one (i.e., anything that follows the third parenthesis) - in this case, an empty string. As a result, the rule produces the string `donor__1__age`.

Formalization. For a given source i, a *Cleaner* is a source-specific method $C_i = \langle \mathrm{T}_i, \mathrm{C}_i, \mathcal{RB}_i \rangle$. For each $T_{i,j} \in \mathrm{T}_i$, for every transformed *sample-pair* $< t_{data}, t_{meta} >$, it builds a cleaned *sample-pair* $< c_{data}, c_{meta} >$ in the dataset $C_{i,j} \in \mathrm{C}_i$, such that:

- c_{data} is a symbolyc link to t_{data}, with the same exact content;
- c_{meta} is a metadata file that contains a cleaned key-value pair $\langle k', v \rangle$ for each corresponding transformed key-value pair $\langle k, v \rangle$ contained in t_{meta}. If k' is empty, a related pair is not produced.

Files in the datasets $C_{i,j} \in \mathrm{C}_i$ are produced by running the rule engine C_i over T_i using the set of rules $\mathcal{RB}_i$.

Method. The description of the method requires the definition of relationships between rules and of rule bases.

Definition 4.1 (Rule Equivalence, Containment, and Partial Overlap) Given two rules $r, r' \in \mathcal{RB}_i$, their antecedents $r.a$ and $r'.a$, and the corresponding generated languages $\mathcal{L}(r.a)$ and $\mathcal{L}(r'.a)$:

[10] `https://github.com/DEIB-GECO/Metadata-Manager/wiki/Rule-Base-Generator`

[11] The replacement strategy specified by a rule is implemented using the `java.util.regex` library (`https://docs.oracle.com/javase/8/docs/api/java/util/regex/package-summary.html`), supporting full regular expressions.

- r is *equivalent* to r' when $\mathcal{L}(r.a) = \mathcal{L}(r'.a)$;
- r is *contained* in r' when $\mathcal{L}(r.a) \subset \mathcal{L}(r'.a)$;
- r *partially overlaps* r' when $\mathcal{L}(r.a) \not\subset \mathcal{L}(r'.a)$, $\mathcal{L}(r'.a) \not\subset \mathcal{L}(r.a)$, and $\mathcal{L}(r.a) \cap \mathcal{L}(r'.a) \neq \emptyset$

Definition 4.2 (Rule Base) The $\mathcal{RB}$ Rule Base is a list of rules such that rule r precedes rule r' in $\mathcal{RB}$ if either 1) r is contained in r', or 2) r partially overlaps r' and the user gives priority to r over r'.

By the effect of the above definitions, rules that are more specific precede more general rules. When the intersection of languages recognized by the rules is non-empty, the user can specify the desired order in which the rules should appear in the *RB*. When the intersection is empty, the rules' order in the *RB* corresponds to the order of insertion.

Algorithm 1 Rule Base Creation

```
 1: function RBCreation(RB, SK, AK)
 2:     UK ← AK − SK
 3:     while UK is not empty do
 4:         newRule ← getRuleFromUser()
 5:         if userApprSimul(RB, newRule) then
 6:             RuleInsertion(RB, newRule)
 7:             matched ← matchAll(RB, UK)
 8:             SK ← SK + matched
 9:             UK ← UK − matched
10:         end if
11:     end while
12: end function
```

Building a *Cleaner* requires building the Rule Base (Algorithm 1), by calling the function to insert a rule in the right order (Algorithm 2), which is based on the comparison between pairs of rules performed by the function Compare (Algorithm 3). When the Rule Base is prepared, it is applied to the transformed files, in particular to the keys from the $\langle key, value\rangle$ pairs in every dataset $T_{i,j}$ of $\mathrm{T}_{i,j}$ (Algorithm 4). After the consolidation of cleaning rules, a rule base can be repeatedly applied to transformed data, until major changes occur at the sources.

Algorithm 1 takes as input *RB*, which stores the information about rules in their order (Def. 4.1), *SK*, the set of seen keys, and *AK*, the set of all keys retrieved from the files of a given source. It first finds the unseen keys *UK* (those that have not been considered for rule creation yet). Then, until all unseen keys have been considered, the user is asked to insert new rules and approve (or not) the simulated effect of the incremented *RB* on all keys. When the user is satisfied with the results, the rule is actually added to the *RB*, and the sets of keys are updated accordingly.

Adding a new rule to the Rule Base means inserting it in the right position with respect to the order defined in Def. 4.2. This is accomplished by Algorithm 2, which iterates over the *RB* list and, based on the comparison between each pre-existing rule

Algorithm 2 Rule addition in Rule Base

```
1: function RULEINSERTION(RB, newRule)
2:     for r in RB do
3:         res ← COMPARE(newRule, r)
4:         if res is EQUIVALENT then
5:             if userPref(newRule, r) = newRule then
6:                 replaceRule(newRule, RB, indexOf(r))
7:             end if
8:             return RB
9:         else if res is CONTAINED then
10:            addRule(newRule, RB, indexOf(r))
11:            return RB
12:        else if res is PARTIALLY_OVERLAPS then
13:            if userPriority(newRule, r) = newRule then
14:                addRule(newRule, RB, indexOf(r))
15:                return RB
16:            end if
17:        end if
18:    end for
19:    addRule(newRule, RB, RB.size)
20:    return RB
21: end function
```

Algorithm 3 Order comparison between rules

1: **function** COMPARE(r, r')
2: $\mathcal{A}_r \leftarrow NFA2DFA(RegEx2NFA(r.a))$
3: $\mathcal{A}'_r \leftarrow NFA2DFA(RegEx2NFA(r'.a))$
4: **if** $\mathcal{L}(\mathcal{A}_r) = \mathcal{L}(\mathcal{A}'_r)$ **then**
5: **return** EQUIVALENT
6: **else if** $\mathcal{L}(\mathcal{A}_r) \subset \mathcal{L}(\mathcal{A}'_r)$ **then**
7: **return** CONTAINED
8: **else if** $\mathcal{L}(\mathcal{A}_r) \not\supset \mathcal{L}(\mathcal{A}'_r) \wedge \mathcal{L}(\mathcal{A}_r \cap \mathcal{A}'_r) \neq \emptyset$ **then**
9: **return** PARTIALLY_OVERLAPS
10: **end if**
11: **end function**

with the one to be added, determines the insertion position (in an "insertion sort" manner).

Comparing rules means evaluating the containment relationship between the languages generated by their antecedents, as described by Algorithm 3. Several procedures exist to convert regular expressions into equivalent Non-deterministic Finite Automata (NFA) [382, 278]; we use the Brics Java library [286] for automata implementations, which is based on Thompson's construction algorithm [382]. Then, NFA need to be converted into equivalent Deterministic Finite State Automata (DFA) $\mathcal{A}_r$ and $\mathcal{A}_{r'}$ – this can be done with the Rabin-Scott powerset construction [323]. Later, the two languages are checked for equivalence, containment, and partial overlapping (by using the automaton constructed from the cross-product of states

Algorithm 4 Application of Rule Base to keys

1: **function** CLEANER(RB,T_i)
2: $C_i \leftarrow []$
3: **for each** $T_{i,j} \in T_i$ **do**
4: $C_{i,j} \leftarrow []$
5: **for each** $\langle t_{data}, t_{meta} \rangle \in T_{i,j}$ **do**
6: $c_{data} \leftarrow createSymbolicLink(t_{data})$
7: $c_{meta} \leftarrow []$
8: **for each** $\langle key, value \rangle \in t_{meta}$ **do**
9: $newKey \leftarrow matchFirst(key, RB)$
10: **if** $nonEmpty(newKey)$ **then**
11: $add(c_{meta}, \langle newKey, value \rangle)$
12: **end if**
13: **end for**
14: $C_{i,j} \leftarrow C_{i,j} + \langle c_{data}, c_{meta} \rangle$
15: **end for**
16: $C_i \leftarrow C_i + C_{i,j}$
17: **end for**
18: **return** C_i
19: **end function**

that accepts the intersection of the languages). Algorithm 4 describes the *Cleaner* as an application of the Rule Base to the input set C_i, iterating over all datasets $C_{i,j} \in C_i$; rules are applied in the order in which they appear in the Rule Base.

Example. Table 4.3 shows the cleaning procedure of a set of transformed ENCODE keys. It assumes an initial set of transformed keys from T_i; for each key, the user produces cleaning rules, driven by Algorithm 1. Eventually, the method produces a rule base made of a list of 7 rules; their application to keys in T_i produces the set of cleaned keys in C_i.

As an example instance of rule base mechanism, we observe that rule (2) deletes the key: `replicates__1__library__biosample__sex`. Instead, rule (3), applied to the key: `replicates__1__library__biosample__biosample_type`, dictates that the key must be rewritten by concatenating the content of the second parenthesis (i.e., `biosample`) with the content of the first (i.e., `1`), and with the content of the fourth (i.e., `type`), obtaining at the end `biosample__1__type`.

4.6 Data Mapping

The *Mapper* module is in charge of the integration at the schema level of a set of cleaned keys produced for each source. The method applies local-to-global mappings using a syntax inspired by Datalog [91]. Mapping rules build relational rows from the key-value pairs output by the *Cleaner* step to achieve the integration of different local schemata into a unique local one, i.e., the Genomic Conceptual Model (see

Transformed keys in T_i

replicates__1__library__biosample__donor__age	32
replicates__1__library__biosample__donor__age_units	year
replicates__1__library__biosample__donor__sex	male
replicates__2__library__biosample__donor__age	4
replicates__2__library__biosample__donor__age_units	year
replicates__2__library__biosample__donor__sex	female
replicates__1__library__biosample__sex	male
replicates__1__library__biosample__biosample_type	tissue
replicates__1__library__biosample__health_status	healthy, CAD
file__biological_replicates	1
file__technical_replicates	1_1
file__assembly	GRCh38
file__file_type	bed narrowPeak
replicates__1__biological_replicate_number	1
replicates__1__technical_replicate_number	1
replicates__2__biological_replicate_number	2
replicates__2__technical_replicate_number	1
assembly	hg19

↓

RuleBase RB_i

(1) replicates(__[0-9]__)library__biosample__(donor)__(age|sex)(.*) ⇒ $2$1$3$4
(2) replicates__[0-9]__library__biosample__sex.* ⇒
(3) replicates(__[0-9]__)library__(biosample)__(biosample_)?(.*) ⇒ $2$1$4
(4) file__(biological|technical)_replicates ⇒
(5) (file__)(file_)?(.*) ⇒ $1$3
(6) (replicate)s(__[0-9]__)(.*) ⇒ $1$2$3
(7) assembly ⇒

↓

Cleaned keys in C_i

donor__1__age	32
donor__1__age_units	year
donor__1__sex	male
donor__2__age	4
donor__2__age_units	year
donor__2__sex	female
biosample__1__type	tissue
biosample__1__health_status	healthy, CAD
file__assembly	GRCh38
file__type	bed narrowPeak
replicate__1__biological_replicate_number	1
replicate__1__technical_replicate_number	1
replicate__2__biological_replicate_number	2
replicate__2__technical_replicate_number	1

Table 4.3 Example of data cleaning process.

Section 3.2). Arbitrary queries on GCM can then be propagated to sources, using the *global-as-view* approach [245].

Note that rules capture only a portion of the data integration semantics, as we allow for exceptions, e.g., attributes that are not in common to most sources (e.g. all clinical diagnosis conditions available for the donor in TCGA), while specific for few experiment types, are modeled as attribute-value pairs; the corresponding data is directly referenced from the Item entity.

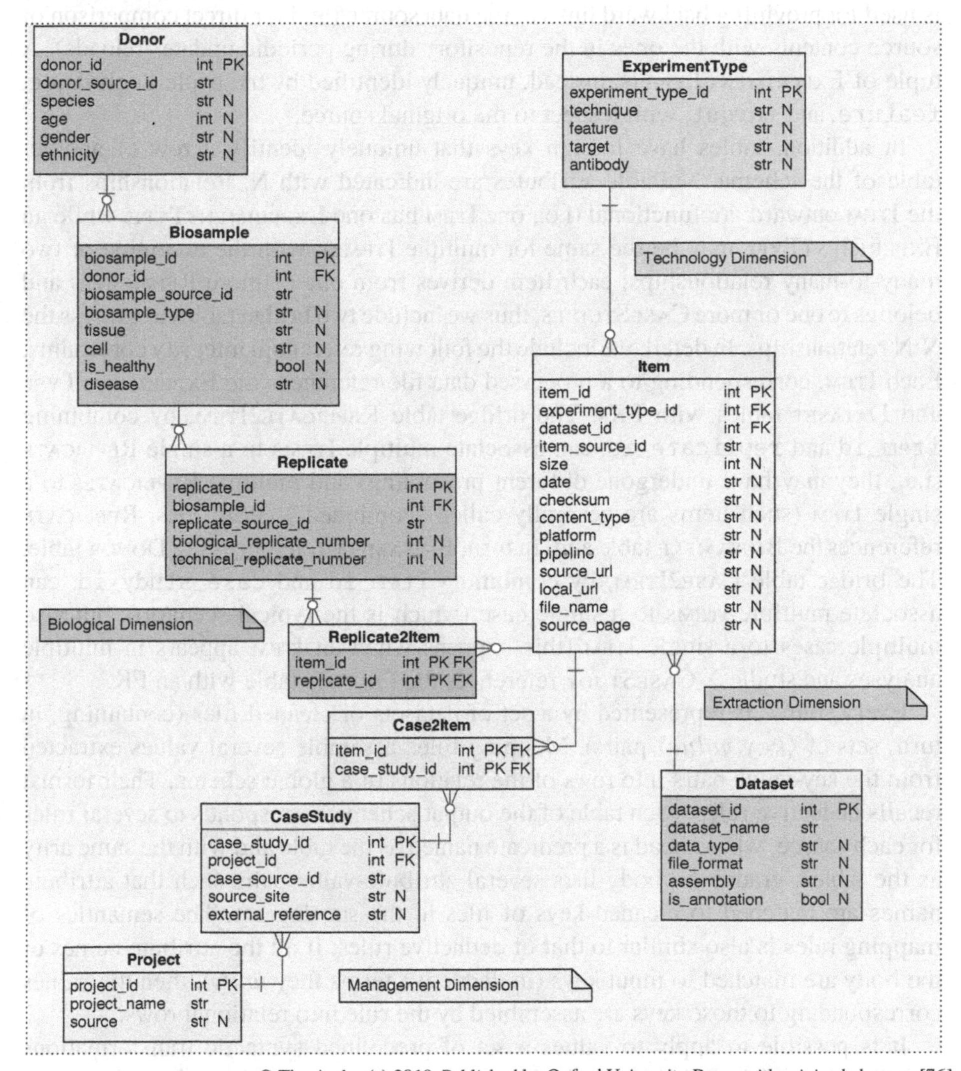

Fig. 4.4 Logical schema of the GCM relational database instance.

The global schema $\mathcal{G}$ shown in Figure 4.4, is obtained as straightforward mapping from the conceptual schema (Figure 3.3 in Section 3.2) and it resembles a classic data mart [67]. It contains the central entity table Item, a set of entity tables Donor, BioSample, Replicate, Project, Case, Dataset, ExperimentType, Item2Replicate, and Item2Case.

All tables have a numerical sequential primary key, conventionally named `<table_name>_id` and indicated as PK in Figure 4.4. Tables Donor, Biosample, Replicate, Item, and CaseStudy have, in addition, a secondary unique key `<table_name>_source_id` that refers to the original source; such secondary key is used for providing backward links to the data source (and for direct comparison of source contents with the ones in the repository during periodic updates/reloads). A tuple of ExperimentType is, instead, uniquely identified by the triple `technique`, `feature`, and `target`, with respect to the original source.

In addition, tables have foreign keys that uniquely identify a row of another table of the schema. Nullable attributes are indicated with N. Relationships from the Item outward are functional (i.e., one Item has one ExperimentType, while an ExperimentType may be the same for multiple Items), with the exception of two many-to-many relationships: each Item derives from one or more Replicates and belongs to one or more CaseStudies, thus we include two bridge tables to express the N:N relationships. In detail, we include the following referential integrity constraints. Each Item, corresponding to a processed data file references the ExperimentType and Dataset tables with FKs. The bridge table Replicate2Item, by combining `item_id` and `replicate_id`, can associate multiple Items to a single Replicate (i.e., they may have undergone different processing) and multiple Replicates to a single Item (such items are generally called "combined"). With FKs, Replicate references the Biosample table and, in turn, Biosample references the Donor table. The bridge table Case2Item, by combining `item_id` and `case_study_id`, can associate multiple Items to a single case (which is the typical scenario), but also multiple cases to a single Item (this happens when an Item appears in multiple analyses and studies). CaseStudy references the Project table with an FK.

Every source is represented by a set of datasets of cleaned files (containing, in turn, sets of $\langle key, value \rangle$ pairs). Mapping rules assemble several values extracted from the key-value pairs into rows of the relations of a global schema. Their format recalls deductive rules: each table of the output schema corresponds to several rules for each source, whose head is a predicate named as the table and with the same arity as the table's grade; the body lists several attribute-value pairs such that attribute names are matched to cleaned keys of files in the starting set. The semantics of mapping rules is also similar to that of deductive rules: if all the attribute names of the body are matched to input keys (in deductive terms they *unify*), then the values corresponding to those keys are assembled by the rule into relational rows.

It is possible to apply to values a set of predefined syntactic transformations (*SynTr*), defined in Table 4.4, which can be freely composed on the left side of mapping rules; transformations can be easily extended. For example, to put into lowercase letters two values that have been first concatenated with a space, the expression `LCase(Conc(`$value_1, value_2,$ " "`))` can be used to generate a value for a specific position of a row.

$\texttt{Conc}(s_1, s_2, c)$: concatenates s_1 and s_2 using c as separation string
$\texttt{Alt}(s_1, s_2)$: outputs s_1 if present and not null, else s_2
$\texttt{Rem}(s_1, s_2)$: removes the occurrences of string s_2 from s_1
$\texttt{Sub}(s_1, s_2, s_3)$: substitutes occurrences of s_2 in s_1 with the new s_3
$\texttt{Eq}(s, p)$: outputs `true` when s is equal to p, else `false`
$\texttt{ATD}(a)$: converts a, a number followed by space and unit of measurement, into the correspondent number of days
$\texttt{LCase}(s)$: converts string s into its lower case version
$\texttt{Int}(n)$: casts number n to its correspondent Integer format
`Id(...)`: generates synthetic *id* for faster indexing of table t from specified arguments

Table 4.4 Example syntactic transformations for mapping rules.

Formalization. A *Mapper* is a source-specific method $\mathcal{M}_i = \langle C_i, \mathcal{G}, \mathcal{MB}_i \rangle$. For every metadata cleaned file c_{meta} in each dataset $C_{i,j} \in C_i$, it assembles several values v present in the pairs $\langle k, v \rangle$ of c_{meta} into rows of the tables of the global schema $\mathcal{G}$. Rows are produced by running the rule engine $\mathcal{M}_i$ over datasets of C_i one by one, using the mapping rules contained in the mapping base $\mathcal{MB}_i$. A *mapping rule* is a declarative rule of the form: $\textsc{Entity}(SynTr(v_1),..., SynTr(v_i),..., SynTr(v_N)) \rightsquigarrow \{\langle k_1, v_1\rangle,..., \langle k_i, v_i\rangle,..., \langle k_N, v_N\rangle\} \subseteq c_{meta}$, where every v_i in the left-hand side of the rule also appears in the rule right-hand side (i.e., rule evaluations are finite), in a positive form (i.e., rules are safe [91]).

Method. Once mapping rules are fully specified, the method consists simply in applying the rules to each file c_{meta} in every dataset $C_{i,j}$ of the set C_i of a data source i, in arbitrary order. Note that every file associated with a data source as produced by the cleaning method may have several versions for the same key, numbered from 1 to n_c (as n is specific for the single file c), e.g., `biosample__1__type` and `biosample__2__type`. Each rule is applied for every version, and associates with each version a distinct row. When a version is present (e.g., in rules for Donor, BioSample and Replicate of Table 4.5), we denote such version by generically naming the keys in the rule's body using k, and then generating a rule for each value of k. For each v_i in the rule left-hand side, if the corresponding $\langle k_i, v_i \rangle$ in the rule right-hand side exists in c_{meta}, then we add $SynTr(v_i)$ to the result, i.e., a new tuple in the Entity table specified in the rule left-hand side.

Example. Table 4.5 illustrates all the rules that are required to build the biological dimension of the relational schema shown in Figure 4.4 for the ENCODE data source.

In Table 4.6 we show two additional rules for the GDC and Roadmap Epigenomics data sources in order to build the Donor table for these two sources. Note that $\texttt{Oid}_t$ is the notation used for the *ObjectIdentifier* of table t, which is a unique accession retrieved from the source, while $\texttt{Id}(\texttt{Oid}_t)$ indicates a numerical synthetic id for faster indexing of table t; PKs and FKs are built on these ids.

$$C_{ENC} = \{C_i \mid i = ENCODE\}, \forall j, c_{meta} \mid \langle c_{data}, c_{meta} \rangle \in C_{ENC,j} \in C_{ENC} \rightarrow k \leq n_c$$

$$\textsc{Donor}(\mathrm{Id}(\mathtt{Oid}_D), \mathtt{Oid}_D, v_1, \mathrm{ATD}(\mathrm{Conc}(v_2, v_3, \text{`` ''})), v_4, v_5) \rightsquigarrow \\ \{\langle \textit{donor\_\_k\_\_accession}, \mathtt{Oid}_D \rangle, \langle \textit{donor\_\_k\_\_organism}, v_1 \rangle, \langle \textit{donor\_\_k\_\_age}, v_2 \rangle, \langle \textit{donor\_\_k\_\_age\_units}, v_3 \rangle, \\ \langle \textit{donor\_\_k\_\_sex}, v_4 \rangle, \langle \textit{donor\_\_k\_\_ethnicity}, v_5 \rangle\} \subseteq c_{meta}$$

$$\textsc{BioSample}(\mathrm{Id}(\mathtt{Oid}_B), \mathrm{Id}(\mathtt{Oid}_D), \mathtt{Oid}_B, \text{``tissue''}, v_2, \textsc{null}, \mathrm{Eq}(v_3, \text{``healthy''}), v_3) \rightsquigarrow \\ \{\langle \textit{biosample\_\_k\_\_accession}, \mathtt{Oid}_B \rangle, \langle \textit{donor\_\_k\_\_accession}, \mathtt{Oid}_D \rangle, \langle \textit{biosample\_\_k\_\_type}, \text{``tissue''} \rangle, \\ \langle \textit{biosample\_\_k\_\_term\_name}, v_2 \rangle \langle \textit{biosample\_\_k\_\_health\_status}, v_3 \rangle\} \subseteq c_{meta}$$

$$\textsc{BioSample}(\mathrm{Id}(\mathtt{Oid}_B), \mathrm{Id}(\mathtt{Oid}_D), \mathtt{Oid}_B, \text{``cell line''}, \textsc{null}, v_2, \mathrm{Eq}(v_3, \text{``healthy''}), v_3) \rightsquigarrow \\ \{\langle \textit{biosample\_\_k\_\_accession}, \mathtt{Oid}_B \rangle, \{\langle \textit{donor\_\_k\_\_accession}, \mathtt{Oid}_D \rangle, \langle \textit{biosample\_\_k\_\_type}, \text{``cell''} \rangle, \\ \langle \textit{biosample\_\_k\_\_term\_name}, v_2 \rangle, \langle \textit{biosample\_\_k\_\_health\_status}, v_3 \rangle\} \subseteq c_{meta}$$

$$\textsc{Replicate}(\mathrm{Id}(\mathtt{Oid}_R), \mathtt{Oid}_R, \mathtt{Oid}_B, v_1, v_2) \rightsquigarrow \\ \{\langle \textit{replicate\_\_k\_\_uuid}, \mathtt{Oid}_R \rangle, \langle \textit{biosample\_\_k\_\_accession}, \mathtt{Oid}_B \rangle, \langle \textit{replicate\_\_k\_\_bio\_rep\_num}, v_1 \rangle, \\ \langle \textit{replicate\_\_k\_\_tech\_rep\_num}, v_2 \rangle\} \subseteq c_{meta}$$

$$\textsc{Item}(\mathrm{Id}(\mathtt{Oid}_I), \mathrm{Id}(v_1, v_2, v_3), \mathrm{Id}(\mathtt{Oid}_{DS}), \mathtt{Oid}_I, v_4, v_5, v_6, v_7, v_8, \mathrm{Conc}(\text{``www.encodeproject.org''}, v_9, \text{``/''}), \\ \mathrm{Conc}(\text{``www.gmql.eu...''}, \mathtt{Oid}_I, \text{``/''})) \rightsquigarrow \\ \{\langle \textit{assay\_term\_name}, v_1 \rangle, \langle \textit{target\_\_investigated\_as}, v_2 \rangle, \langle \textit{target\_\_label}, v_3 \rangle, \langle \textit{dataset\_name}, \mathtt{Oid}_{DS} \rangle, \\ \langle \textit{file\_\_accession}, \mathtt{Oid}_I \rangle, \langle \textit{file\_\_size}, v_4 \rangle, \langle \textit{file\_\_date\_created}, v_5 \rangle, \langle \textit{file\_\_md5sum}, v_6 \rangle, \langle \textit{file\_\_pipeline}, v_7 \rangle, \\ \langle \textit{file\_\_platform}, v_8 \rangle, \langle \textit{file\_\_href}, v_9 \rangle\} \subseteq c_{meta}$$

$$\textsc{Item2Replicate}(\mathrm{Id}(\mathtt{Oid}_I), \mathrm{Id}(\mathtt{Oid}_R)) \rightsquigarrow \\ \{\langle \textit{file\_\_accession}, \mathtt{Oid}_I \rangle, \langle \textit{replicate\_\_k\_\_uuid}, \mathtt{Oid}_R \rangle\} \subseteq c_{meta}$$

Table 4.5 Mapping rules for biological view of ENCODE source.

$$C_{GDC} = \{C_i \mid i = GDC\}$$

$$\textsc{Donor}(\mathrm{Id}(\mathtt{Oid}_D), \mathtt{Oid}_D, \text{``Homo sapiens''}, v_1, v_2, \mathrm{LCase}(\mathrm{Conc}(v_3, v_4, \text{`` ''})) \rightsquigarrow \\ \{\langle \textit{bcr\_patient\_uid}, \mathtt{Oid}_D \rangle, \langle \textit{demographic\_\_days\_to\_birth}, v_1 \rangle, \langle \textit{demographic\_\_gender}, v_2 \rangle, \\ \langle \textit{demographic\_\_race}, v_3 \rangle, \langle \textit{demographic\_\_ethnicity}, v_4 \rangle\} \subseteq c_{meta}$$

$$C_{REP} = \{C_i \mid i = RoadmapEpigenomics\}, \forall c_{meta} \forall j \mid \langle c_{data}, c_{meta} \rangle \in C_{REP,j} \in C_{REP} \rightarrow k \leq n_c$$

$$\textsc{Donor}(\mathrm{Id}(\mathtt{Oid}_D), \mathtt{Oid}_D, \text{``Homo sapiens''}, \mathrm{ATD}(v_1), v_2, v_3) \rightsquigarrow \\ \{\langle \textit{donor\_\_k\_\_id}, \mathtt{Oid}_D \rangle, \langle \textit{donor\_\_k\_\_age\_weeks}, v_1 \rangle, \langle \textit{donor\_\_sex}, v_2 \rangle, \langle \textit{donor\_\_k\_\_ethnicity}, v_3 \rangle\} \subseteq c_{meta}$$

Table 4.6 Mapping rules for Donor table from sources GDC and Roadmap Epigenomics

4.7 Data Normalization and Enrichment

During this step, specific values of the global schema are associated with controlled terms, lists of synonyms and hyperonyms, and external references to reference ontologies. We consider ten *semantically enrichable* attributes of the global schema: *Technique, Feature* and *Target* of experiment types, *Disease, Tissue* and *Cell* of biosamples, *Ethnicity* and *Species* of donors, *ContentType* and *Platform* of items.

The adoption of a specific knowledge base for each semantically enrichable attribute provides us with *value normalization*, as we transform the values of reference knowledge bases into restricted vocabularies. Using external knowledge bases (rather than creating a new one) is essential in the biomedical domain, where specialized ontologies are already available and well-recognized and their use boosts interoperability.

This process is supervised and requires a preliminary selection of the most suitable ontologies to describe each semantically enrichable attribute of the global schema: Section 4.7.1 presents our solution to the problem of selecting appropriate search services and ontologies to annotate metadata. Section 4.7.2 describes how the enrichment procedure works.

4.7.1 Search Service and Ontology Selection

First, we present the four most used and well-known ontology search services in literature, and how we score them in order to select the most appropriate one for our purpose. Next, we compare the ontologies provided by that search service and select the specific ontology that is most suitable to annotate values for each ontological attribute.

4.7.1.1 Ontology Search Services

Ontological access to genomic data is well supported by several search services, which are capable in turn to integrate a high number of ontologies. Therefore, we are initially concerned with choosing the best search service that will be used within our system as a broker to the underlying ontologies. We consider four different search services.

BioPortal [399] is a repository of biomedical ontologies and terminologies whose access is provided through a Web portal and Web services. We exploit its *term search* service, an endpoint that takes a free text input and provides a result in JSON format, listing a (configurable) number of annotations to ontological terms, showing different degrees of matching with the free text. These can be considered as possible annotations for the input text. Each term is identified by the pair $\langle ontology, id \rangle$, describing the code which references the ontology inside the BioPortal system and an identification number that references the term inside the ontology. A term also contains a single preferred label and its synonyms. An annotation is composed of a term and a match type: "PREF" if the match with the term is established with the preferred label or "SYN" if the match is with one of the term synonyms.

Ontology Recommender [270] is a BioPortal service that receives a free text or a list of keywords and suggests a set of ontologies appropriate for annotating the indicated terms, considered all together. The structure of annotations is identical to BioPortal's. Additionally, Recommender provides four scores that reflect how well the ontology (set) annotates the input data:

- *Coverage*: measures to which extent the ontology represents the input data by matching (either exactly or fuzzily, e.g., with containment or stemming) the query terms to the class and property names of the ontology. In literature, coverage is also referred to as term matching [213], class match measure [8], or topic coverage [342].
- *Acceptance*: indicates how well-known and trusted the ontology is by the community of interest, i.e., the biomedical one. Criteria related to acceptance have to do with popularity (how much the community supports the ontol-

ogy) [273, 342], which relies on the assumption that relevant ontologies are referenced by many other ontologies (thus, they have a high connectivity [213] or connectedness [71]—when also the quality of connections is assessed).
- *Detail*: shows the level of specification provided by the ontology for the input data. Definitions of detail refer to semantic richness [273], structure [71] (measuring the number of properties over the number of classes of the ontology, under the assumption that more advanced ontologies generally have a large number of properties), and granularity [266] (where the perspective is the one of Keet [221], based on categories of non-scale dependency, providing levels ordered through primitive relations like *is_a* and *part_of*).
- *Specialization*: indicates how specialized the ontology is with respect to the domain of input data. This measure is impacted by the fact that some biomedical ontologies aim to focus on specific sub-domains or particular tasks; in this case, they may be particularly specialized on a specific input, if this can be fit into the same sub-domain.

Ontology Lookup Service (OLS) [215]) provides ontology search, visualization, and ontology-based services. The accepted input is a keyword, and the provided result is a list of annotations, similar to the other services but not including a match type. In the API request, a *fieldList* parameter can be used to specify the specific elements to be included in the output along with other formatting preferences.

Zooma[12] is a service from OLS which provides mappings between textual input and a manually curated repository of text-to-ontology-term mappings. If no mappings are found, it uses the basic OLS search. In addition to the usual annotation information, Zooma also returns a confidence label associated with the annotation, ranging from HIGH to LOW.

We exclude other ontology search portals such as HeTOP [175] and UMLS [63], as they are more focused on multilingual support and medical terminologies, therefore do not include many ontologies that are important to annotate our values. Also the NCBO Annotator [214] is not considered since its functionalities are completely covered by the Ontology Recommender.

4.7.1.2 Scoring

Every search service provides a search API, which is repeatedly used for the score evaluation. For each API call we store: the used service; the attribute from GCM characterizing the values (the "type" of the values); the original `raw` value deriving from the GCM, imported through the mapping phase; possible `parsed` values deriving from a simple syntactic pre-processing of `raw` values (e.g., removal of punctuation, split of long expressions...); the ⟨`ontology,ontology_id`⟩ pair, uniquely identifying an ontological term within a service; `pref_label` and `synonym`, respectively the primary textual expression used for the term and its alternative version; `score`,

[12] `https://www.ebi.ac.uk/spot/zooma/`

textual information regarding the goodness of a match, directly retrieved from the services, if available.

In total, we performed 1,783 API calls to each of the four services, corresponding to 1,299 original values to be enriched; some of these were split during a pre-processing phase. As a result, we retrieved 1,783 interesting matches from BioPortal, 885 from Recommender, 1,782 from OLS, and 1,779 from ZOOMA, all of which were used for the following processing after calculating our scores.[13]

Starting from the retrieved information, we calculate the *match_score* as a measure of how well a term matches a value, by using a scoring system that is specifically designed for the task: the general formula returning the *match_score* value, shown in Equation 4.1, subtracts from an initial maximum number (10, when there is a perfect match with a `pref_label`, 9 with a `synonym`) a penalty measuring how the raw value differs from the label retrieved from the services:

$$match\_score(raw, label) = \{10, 9\} - distance(raw, label) \tag{4.1}$$

To compute the distance, we originally use a modified version of Needleman-Wunsch algorithm [298], a protein and nucleotide sequence alignment algorithm that is widely used in bioinformatics. In the original algorithm, the input is represented by two strings whose letters need to be aligned. The letters may have a "match", a "mismatch" or an "indel" (i.e., adding a gap in one of the strings). In our modified version, we define each word as a distinct letter of the original algorithm, and we a specific type of *mismatch*, i.e., the *swap*. All in all, the total distance is calculated as a sum of distances between words:

- **Match:** Two words are the same, then their distance is 0
- **Swap:** Two consecutive words traded places, then their distance is 0.5
- **Insert:** A new word is added to the *raw*, then their distance distance is 1
- **Delete:** One word is deleted from the *raw*, then their distance is 2
- **Mismatch:** Two words are different, then their distance is 2.5

The indicated distance values are chosen in such a way that the number of deletions is minimized (i.e., we penalize a *label* which does not include a word present in *raw*) and the swap is preferred to indel and mismatch. For example, for the *raw* "breast invasive carcinoma", the *label* "invasive breast carcinoma" (i.e., one *swap*) is considered better than "breast carcinoma" (i.e., one *deletion*).

Then, we assign different scores to each used ontology: i) the *onto_acceptance*, a measure of how well-known and trusted the ontology is by the biomedical community; ii) the *onto_suitability*, a measure of how much an ontology is adequate for a given attribute (thus one ontology obtains multiple scores, one for each attribute).

Acceptance Score. The acceptance score is a metric that reflects the appreciation of the ontology give by the biomedical community. We retrieve it through Recommender Web Services [270], where it is derived from the number of visits to the ontology page in BioPortal and the presence or absence of the ontology in UMLS [63].

[13] Count corresponds to the results retrieved at the time of experiments, i.e., October 2018.

Suitability Score. The suitability score represents how well the ontology annotates the terms of a given attribute. Two measures are considered: 1. how many terms the ontology can annotate with respect to the total number of terms in input (i.e., coverage); 2. how many terms the ontology can annotate with a better match score: the score will be higher if the ontology annotates more terms with `pref_labels` rather than with `synonyms`.

For a given ontology *o* and a given attribute *a*, the coverage is calculated by dividing the number of obtained annotations by the number of input *raw* values belonging to the attribute *a*:

$$coverage(o, a) = \frac{\#annotations(o, a)}{\#inputRawValues(a)} \tag{4.2}$$

To calculate suitability, for each attribute *a* and ontology *o*, we sum the match scores associated with all the annotations obtained from the ontology *o*, normalize by the number of total annotations, and multiply the obtained value by the coverage obtained from Equation 4.2:

$$onto\_suitability(o, a) = \frac{\sum_{n \in annotations(o,a)} match\_score(n)}{\#annotations(o, a)} * coverage(o, a) \tag{4.3}$$

Overall Score Computation. For each annotation, i.e., the mapping between a *raw* value (belonging to a specific attribute *a*) and a *label* performed by one ontology *o*, we compute a score that summarizes all the others available. The *overall_score* is obtained by multiplying each *raw* value's *match_score* by a weighted average of the *onto_suitability* and *onto_acceptance*, as shown in Equation 4.4:

$$overall\_score(raw, label, o, a) = \frac{match\_score(raw, label, o)}{5} * \left(\frac{2 * onto\_suitability(o, a)}{5} + onto\_accept.(o)\right) \tag{4.4}$$

Note that coefficients have been chosen as they guaranteed a reasonable distribution of scores over our dataset. Based on the *overall_scores*, for each pair of attribute and *parsed* value (derived by syntactic pre-processing of *raw* values), we informed the service evaluation phase, described next.

4.7.1.3 Service Evaluation

Table 4.7 describes the obtained results by applying Equation 4.4 over the annotations on all raw values performed by the ontologies in the four services. The "Service Properties" part contains an overview of service properties. BioPortal and Recommender provide a *match_type* (MT) in their APIs response, which means that they specify if the input text is more similar to the preferred label rather than to one of the synonyms associated with a term. Recommender offers the additional function of searching for multiple keywords at the same time (MK) and consequently suggests a minimal set of ontologies suitable for annotating the maximum possible number of keywords. This function is also offered by ZOOMA which, however, in practice just performs multiple single keyword requests and lists all results at the same time. Only Recommender executes a good attempt at annotating free texts (FT). BioPortal's set of ontologies is much broader than OLS' since minor efforts are also included. ZOOMA exploits search results from OLS but also provides results coming from previous manual curation works as an additional service to the user.

		BioPortal	Recommender	OLS	ZOOMA
Service Properties	Search properties	MT	MT,MK,FT	-	MK
	Num. of ontologies	728	728	214	214
	Previous curation	no	no	no	yes
Example scoring of "cervical adenocarcinoma"	1st best match	ncit_c4029	ncit_c4029	ncit_c4029	efo_0001416
	2nd best match	efo_0001416	None	efo_0001416	None
	3rd best match	doid_3702	None	ncit_c136651	None
	Occurrence score	1	0.5	1	0.5
	Coverage score	1	1	1	1
Aggregated scores	Occurrence	83.17%	46.97%	**90.54%**	75.96%
	Coverage	100.00%	49.88%	**99.94%**	99.78%

Table 4.7 Summary of Ontology Search Services.

The "Example scoring" part contains an example of how services are rewarded based on the matching terms they find. To evaluate the match, we use the *overall_score* described above. When the disease-related text "cervical adenocarcinoma" is searched, BioPortal suggests, on top of others, the three terms "ncit_c4029", "efo_0001416", and "doid_3702", while Recommender just provides one result, "ncit_c4029". Our algorithm for *Occurrence* computes the set of terms that occur the highest amount of times in the top three matches of the services (in this case ["ncit_c4029","efo_0001416"]) and assigns a weighted reward (1 if the set only contains one entry, 0.5 if it contains 2, and so on) to the services which include that term in the top results. Indeed BioPortal scores 1 since it contains both top results, while Recommender scores 0.5 since it contains just one. *Coverage* is 1 when the service provides at least one result, 0 otherwise.

As scores for service selection, we use the average *Occurrence* and *Coverage* over all the searched raw values. On this basis, OLS is selected as the best-suited search service to pursue the enrichment annotations in our system (see results highlighted with bold font in Table 4.7).

4.7.1.4 Ontology Selection

Based on the *overall_score* described above, we aggregate results over specific attributes and ontologies. This calculation produces, as a result, one top ontology for each attribute. Since most of the time only one ontology does not provide an acceptable coverage for all the values belonging to that attribute, we use an algorithm to compute a small set of ontologies to annotate values from an attribute. Such an algorithm first tries to match values only with the first ontology, then tries to match only the ones left unmatched with the following ontologies, until a fixed point for coverage is found. If the computational costs become too high, the algorithm can be stopped at a predefined threshold coverage, considered acceptable. In our case, we set the threshold equal to 95%.

Entity	Attribute	Pref. ontologies	Coverage	Score	Suitability
ExperimentType	*Technique*	OBI, EFO	0.857	0.486	0.490
ExperimentType	*Feature*	NCIT	1.000	0.854	0.893
ExperimentType	*Target*	OGG	0.950	0.747	0.948
BioSample	*Disease*	NCIT	0.978	0.784	0.802
BioSample	*Tissue*	UBERON	0.957	0.753	0.937
BioSample	*Cell*	EFO, CL	0.953	0.644	0.577
Item	*ContentType*	NCIT, SO	0.950	0.510	0.743
Item	*Platform*	NCIT	1.000	0.909	0.950
Donor	*Ethnicity*	NCIT	0.962	0.907	0.912
Donor	*Species*	NCBITaxon	1.000	0.667	1.000

Table 4.8 Choice of reference ontologies for semantically enrichable attributes. Definition of ontology acronyms: Ontology for Biomedical Investigations (OBI) [25], Experimental Factor Ontology (EFO) [268], Uber Anatomy Ontology (UBERON) [291], Cell Ontology (CL) [280], National Cancer Institute Thesaurus (NCIT) [121], NCBI Taxonomy Database (NCBITaxon) [144], Sequence Ontology (SO) [135], Ontology of Genes and Genomes (OGG) [193].

The results of our selection are shown in Table 4.8, where, for each semantically enrichable attribute, we indicate the preferred ontology and three normalized indicators. Coverage indicates the percentage of attribute values that are found in the ontologies. Score is an average matching score of all the annotated attribute values weighted by ontology acceptance. Suitability is a measure of how much an ontology set is adequate for an attribute. Note that a second preferred ontology is added when the first one did not reach 0.85 coverage; in this case, indicators refer to the union of the ontologies In the specific case of *Technique*, unfortunately, we were not able to achieve the target coverage threshold of 0.95 because a third added ontology did not add any matches with respect to the second ontology.

4.7.2 Enrichment Process

The *Enricher* is supported by an interactive tool[14] that: i) calls external services to annotate values with concepts from controlled vocabularies or dedicated ontologies; ii) asks for user feedback when annotations have a low matching score; users can either accept one of the proposed solutions or manually specify new annotations. After selecting such sets, we proceed with the enrichment of the values contained in the ontological attributes of the GCM.

The result of the normalization is contained within the relational database $\mathcal{K}$, called Local Knowledge Base (LKB). Figure 4.5 describes the logical schema of the relational database, whose orange part is populated from ontologies and referenced from the global schema $\mathcal{G}$. The "Knowledge Base" frame stores all the information retrieved from OLS services and is relevant to annotate our values. The main tables are:

- VOCABULARY, whose PK term identifier `tid` is referenced from all the GCM tables that contain semantically enriched attributes, has the acronym of the ontology providing the term (`source`, e.g., NCIT), the `code` used for the term in that ontology (e.g., NCIT_C4872) and its label (`pref_label`, e.g., Breast Carcinoma), in addition to an optional `description` and `iri` (i.e., International Resource Identifier);
- SYNONYM, containing alternative labels that can be used as synonyms of the preferred label along with their type (e.g., alternative syntax, related nomenclature, related adjectives) – referencing the term in the vocabulary table;
- REFERENCE, containing references to equivalent terms from other ontologies (in the form of a `<source, code>` pair) – referencing, with the FK `tid`, the corresponding term in the vocabulary table;
- ONTOLOGY, a dimension table presenting details on the specialized ontologies contained (even partially) in the knowledge base – referenced with an FK from the vocabulary table;
- RELATIONSHIP, containing ontological hierarchies between terms and the type of the relationships (either generalization *is_a* or containment *part_of*[15]) – the primary key is composed of parent, child, and type of the relationship; the first two reference the vocabulary table with FKs.[16]

The "Expert Support" frame includes the tables used to contain information for expert users.

The "Genomic Conceptual Model" frame contains the tables from the GCM (of which we only show in detail the ones which have ontological attributes).

[14] The GitHub repository of the tool is `https://github.com/DEIB-GECO/Metadata-Enricher`

[15] We restricted to *generalization* and *containment* as these are the most common relationships in the involved ontologies – typical in all bio-ontologies. This choice reflects a trade-off between performance and richness of information.

[16] The system provides the unfolding of the hierarchies as an internal materialized view over the table `relationship`, used for faster query processing.

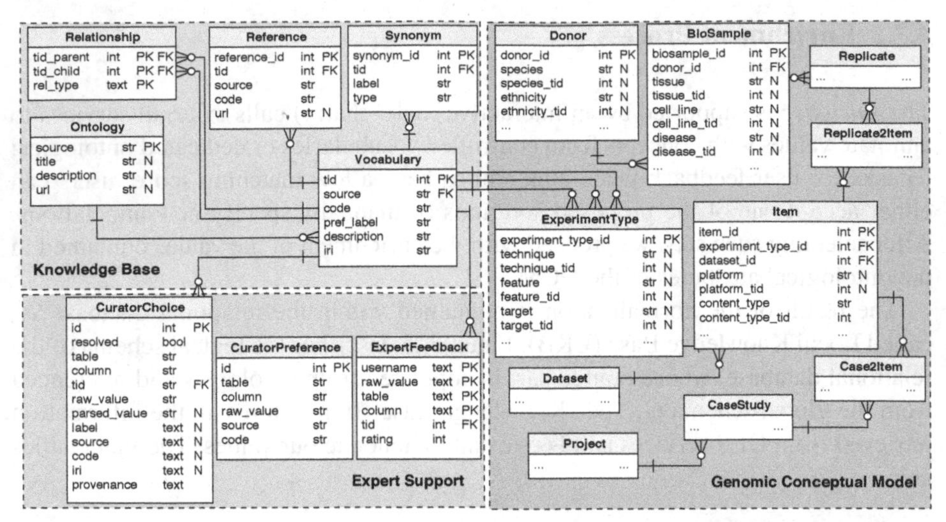

Fig. 4.5 Relational schema for tables of the GCM, LKB, and user feedback routines.

Each GCM ⟨`ontological_attribute`⟩ is equipped with a companion-attribute ⟨`ontological_attribute`⟩`_tid`, which references the ontological term in the vocabulary table (e.g., *Platform* with value "Illumina Human Methylation 450" is associated to *Platform_tid* = 10, representing the vocabulary object OBI_0001870, taken from the OBI ontology). The Vocabulary table is the central entity of the LKB schema. The *tid* column is the primary key which is referenced by all other tables in LKB and from the tables in GCM. Also tables from the LKB and from the Expert tables are linked using *tids*.

Formalization. The *Enricher* is a source-independent method $\mathcal{E} = \langle A, O, \mathcal{G}, \mathcal{K} \rangle$. A is the set of semantically enrichable attributes of the global schema $\mathcal{G}$. For each attribute a in A and each possible value of a, $\mathcal{E}$ generates the corresponding entries in the Local Knowledge Base $\mathcal{K}$, extracted from the preferred ontologies of a in O, i.e., the set of reference ontologies defined in Table 4.8.

Method. Value normalization and enrichment is a supervised procedure illustrated in Figure 4.6. The workflow is executed for all values of semantically enrichable attributes, and consists of two parts: 1) For each such value associated with a *_tid* column, the system initially looks for a suitable term in the vocabulary of the LKB; if a match is available, and the term was already annotated in the past, the procedure is completed. When the match is successful but annotations are lacking, a user's feedback is requested. 2) Terms that do not match with the vocabulary, or whose annotations are not approved by the user, are then searched within the specific ontologies associated with the attribute, as defined in Table 4.8. If matches are of high confidence (i.e., `match_score` – calculated according to Equation 4.1 greater than 5), we select the corresponding term and proceed with the annotation; if the

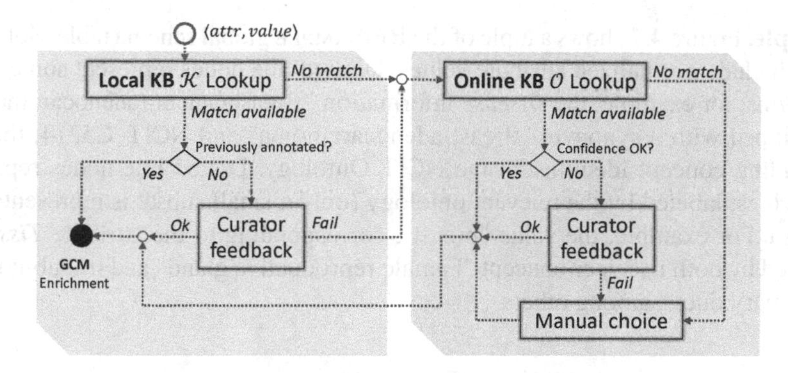

Fig. 4.6 Iterative supervised normalization and enrichment procedure.

confidence is low, user feedback is requested. When feedback is negative or there is no match, users are asked to provide a new vocabulary term.

Once the term has been selected, we populate the tables of the LKB with all the information derived from OLS regarding the term: description, iri, synonyms, xrefs, hyperonyms, and hyponyms (both of IS_A and PART_OF kinds). The depths of ancestors and descendants retrieved from the ontology are configurable by constant specification. With the current implementation and data, the automatic enrichment process successfully annotates about 83% of the total raw values, meaning that this fraction of the input values is annotated with ontological terms that reach a *match_score* of at least 5 (out of 10, i.e., perfect match with a preferred label).[17] The remaining non-annotated values are handled using a manual curation procedure. Indeed, we prepared two procedures that allow expert curators to support the annotation algorithm; we assume them to be knowledgeable about biological data management and to be experts in genomic data curation.

In the first procedure, a curator can examine all cases in which the algorithm is not able to provide a high-quality match (i.e., the service provides either partial matches with low score or no result). The low scores matches are proposed as suggestions so that the curator may select one of them. In any case, a manual annotation can always be provided. The procedure can be configured so that it also shows the cases with the same score.

The second procedure is started when a pre-existing annotation is not adequate (i.e., a *_tid* column has been filled with a wrong vocabulary term). In this case, the curator can invalidate the annotation and provide an alternative.

So far, we enriched attribute values by linking them to 1,629 terms in the 8 specified ontologies. In addition to terms that directly annotate values, we included all terms that could be reached by traversing up to three ontology levels from the base term (12,087 concepts in total).

[17] Note that we are not considering accuracy at this stage, as that will only be evaluated through an expert validation in Section 4.10.2.

Example. Figure 4.7 shows a tuple of the BioSample global schema table. Solid line nodes include normalized attribute values. Dashed line nodes represent some of the synonyms; for example, the *Disease* information "Breast cancer (adenocarcinoma)" is equipped with a synonym "Breast adenocarcinoma" and NCIT_C5214, the corresponding concept identifier in the NCIT Ontology. Dotted line nodes represent hierarchies, labeled by the relevant ontology (only a small subset is represented for brevity). For example, the value "breast", corresponding to the attribute *Tissue*, is enriched by both its super-concept "Female reproductive gland" and its sub-concept "Mammary duct", among others.

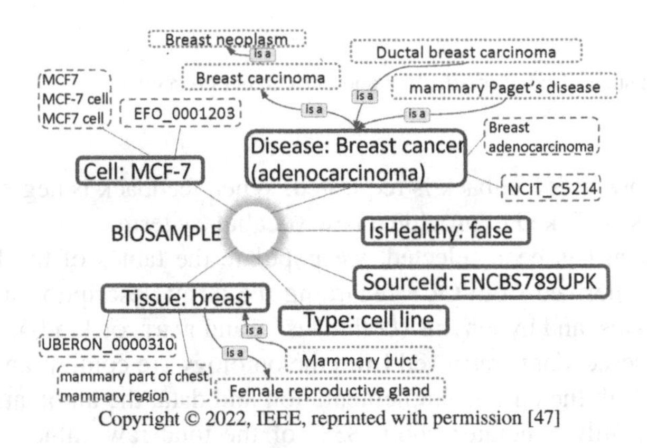

Fig. 4.7 Normalization and enrichment of a BioSample tuple.

4.8 Data Constraint Checking

At the end of the integration process, we introduce integrity constraints, which define dependencies between values of the global schema $\mathcal{G}$. These implement the contextual and dependent features assigned to GCM attributes, which were introduced in Section 3.2. We consider pairs of attributes ($A_S \in R_S$ and $A_E \in R_E$), where R_S and R_E denote the starting and ending tables in the $\mathcal{G}$ global schema, connected by a join path in $\mathcal{G}$. Given that $\mathcal{G}$ is an acyclic schema, there is just one join path between any two tables in $\mathcal{G}$.

Definition 4.3 (Attribute Dependency Rule) A *dependency rule* between attributes $R_S.A_S$ and $R_E.A_E$ of $\mathcal{G}$ is an expression of the form: $Boolean(R_S.A_S) \rightarrow Boolean(R_E.A_E)$, where $Boolean(A)$ is a Boolean expression over an attribute A of $\mathcal{G}$. The interpretation of the dependency rule is that: 1) **when** the Boolean expression in the left part of the rule is true for a value $v_S \in A_S$, 2) **if** there exists one value $v_E \in A_E$ such that $\langle v_S, v_E \rangle$ are connected by the join path between R_S and

R_E, 3) **then** the Boolean expression on the right part of the rule must be true for v_E. Boolean expressions include as special cases the predicates IS NULL or IS NOT NULL.

Dependencies can be manually defined during the lifetime of the META-BASE repository; they are manually defined and their identification is not assisted by a tool, even if we are evaluating to aid the production of such rules using an Association Rules Mining approach inspired by [271].[18] Table 4.9 shows some examples of dependency rules. For example, the first rule indicates that if the *Species* of a DONOR is "Homo sapiens" and the donor is connected to a DATASET through the only possible path in $\mathcal{G}$, then the *Assembly* of the dataset must be one of "hg19", "hg38", or "GRCh38". Dependency rules allow including in the GCM relevant attributes that are not common to all data types. For example, attributes *Target* and *Antibody* of EXPERIMENTTYPE are of great interest in ChIP-seq experiments, but are not significant in other experiments. Thus, a rule can specify that when *Technique* is not "ChIP-seq", then these attributes are null.

$\langle e_S, e_E \rangle$ in (DONOR·BIOSAMPLE·REPLICATE·ITEM·DATASET)
e_S.*Species* = "Homo sapiens" $\rightarrow$ e_E.*Assembly* $\in$ ["hg19", "hg38", "GRCh38"]

$\langle e_S, e_E \rangle$ in (DONOR·BIOSAMPLE)
e_S.*Gender* = "Male" $\rightarrow$ e_E.*Disease* $\neq$ "Ovarian cancer"
e_E.*Tissue* $\neq$ "Uterus"
e_S.*Gender* = "Female" $\rightarrow$ e_E.*Disease* $\neq$ "Prostate cancer"

$\langle e_S, e_E \rangle$ in (PROJECT·CASESTUDY)
e_S.*Source* = "ENCODE" $\rightarrow$ e_E.*SourceId* = "ENCSR.*"

$\langle e_S, e_E \rangle$ in (PROJECT·CASESTUDY·ITEM)
e_S.*Source* = "ENCODE" $\rightarrow$ e_E.*SourceId* = "ENCFF.*"
e_S.*Source* = "TCGA" $\rightarrow$ e_E.*SourceId* = "^[0-9a-z]{8}-([0-9a-z]{4}-){3}[0-9a-z]{12}$"
e_S.*Source* = "ENCODE" $\rightarrow$ e_E.*SourceUrl* is not null
e_S.*Source* = "TCGA" $\rightarrow$ e_E.*SourceUrl* is not null

$\langle e_S, e_E \rangle$ in (BIOSAMPLE)
e_S.*BiosampleType* = "tissue" $\rightarrow$ e_E.*Tissue* is not null
e_S.*BiosampleType* = "cell line" $\rightarrow$ e_E.*Cell* is not null
e_S.*isHealthy* = `false` $\rightarrow$ e_E.*Disease* is not null
e_S.*Disease* = "B-cell lymphoma" $\rightarrow$ e_E.*Tissue* = "Blood"
e_S.*Disease* = "Colon Adenocarcinoma" $\rightarrow$ e_E.*Tissue* $\in$ ["Colon", "Rectosigmoid junction"]

$\langle e_S, e_E \rangle$ in (DATASET·ITEM)
e_S.*IsAnn* = `true` $\rightarrow$ e_E.*ContentType* is not null

$\langle e_S, e_E \rangle$ in (PROJECT·CASESTUDY·ITEM·DATASET)
e_S.*Source* = "ENCODE" $\rightarrow$ e_E.*DatasetName* = ".*ENCODE.*"
e_S.*Source* = "Roadmap Epigenomics" $\rightarrow$ e_E.*DatasetName* = ".*ROADMAP_EPIGENOMICS.*"

$\langle e_S, e_E \rangle$ in (EXPERIMENTTYPE)
e_S.*Technique* $\neq$ "Chip-seq" $\rightarrow$ e_E.*Target* is null
e_S.*Technique* $\neq$ "Chip-seq" $\rightarrow$ e_E.*Antibody* is null

Table 4.9 Examples of dependency rules, including a description of the join path connecting the two attributes used in the left and right parts of the rule.

[18] In [271] authors try to enhance the metadata authoring process by supporting it with association rules-based suggestions; we would like to use a similar automatic generation of rules to support the content check of our database *a posteriori*

Dependencies complement the conceptual model specification; when the dependencies are specified for attributes belonging to different entities, they hold for all the instance pairs connected with an arbitrary join path connecting the two entities (this is not ambiguous because the conceptual model is acyclic). Note that dependency rules can work on attributes of the same table (intra-tabular) or across different tables of the $\mathcal{G}$ (inter-tabular).

Formalization. The *Constraint Checker* is a source-independent method $\mathcal{I} = \langle \mathcal{G}, \mathcal{IB} \rangle$. It is run on the whole content of the global schema $\mathcal{G}$, to which the integrity check rule base $\mathcal{IB}$ is applied.

Method. The *Checker* is applied at the end of data mapping and enriching phases. It produces a report of inconsistencies found in the database, with the indication of the violated constraints. Our concept of dependency rule application is similar to the integrity checks introduced by the ENCODE DCC [267].

4.9 Architecture Implementation

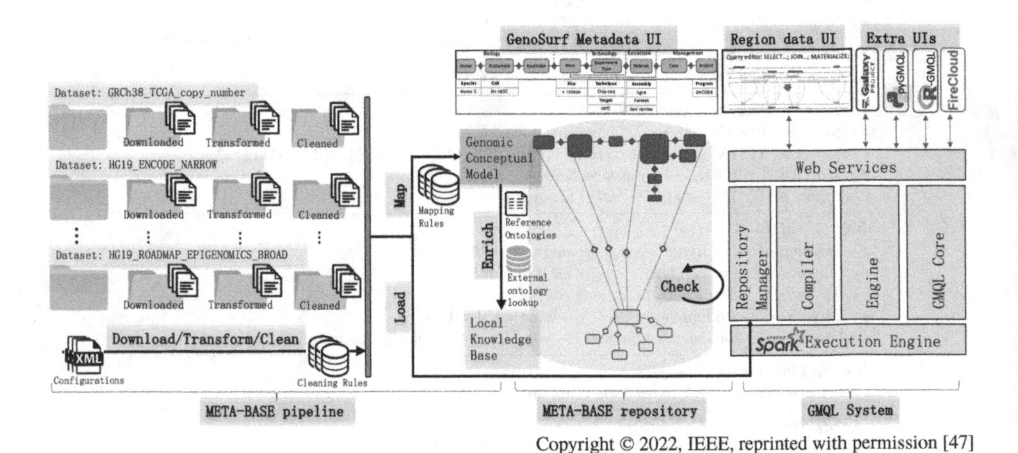

Fig. 4.8 Overall architecture for genomic data processing.

META-BASE is part of a broad architecture, whose main purpose is providing a cloud-based environment for genomic data processing. The overall system architecture is presented in Figure 4.8. In the left part of the figure we show the META-BASE pipeline discussed in Sections 4.3-4.8; the whole pipeline is configured using parameters provided as a single XML configuration file. Details are provided in Appendix A. Each dataset (on the left) is progressively downloaded, transformed, and cleaned. The data mapping method transforms cleaned attribute-value pairs into the global database GCM. The normalization and enrichment method adds references from the semantically enrichable attributes to the Local Knowledge Base, which is implemented by relational tables. Interactive access to the META-BASE repository is provided by a user-friendly interface (called GenoSurf, discussed in Chapter 6), that exploits the acyclic structure of the global schema to support simple conjunctive queries at the center of Figure 4.8).

As shown in the right part of the figure, the META-BASE repository can also be queried using the GMQL System [275], which supports integrated data management on the cloud; the system is accessed through Web Services as a common point of access from a variety of interfaces, including a visual user interface, programmatic interfaces for Python [295] and R/Bioconductor, and workflow-based interfaces for Galaxy and FireCloud. The implementation is executed using the Apache Spark engine, deployed either on a single server or a cloud-based system.

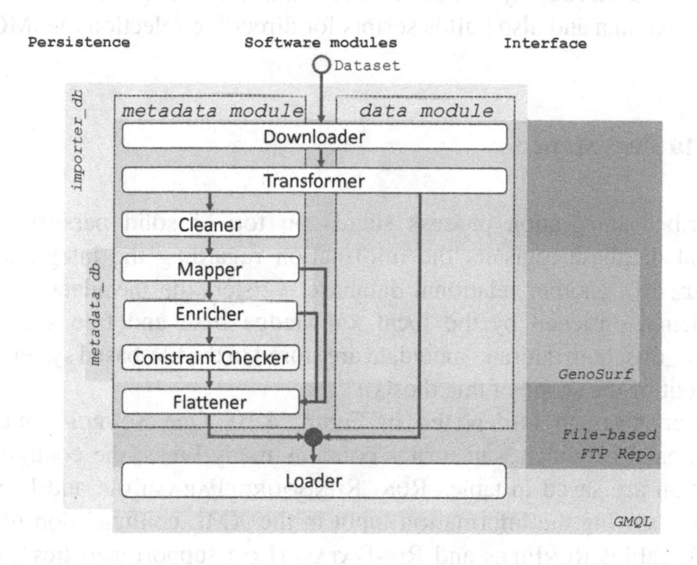

Fig. 4.9 META-BASE software integration process, from the *download* of a source partition (based on the prior definition of a GDM dataset) to its *transformation* and all following phases. *Cleaning*, *mapping*, *enriching* and *checking* are only performed on the metadata, while data are left unchanged. Metadata are *flattened* from the GCM relational implementation back to the file-based GeCo repository. Data and corresponding metadata of each dataset are *loaded* into the `http://gmql.eu/` system at the end.

META-BASE is a software architecture implemented in Scala,[19] available open-source at `https://github.com/DEIB-GECO/Metadata-Manager/`. The software process is depicted in Figure 4.9. The *Downloader* and *Transformer* modules act both on data and on metadata at the same time. The phases of cleaning, mapping, enriching, and constraint checking, instead, only concern metadata. After completion of the *Mapper* or of the *Enricher*, the *Flattener* is triggered, to re-generate

[19] `https://www.scala-lang.org/`

GDM attribute-value metadata files enhanced with GCM structured information; this is done to exploit the integration effort performed on metadata back also in the file-based representation of the repository. The *Loader* consequently prepares the datasets for the GMQL interface.

All phases are recorded in the `importer_db`, while the `metadata_db` contains all metadata of the repository samples in a relational format, referenced by sample IDs. On the right side of Figure 4.9 we can appreciate the three main access points to our repository:

i) FTP of the file-based GeCo Repository (`http://www.gmql.eu/datasets/`);
ii) GMQL interface for querying datasets using our tertiary analysis data processing environment (`http://www.gmql.eu/gmql-rest/`);
iii) GenoSurf metadata-based search engine (`http://www.gmql.eu/genosurf/`) – described thoroughly in Section 6.4 – that allow user-friendly surfing upon integrated data and also builds scripts for direct file selection in GMQL.

4.9.1 Data Persistence

The described integration process stands on top of solid persisting supports: a relational database registers the information regarding the integration process (`importer_db`); another relational database registers the metadata content in the GCM schema, enriched by the local knowledge base and free key-value pairs (`metadata_db`); both data and metadata are stored in the file-based system of GMQL (which is out of the scope of this thesis).

The `importer_db` is depicted in Figure 4.10. One Source contains many Datasets, each of which – in turn – contains many Files; the configurations for the execution are saved in tables Run, RunSourceParameter and RunDataset-Parameter, storing the information input in the XML configuration file (see Appendix A). Tables RunFiles and RunDatasetLog support statistics computation and reporting to the user (such as total files to download, as well as failed, outdated, and updated files after an execution). Possible values in the attribute `status` of a file are [Updated, Failed, Outdated, Compare].

The `metadata_db` is illustrated in Figure 4.11. The core of the schema represents the Genomic Conceptual Model; it is extended by two sub-schemata representing, respectively: the semantic enrichment for specific attributes of four tables, i.e., the Knowledge Base, and the original unstructured metadata – in the form of key-value pairs. Note that, out of all metadata extracted from sources, many attributes and their respective values cannot be mapped to the GCM. In these cases, in continuity with the metadata format of GDM, we store such extra attributes in an unstructured format, i.e., a table Pair containing pairs of a `key` and a `value`, extended with the `item_id` of the Item which they refer to; all attributes together form the PK, while the `item_id` also acts as FK.

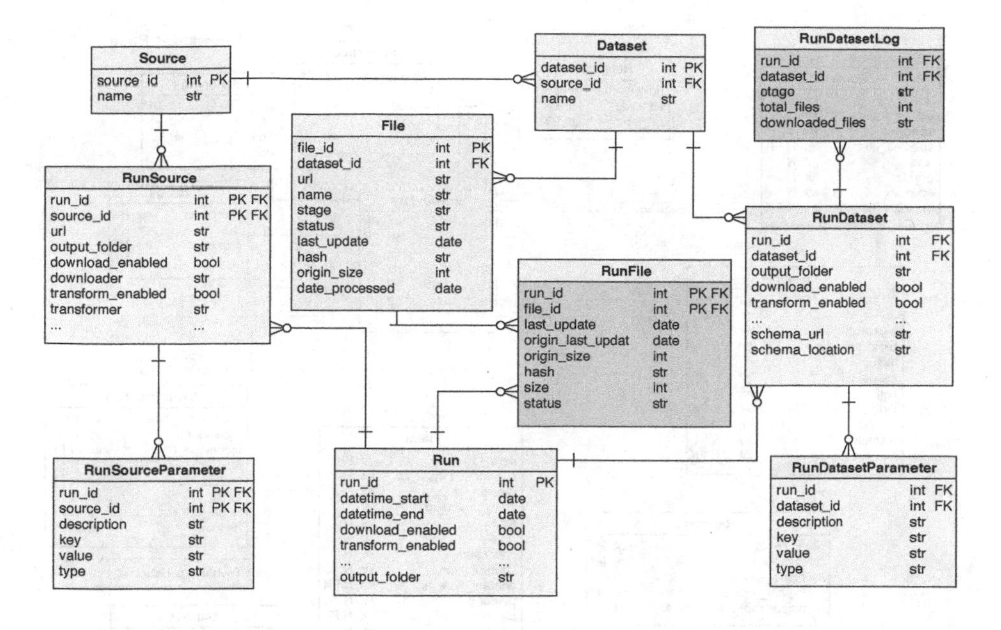

Fig. 4.10 Logical database schema of the `importer_db`.

4.10 Architecture Validation

We assessed the META-BASE architecture from different perspectives:

1. *Lossless integration*: the import process – operated by the *Downloader*, *Transformer* and *Cleaner* – does not miss information available in the original sources and the schema integration effort – operated by the *Mapper* – correctly reports the information from the sources into a global schema.
2. *Semantic enrichment*: the semantic information complementing original metadata (annotated by the *Enricher*) effectively identifies mapping to ontological concepts.
3. *Use evaluation*: the repository is useful for the targeted bioinformatics community and contributes to making their work more efficient.

4.10.1 Lossless Integration

We propose to consider how the integration process was lossless based on two perspectives, i.e., extensional and intentional. First, by *extensional* terms, i.e., we define our process as a process that satisfies its objective for all the imported data, by assuming that this could be tried and checked for all the inputs (while we only show it for a small manually checked sample).

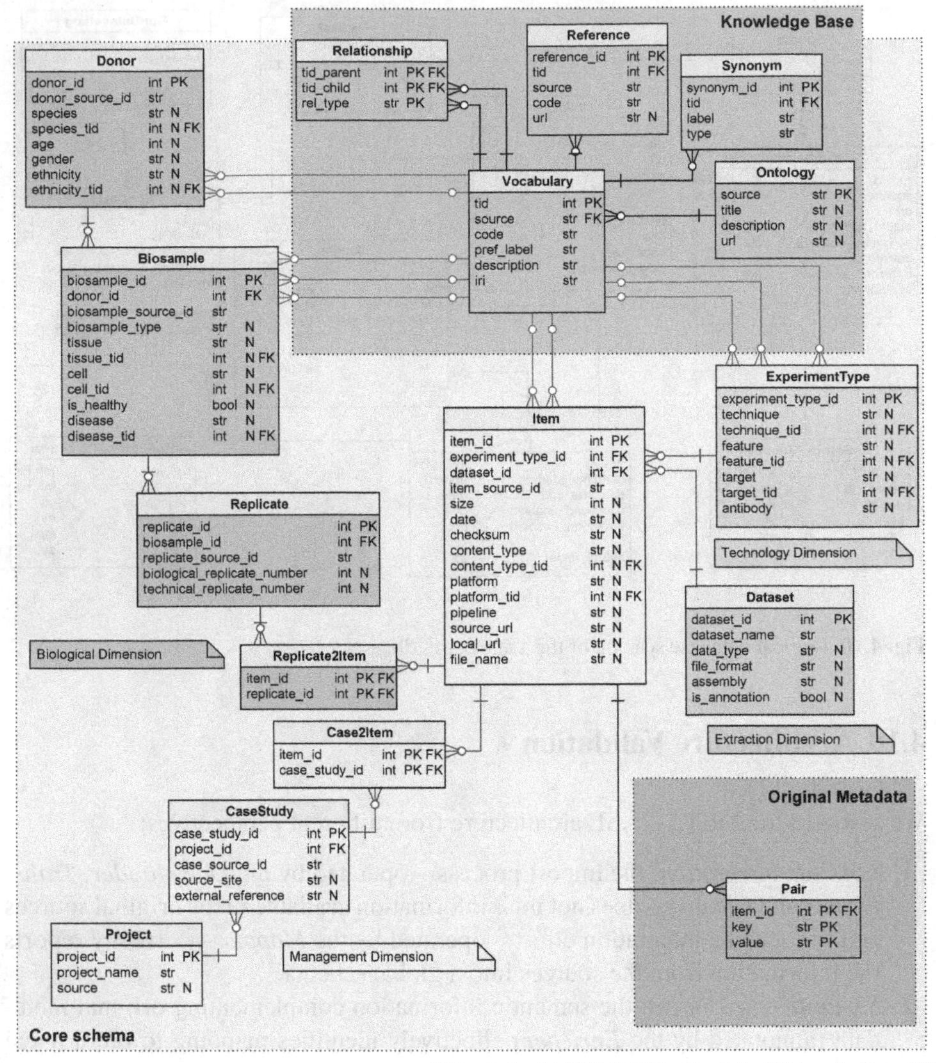

Fig. 4.11 Logical schema of `metadata_db`. Red relations represent foreign keys (FK) between core schema tables; blue relations link core schema values to corresponding ontology vocabulary terms. Data types are shortened: str for character varying, int for integer, and bool for Boolean. PK identifies primary keys; N marks nullable attributes.

Definition 4.4 (Extensional Lossless Integration) For each data source i, for each retrieved partition P_i, for each file f in P_i, for each pair $\langle key, value\rangle$, there exists a key' corresponding to key and a $value'$ corresponding to $value$, such that $\langle key', value'\rangle$ is also a pair and it is stored in the META-BASE repository.

Attribute	Query	META-BASE	ENCODE	GDC	Roadmap Ep.	Cistrome	1000 Genomes
Target	H3K27me3	1,990	814	-	381	795	-
Cell	MCF-7	1,447	6,442	-	-	125	-
Tissue	fat	57	-	-	57	-	-
Tissue	breast	23,729	92	24,788	94	264	-
Disease	breast cancer	114	-	45	-	163	-
Platform	Illumina Genome Analyzer II	981	723	-	-	-	258
Technique	RNA-seq	56,047	5,503	55,650	399	-	-

Table 4.10 Comparison of the number of items from exact-match queries in META-BASE vs. other sources.

Exploiting this definition, we tried to evaluate the goodness of the ETL process operated by the *Downloader*, *Transformer*, and *Cleaner* phases of META-BASE.

This evaluation is hard due to a number of issues, including among others the following ones: i) reproducing queries on the source just within a partition of interest is not always possible; ii) query interfaces at sources are different from ours (e.g., free-text search vs. attribute-based search); iii) sources assign metadata to different entities (e.g., experiments vs. data files included in experiments); iv) we cannot generate all possible queries and manual check of results is time-consuming.

We carefully considered all these issues: we studied the source query mechanisms at our best and favored an exact-match strategy for the benefit of comparison. As the generation of all possible queries and their manual check is not feasible, we performed the evaluation on a restricted number of meaningful example queries that show the effectiveness of our approach.

Table 4.10 reports the number of items resulting from seven exact-match queries as found either in the META-BASE repository or in individual sources. Note that the queries "H3K27me3", "fat", and "Illumina Genome Analyzer II" returned a number of matches in our system equal to the sum of matches in the integrated sources. However, Table 4.10 shows also some unavoidable difficulties, such as *semantic mismatch* (i.e., items are *false positives* when they match the searched string but their attribute's semantic is different from the one intended by the search). Indeed, the query "MCF-7" suffers from a problem with ENCODE matches: out of 6,442, only 1,322 (the ones found by our system for the same source) actually refer to the value describing the cell line of the item, while the remaining ones match with alternative information, regarding *possible controls*, *revoked files*, or *summary* of the experiment. Likewise, in ENCODE the query "RNA-seq" matches 5,503 items only because this information is contained in *related series* attributes, thus not describing the technique of the item. Similarly, "breast" and "breast cancer" (to a lesser extent) present a considerable number of false positives in GDC: in the first case, 1,509 items are wrongly matched due to information regarding family members (e.g., tagged with XML key "clinical_patient.family_history_cancer_type"), or information that does not directly describe the sample. Note that each query to the META-BASE repository is targeted to a specific attribute; thus, it finds items that are correctly related to the query, as we checked.

Then, we propose a definition of lossless integration in *intensional* terms, i.e., we define the necessary and sufficient conditions for our integration process to be lossless.

Definition 4.5 (Intensional Lossless Integration) Given a set of data sources I to be integrated into a global schema $\mathcal{G}$: i) for every attribute a in $\mathcal{G}$, a is mapped to all and only the keys in I that express the same semantics; ii) for every value contained in the database with schema $\mathcal{G}$, this is either set manually or derived from one or multiple values contained in I, where the corresponding $keys$ are mapped to one attribute a of $\mathcal{G}$, as defined by i).

Using this definition, we exploit the fact that our schema mapping operation uses global-as-view-like queries to import data into a warehouse. As we employ the same principles of the global-as-view theory [245], we also inherit the soundness and completeness of that approach: our mapping function (as shown in Tables 4.5 and 4.6) associates to each attribute in the GCM one query over each of the sources.

4.10.2 Semantic Enrichment

To inform our assessment of the effectiveness of the semantic enrichment process of META-BASE, we conducted a validation by engaging six experts with proven biological knowledge. Manual expert validation is the most common technique used in current biomedical literature [146, 79, 96], both because experts can provide the "optimal" outcome, and because the task of evaluating the semantic matching goodness is not easy to automatize, particularly in a specialized field such as genomics.

For each considered attribute, we presented to them a random set of annotations (i.e., matches between an original value and an ontological term, equipped with synonyms and its descriptions) automatically produced by the enrichment procedure. We asked them to rate the associations according to how accurate they are with respect to their own knowledge.

Attribute	*Platform*	*Ethnicity*	*Species*	*Disease*	*Tissue*	*CellLine*	*Technique*	*Feature*	*Target*
#annot/total	3/4	20/33	4/4	76/97	82/121	191/282	10/14	9/22	738/787
GOOD	33.34%	74.17%	100.00%	70.84%	95.00%	88.34%	81.66%	70.37%	100.00%
FAIR	50.00%	19.17%	-	10.83%	2.50%	4.17%	6.67%	12.96%	-
WRONG	16.67%	5.83%	-	17.50%	1.67%	6.67%	6.67%	14.81%	-
DO NOT KNOW	-	0.83%	-	0.83%	0.83%	0.83%	5.00%	1.85%	-

Table 4.11 Expert validation results.

The questionnaire contains up to 20 matches for each attribute (or less in the case of *Platform*, *Species*, *Technique*, and *Feature*, for which less distinct matches were found), selected randomly from their value pools, therefore considered representative of the sets. The test allows four choices: 1. GOOD, 2. FAIR, 3. WRONG, 4. DO NOT KNOW.

In Table 4.11, in the first row we indicate, for each attribute, the ratio between the number of automatically annotated values and the number of their total distinct values; the ones that are not annotated by the process are directly assigned to the expert annotation process. In the following rows of the table, we show in detail the results from the attributes presented to experts. The averaged results highlight that in 83.06% of cases the experts marked as good the examined matches, in 8.67% fair, and only the remaining 7.18% were marked as wrong. In the 1.08% of cases the experts declared they were not able to evaluate the match.

This small experiment showed that, in the great majority of cases, we correctly matched original values with an appropriate ontological term, therefore also with its synonyms and hierarchies. This brings important advantages, such as enabling the semantic search process that will be described in Chapter 6.

4.10.3 User Evaluation

The usefulness of META-BASE can be assessed on the basis of the interfaces that access its content and benefit from its sound integration approach. In particular, the META-BASE integrated repository is employed – through the GMQL Engine interface[20] – by external users and within the GeCo project for several genomic studies, some recently published,[21] others still in the process of publication preparation or review. All such works use the integrated metadata repository for locating data and the integrated genomic data repository for processing genomic data; this use will continue throughout the ERC project and beyond. The integrated metadata repository can be accessed also using GenoSurf [76], as it will be described in Chapter 6, where we also discuss validation of the interface by means of an empirical study involving 40 experts using the system and assessing their understanding of it with a questionnaire of composite queries. Detailed results are shown in Section 6.4.7, overall indicating positive feedback. Additionally, we will show that the queries provided in Section 6.4.6 allow META-BASE users perform a more efficient search process when compared to the same process on the original sources.

4.11 Related Works

Genomic repositories are growing in number, diversity, and complexity (see last Nucleic Acid Research yearly report [337]). In this context, data integration challenges in the *omics* domain are many, as reviewed in [170, 167, 238].

---

[20] http://gmql.eu/gmql-rest

[21] See 'Genomic Applications' sections at http://www.bioinformatics.deib.polimi.it/geco/?publications

There have been efforts to integrate multiple projects in single initiatives or portals, proposing a unifying data model or strategy to offer integrated access and management of genomic data. Several general projects are focused on offering integrated access to biomedical data and knowledge extracted from heterogeneous sources, including BioMart [372] (for biomedical databases), NIF [182] (in the field of neuroscience), Pathway Commons [339] (collecting pathways from different databases into a shared repository), and ExPASy [18] (linking many resources in life sciences).

Instead of addressing the problem of automatic integration, some works promote methodologies to improve the process of *metadata authoring* (i.e., preliminary preparation and submission) or the manual curation of metadata. Among these, we mention: CEDAR [293], a system for the development, evaluation, use, and refinement of genomics and biomedical metadata, using ontology-based recommendations from BioPortal [399]; BioSchemas.org [173], which applies schemata to online resources making them easily searchable; DNAdigest [231], promoting efficient sharing of human genomic datasets; DATS [346] boosting datasets' discoverability.

In general, we have observed a trend of initiatives that gather tools and data structures to support interoperability among highly heterogeneous systems, to help bioinformaticians perform a set of curation and annotation operations. These include community-driven efforts such as bio.tools [208] (anchored within ELIXIR[22]), service providers (EBI [309]), software suites (Bioconductor [204]), or lists (`http://msutils.org/`). By using the EDAM ontology [209], single initiatives can build bridges among resources while conforming to well-established operations, types/formats of data, and application domains.

Semantic integration. Many works in the literature consider the problem of recognizing ontological concepts to perform semantic annotation of data. Bodenreider proposes a (dated) survey on the use of ontologies in biomedical data management and integration [64] and, together with Fung, in [154] surveys a more recent use of ontologies in biomedical data management and integration. Grosjean *et al.* study differences among portals for bio-ontologies [176]. Oliveira *et al.* identify key search factors for biomedical ontologies to help biomedical experts in selecting the best-suited ones in their search cases [303]. Malone *et al.* in [269] suggest ten rules for selecting a bio-ontology stressing, in particular, coverage, acceptance (which is higher for ontologies that are under active community development), and detail. Textual definitions in classes are a positive point, as they allow human experts to evaluate if concepts used for specific annotations are appropriate.

In this Chapter we have presented BioPortal [399], Ontology Recommender [270], Ontology Lookup Service [215] and Zooma, as tools that can be used to annotate text or structured metadata with ontological terms. Other tools (not suited for our purpose, but worth mentioning) are UMLS [63], HeTop [175], and Annotator [214]. Several works [212, 366, 272, 134] capitalize on these tools and debate solutions devoted to data integration, by using ontology-based recommendations.

[22] `https://www.elixir-europe.org/`

We did not use any of these ready-to-use approaches; instead, as described in Section 4.7, the enrichment of our repository required a method tailored to its specific content.

Consortia's efforts Integrative efforts have also involved big consortia, as we have previously shown in Chapter 2. Among cancer-related consortia, we mention the standardization work of GDC, which produced a metadata model and several data harmonization pipelines, and ICGC [415] also provides a sophisticated metadata structure and services. In the epigenomics domain, we mention the work of the International Human Epigenome Consortium [72], of the Blueprint Consortium (with the curation effort of DeepBlue [9]), and of ENCODE, whose Data Coordination Center published interesting results regarding its data curation and integration achievements [199, 267, 196, 155], respectively reporting on ontologies used for annotation, metadata organization, storage system, and duplication prevention.

In addition, GWAS Catalog includes metadata enrichment for the information regarding traits,[23] and some works, which further improve the single consortia's efforts, have also been produced ([165] for GEO, [146] for ENCODE).

However, all the mentioned works provide so far neither models nor integration frameworks that are general enough to cover aspects falling outside the specific focus of their scope. Differently, the META-BASE approach described in this chapter is independent of specific sub-branches of genomics and can be applied to a large number of heterogeneous sources by any integration designer, in *a-posteriori* fashion, i.e., without having to follow any guidelines in the preliminary production of metadata. To the best of our knowledge, no one before has formally described an integration process for generic genomic data sources as we did with META-BASE.

[23] `https://www.ebi.ac.uk/gwas/docs/ontology`

Chapter 5
Snapshot of the Data Repository

"The most serious outbreak on the planet earth is that of the species Homo sapiens."
— David Quammen, Spillover: Animal Infections and the Next Human Pandemic

In July 2017 the genomic data repository of GeCo contained 16 datasets, about 80,000 GDM samples, and less than 0.5 TB of data. It stored experimental datasets and annotations collected from ENCODE (human broad and narrow peak signals, relevant to epigenomic research), Roadmap Epigenomics (human epigenomic datasets for stem cells and ex-vivo tissues), and TCGA (processed datasets for more than 30 cancer types, including mutations, copy number variations, gene and miRNA expressions, methylations); detailed statistics can be seen in Table 5.1. In that instance, metadata were stored as attribute-value pairs. Such an arrangement was just a preliminary attempt to provide a generic solution for metadata management.

Data source	Included datasets	#samples	Size (MB)
ENCODE	HG19_ENCODE_BROAD	1970	23552
	HG19_ENCODE_NARROW	1999	7168
Roadmap Epigenomics	HG19_EPIGENOMICS_ROADMAP_BED	78	595
	HG19_EPIGENOMICS_ROADMAP_BROAD	979	23244
TCGA	HG19_TCGA_Cnv	2623	117
	HG19_TCGA_DnaSeq	6361	276
	HG19_TCGA_Dnamethylation	1384	29696
	HG19_TCGA_Mirna_Isoform	9227	3379
	HG19_TCGA_Mirna_Mirnaseq	9227	569
	HG19_TCGA_RnaSeq_Exon	2544	31744
	HG19_TCGA_RnaSeq_Gene	2544	3584
	HG19_TCGA_RnaSeq_Spljxn	2544	30720
	HG19_TCGA_RnaSeqV2_Exon	9217	114688
	HG19_TCGA_RnaSeqV2_Gene	9217	20480
	HG19_TCGA_RnaSeqV2_Spljxn	9217	105472
	HG19_TCGA_RnaSeqV2_Isoform	9217	49152
Grand total	**16 datasets**	**78348**	**0.44 TB**

Table 5.1 Description of datasets content in the old repository in July 2017.

A. Bernasconi: *Model, Integrate, Search... Repeat*, LNBIP 496, pp. 99–118, 2023.
https://doi.org/10.1007/978-3-031-44907-9_5

Since then we have worked towards a much richer repository. In addition to including other relevant sources and describing a general framework – that allows adding more information with minimal effort – this thesis work has been concerned with giving more relevance, structure, and semantics to genomic datasets metadata. In these years we produced the Genomic Conceptual Model described in Chapter 3 and included it in the integration and enrichment pipeline described in Chapter 4.

In the landscape of genomic data platforms that integrate datasets, in 2020 we propose the first solution that joins together such a broad range of heterogeneous data, spanning from epigenomics to all data types typical of cancer genomics (e.g., mutation, variation, expression, etc.), until annotations. The framework was designed starting from three important data sources: ENCODE, Roadmap Epigenomics, and TCGA, which provided us with the most complex integration scenarios that can be faced in genomic metadata integration. Later, exploiting the generality of the integration framework, we easily extended our repository by adding a more updated version of TCGA (as exposed by Genomic Data Commons), 1000 Genomes, Cistrome, GENCODE, and RefSeq. We are currently finalizing the inclusion of GWAS Catalog, FinnGen, and working towards relevant subsets of GEO. A full list of datasets is available in Table 5.2, divided by source and assembly. Some sources require periodical updates of the imported content,[1] therefore we distinguish versions with a date (year_month) and we mark with × the last release included by META-BASE. Maintaining old versions accessible is important because users may need to reproduce their past research and computations on previously used datasets (so we make them available through the GMQL interface (`http://gmql.eu/gmql-rest/`); this is a very common practice in bioinformatics. From hereon, we will refer to the subset of the repository that includes only the latest release of each source; the search system (described in Chapter 6) only allows to locate the last versions of the same item, e.g., the ENCODE file ENCFF429VMY is only available in the GRCh38_ENCODE_NARROW_2020_01, while it cannot be accessed in the previous datasets.

Overall the repository includes now more than 9 TB of data, with more than 550,000 GDM samples. The version of the repository that considers only the last versions of datasets currently contains more than 250k processed data items (with an average size in the order of Megabytes), fully interoperable among each other, ranging from very small experimental files containing few regions corresponding to datasets linked to published papers to huge files of about 300k region lines. Distinct metadata attribute-value pairs reach about 50 million. Figure 5.1 provides a quantitative description; on the left, we appreciate the distribution of MB occupied by the different sources – considering only the most recent datasets of each one – and on the right the distribution of MB occupied by files of different assemblies.

Chapter organization. In Section 5.1 we overview one by one the sources included in the repository so far, specifying which efforts have been made and which datasets have been produced. Genomic Data Commons – in particular its data from The Cancer Genome Atlas – deserves particular consideration as it required the

[1] The updates are operated in an incremental way, by selecting specific partitions at the source and by replicating the procedure at different time points, as explained in Section 4.3.

Data source	Assembly	Included datasets	Last rel.	#samples	Size (MB)
ENCODE	GRCh38	GRCh38_ENCODE_BROAD_2017_08		366	2908.17
		GRCh38_ENCODE_BROAD_2017_11		850	6869.1
		GRCh38_ENCODE_BROAD_2019_01		642	5330.21
		GRCh38_ENCODE_BROAD_2019_07		633	5278.92
		GRCh38_ENCODE_BROAD_2020_01	×	641	4990.33
		GRCh38_ENCODE_NARROW_2017_08		10222	119110.9
		GRCh38_ENCODE_NARROW_2017_11		11573	128315.69
		GRCh38_ENCODE_NARROW_2019_01		12559	141731.77
		GRCh38_ENCODE_NARROW_2019_07		12913	142778.98
		GRCh38_ENCODE_NARROW_2020_01	×	13466	146800.22
	hg19	HG19_ENCODE_BROAD_2017_08		2136	25597.83
		HG19_ENCODE_BROAD_2017_11		844	18382.27
		HG19_ENCODE_BROAD_2019_01		909	19158.74
		HG19_ENCODE_BROAD_2019_07		898	19113.27
		HG19_ENCODE_BROAD_2020_01	×	859	18656.86
		HG19_ENCODE_NARROW_2017_08		11468	112441.28
		HG19_ENCODE_NARROW_2017_11		10342	111925.33
		HG19_ENCODE_NARROW_2019_01		12601	140019.11
		HG19_ENCODE_NARROW_2019_07		12733	141024.88
		HG19_ENCODE_NARROW_2020_01	×	13232	144895.45
Roadmap Epigenomics	hg19	HG19_ROADMAP_EPIGENOMICS_BED		156	967.95
		HG19_ROADMAP_EPIGENOMICS_BROAD		979	24332.19
		HG19_ROADMAP_EPIGENOMICS_DMR		66	3060.05
		HG19_ROADMAP_EPIGENOMICS_GAPPED		979	6875.22
		HG19_ROADMAP_EPIGENOMICS_NARROW		1032	11787.53
		HG19_ROADMAP_EPIGENOMICS_RNA_expression		399	2452.9
TCGA GDC	GRCh38	GRCh38_TCGA_copy_number		22374	686.17
		GRCh38_TCGA_copy_number_2018_12		22371	701.55
		GRCh38_TCGA_copy_number_2019_10	×	22627	780.76
		GRCh38_TCGA_copy_number_masked		22375	337.13
		GRCh38_TCGA_copy_number_masked_2018_12		22374	368.43
		GRCh38_TCGA_copy_number_masked_2019_10	×	22616	453.67
		GRCh38_TCGA_gene_expression		11091	56542.18
		GRCh38_TCGA_gene_expression_2018_12		11092	57112.96
		GRCh38_TCGA_gene_expression_2019_10	×	11092	57148.36
		GRCh38_TCGA_methylation		12218	1348516.13
		GRCh38_TCGA_methylation_2018_12		12218	1348528.88
		GRCh38_TCGA_methylation_2019_10	×	12215	1347776
		GRCh38_TCGA_miRNA_expression		10947	1502.32
		GRCh38_TCGA_miRNA_expression_2018_12		10996	1529.16
		GRCh38_TCGA_miRNA_expression_2019_10	×	11081	1572.81
		GRCh38_TCGA_miRNA_isoform_expression		10999	5003.92
		GRCh38_TCGA_miRNA_isoform_expression_2018_12		10996	4874.87
		GRCh38_TCGA_miRNA_isoform_expression_2019_10	×	11081	4934.77
		GRCh38_TCGA_somatic_mutation_masked		10188	2279.86
		GRCh38_TCGA_somatic_mutation_masked_2018_12		10187	648.59
		GRCh38_TCGA_somatic_mutation_masked_2019_10	×	10187	648.59
	hg19	HG19_TCGA_cnv	×	22632	796.82
		HG19_TCGA_dnamethylation	×	12860	247742.08
		HG19_TCGA_dnaseq	×	6914	285.94
		HG19_TCGA_mirnaseq_isoform	×	9909	4206.73
		HG19_TCGA_mirnaseq_mirna	×	9909	746.48
		HG19_TCGA_rnaseq_exon	×	3675	47667.84
		HG19_TCGA_rnaseq_gene	×	3675	5327.29
		HG19_TCGA_rnaseq_spljxn	×	3675	44376.74
		HG19_TCGA_rnaseqv2_exon	×	9825	124343.29
		HG19_TCGA_rnaseqv2_gene	×	9825	21861.72
		HG19_TCGA_rnaseqv2_isoform	×	4056	53081.88
		HG19_TCGA_rnaseqv2_spljxn	×	9825	115087.7
1000 Genomes	GRCh38	GRCh38_1000GENOMES_2020_01	×	2548	1101897.38
	hg19	HG19_1000GENOMES_2020_01	×	2535	1591854.63
Cistrome	GRCh38	GRCh38_CISTROME_broadPeak	×	1382	5406.46
		GRCh38_CISTROME_narrowPeak	×	4683	15875.22
GENCODE	GRCh38	GRCh38_ANNOTATION_GENCODE	×	24	1797.67
	hg19	HG19_ANNOTATION_GENCODE	×	20	739.67
RefSeq	GRCh38	GRCh38_ANNOTATION_REFSEQ	×	31	1324.24
	hg19	HG19_ANNOTATION_REFSEQ	×	30	274.61
Grand total		**67 datasets**		**564422**	**9.23 TB**

Table 5.2 Description of datasets content of new META-BASE repository as of October 2020.

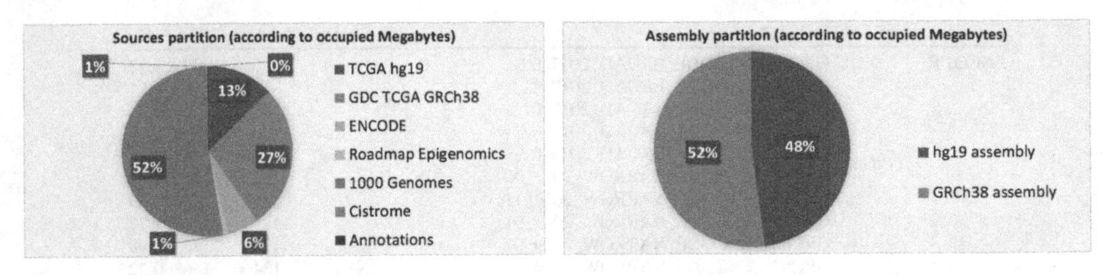

Fig. 5.1 Partition of the integrated repository according to MB occupied by each source and belonging to files of different assemblies.

development of a separate framework (described in Section 5.2) prior to inclusion into META-BASE. Gene Expression Omnibus is a very complex source (not yet included in META-BASE), as it is very big, and heterogeneous in data types and experiment types. We performed a first successful exercise to systematize the describing metadata of their samples, as outlined in Section 5.3.

5.1 Included Data Sources

In the following, we describe more in detail the content of the repository divided by source. Descriptions are guided by Figure 5.2; each pie chart shows how GDM samples (belonging to the most updated versions of the datasets) are distributed among different datasets within each source.

ENCODE. It is the result of an international collaboration of research groups, funded by the National Human Genome Research Institute (NHGRI); it collects projects regarding functional DNA sequences that intervene at the protein/RNA levels. From ENCODE we have included in our data repository all available processed human data files in narrowPeak and broadPeak format, for both GRCh38 and hg19 assemblies (see 4 datasets in Figure 5.2a); archived/revoked data are avoided. We extract datasets from the ENCODE portal;[2] this is a rapidly changing source so we periodically produce a new download, applying an API request for the same partition. The repository currently contains 5 versions since mid-2017. Size and number of files from one version to another change because the content changes at the sources (for example ENCODE progressively moved from producing more broadPeak files to preferring narrowPeak ones); it is also possible that – at the specific time of our query to their system – some files are undergoing maintenance and manual curation therefore are temporarily unavailable.

Roadmap Epigenomics. It is a public resource of human epigenomic data; as indicated in Figure 5.2b, we included in our repository 6 datasets regarding NGS-based consolidated processed data about: 1) broad and narrow regions associated with pro-

[2] `https://www.encodeproject.org/`

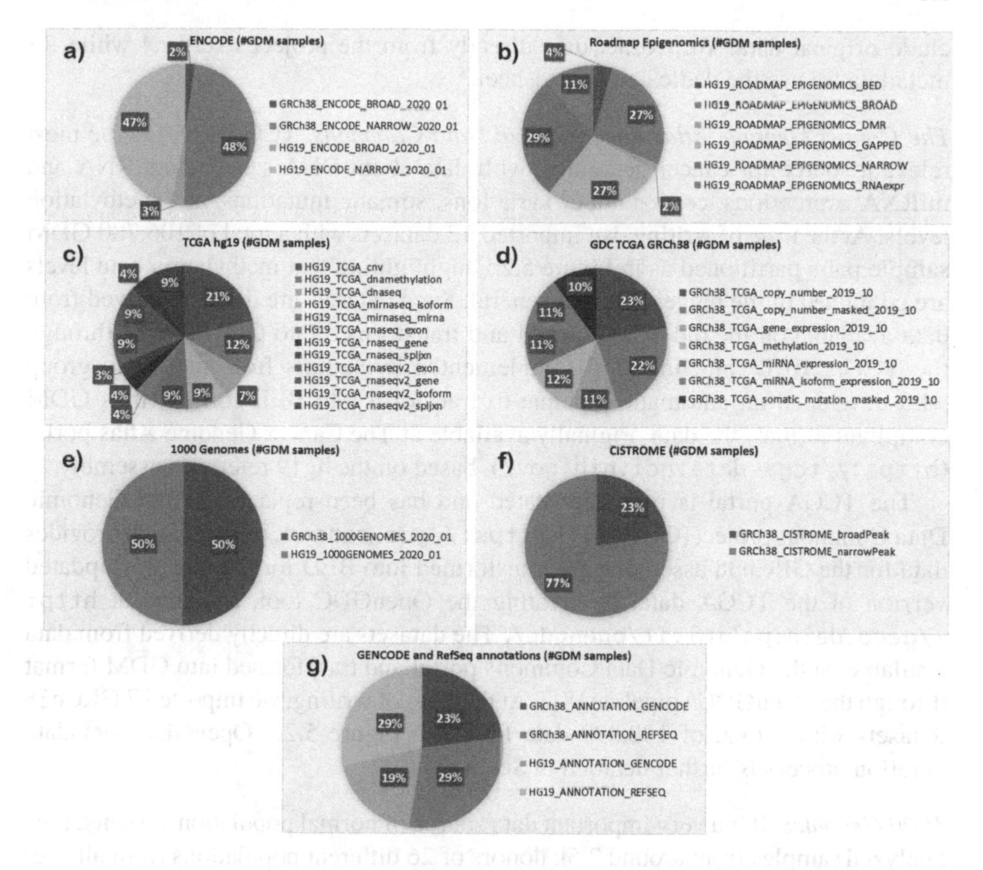

Fig. 5.2 From top to bottom and from left to right, a partition of the integrated repository according to the number of GDM samples per each dataset of ENCODE, Roadmap Epigenomics, TCGA, TCGA-GDC, 1000 Genomes, Cistrome, GENCODE/RefSeq annotations.

tein binding sites, histone modifications, or open chromatin areas identified using ChIP-seq (called using MACS2 and producing respectively broadPeak/narrowPeak/-gappedPeak format files) or DNase-seq experiments (BED, both narrow and broad regions using HOTSPOT peak caller); 2) differentially methylated regions (dataset DMR in BED format); 3) expression quantification of regions associated with genes, exons, and other known genomic regions of interest (dataset RNA Expression).

The Roadmap Epigenomics Project has been concluded; we do not retrieve datasets periodically, as they are not updated at the source.[3] The data and documentation we refer to is described in [233]. Later, ENCODE included part of the data in its portal, by reprocessing it using its normalized pipelines. Instead, we in-

[3] `http://www.roadmapepigenomics.org/`

clude original data. We extract data directly from the project's server,[4] while for metadata we use the dedicated spreadsheet.[5]

The Cancer Genome Atlas and Genomic Data Commons. TCGA [397] is the most relevant source for cancer genomics, with data about DNA sequencing, RNA and miRNA expressions, copy number variations, somatic mutations, and methylation levels. At the time of writing, we imported 12 datasets with a total of 106,780 GDM sample pairs partitioned as in Figure 5.2c, highlighting that methylation beta levels are contained in the biggest files, in general. Such datasets are directly derived from data available on the old TCGA portal and transformed into GDM format through the TCGA2BED pipeline [115], implemented by authors from the GeCo group who developed an automatic pipeline to transform into BED format (thus GDM compliant format) the data originally available at The Cancer Genome Atlas portal (`https://tcga-data.nci.nih.gov/`), based on the hg19 reference assembly.

The TCGA portal is now deprecated and has been replaced by the Genomic Data Commons project (GDC [177], `https://gdc.cancer.gov/`), which provides data for the GRCh38 assembly; we transformed into BED format also this updated version of the TCGA data, by creating the OpenGDC tool, available at `http://geco.deib.polimi.it/opengdc/`. The datasets are directly derived from data available on the Genomic Data Commons portal and transformed into GDM format through the OpenGDC[6] pipeline [81]. At the time of writing, we imported 7 GRCh38 datasets with a total of 100,899 data files, see Figure 5.2d. OpenGDC metadata curation process is further detailed in Section 5.2.

1000 Genomes. It is a very important data source of normal population variants; they analyzed samples from around 2.5k donors of 26 different populations from all over the world (such as, e.g., Japanese in Tokyo, Japan, Esan in Nigeria, Toscani in Italia). 1000 Genomes source files are retrieved from their FTP repository.[7] As of today, we include two very big datasets – around 1 TB each – one for hg19 assembly (2,535 GDM samples) and one for GRCh38 assembly (2,548 GDM samples). Figure 5.2e shows that the two datasets are very well balanced, however, the next phases of the project will only update the GRCh38 version. The process of integration of 1000 Genomes within META-BASE is described in [12].

Cistrome. We imported data relevant to ChIP-seq experiments targeted to histone modifications (while we will include the next transcription factor, chromatin regulator binding sites, and chromatin accessibility datasets). The included data is originally from ENCODE, Roadmap Epigenomics, and GEO experiments, but it has been

[4] `https://egg2.wustl.edu/roadmap/data/`

[5] `https://egg2.wustl.edu/roadmap/web_portal/meta.html`

[6] Documentation is available at `http://geco.deib.polimi.it/opengdc/data/OpenGDC_format_definition.pdf`

[7] `https://www.internationalgenome.org/`

thoroughly curated and re-processed homogenous pipelines.[8] The datasets we show in Figure 5.2f are contained in our private repository but are not publicly available on the GMQL interface as they are not released publicly. However, we make their data searchable over metadata in GenoSurf. Original files can be retrieved by interested users directly from `http://cistrome.org/db/#/bdown`, upon signing their usage agreement.

Genomic Annotations from GENCODE and RefSeq. GENCODE aims to create a comprehensive set of annotations, including genes, transcripts, exons, protein-coding and non-coding loci, as well as variants. For the hg19 assembly, we included releases 10 (December 2011) and 19 (December 2013), whereas for GRCh38 we imported versions 22 (March 2015), 24 (December 2015), and 27 (August 2017).[9] RefSeq is a stable reference for genome annotation, analysis of mutations, and studies on gene expression. We imported annotation files of GRCh38 v10 and hg19 v13 releases.[10] Figure 5.2g shows the number of samples available for different genomic elements; each file has all possible annotations for genes, exons, CDS, UTR, etc.

Next Sources. We are currently proceeding with the analysis of other sources to be added next. ICGC is an important addition to enrich the cancer genomics contribution of our work. It can be used together with TCGA samples to produce interesting pan-cancer analyses [317]. We already implemented the download and transform phase of GWAS-related sources (i.e., GWAS Catalog and FinnGen project[11]); this work is described in [46]. While modeling the metadata, we understood that the concept of "study" should be mapped on our Item (containing all regions corresponding to variants reported in a single GWAS), and their "publication" is our CaseStudy, a "trait" can be expressed using our *Disease* attribute. The biological view of the GCM should be slightly modified in order to capture the concepts of "cohort" (a population that is divided into case and control individuals, either exhibiting or not a certain trait) and "ancestry" (of donor individuals) that are relevant for GWAS. Such need highlighted the insufficient flexibility of our framework based on the GCM global schema. However, we accounted for this additional information using $\langle key, value \rangle$ pairs, which our current search infrastructure (especially GenoSurf) allows to search, match, and combine with other data.

The Gene Expression Omnibus is certainly a very attractive data source, possibly the biggest and most varied one in genomics, but poor in data curation and structure. This hinders the possibility of easily defining partitions of its content. We have made a first attempt to systematize their metadata, by extracting structured information from their experiments' unstructured descriptions. The results of such effort are reported in Section 5.3.

---

[8] The pipelines are detailed at `http://cistrome.org/db/#/about/`.

[9] The GENCODE datasets were retrieved from `https://www.gencodegenes.org/human/releases.html`

[10] The RefSeq datasets were retrieved from `https://www.ncbi.nlm.nih.gov/projects/genome/guide/human/`.

[11] `https://www.finngen.fi/en`

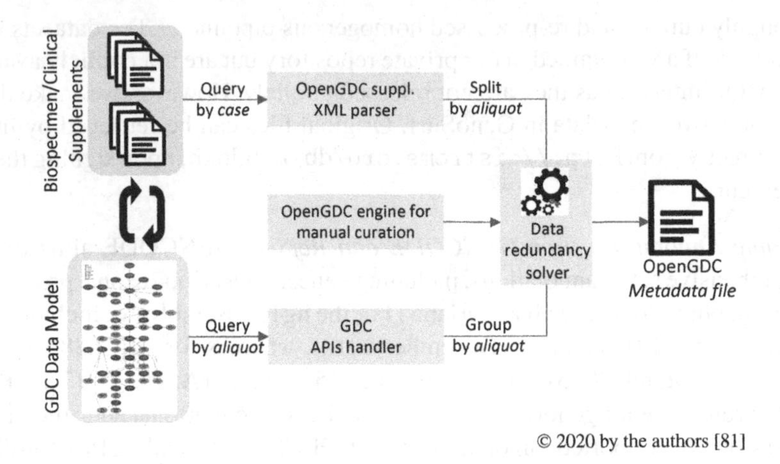

Fig. 5.3 OpenGDC metadata pipeline overview.

5.2 OpenGDC: Integrating Cancer Genomic Data and Metadata

In 2019 TCGA data has been included in the repository following a data curation process applied to the datasets retrieved from the Genomic Data Commons portal, which currently exposes the official and most updated TCGA data version.

The region data files were enriched thoroughly and transformed from various formats into the unique GDM format; however, the focus of this section is on metadata (which is my personal contribution to the work). To populate the OpenGDC metadata files (indexed with an `opengdc_id`, which identifies the GDM sample-pair), we retrieve clinical/biospecimen information from the GDC data type called Clinical and Biospecimen Supplements;[12] They are a special data type that contains data documentation; this information is stored in two different XML format files, originally provided by Biospecimen Core Repository (BCR) under contract of the NCI. In addition, we consider other properties retrieved using the GDC APIs. Finally, we add manually curated attributes computed within our standardization pipelines.

Given a converted experimental data file in GDM format, identified by an `opengdc_id`, the corresponding metadata file is generated according to the pipeline shown in Figure 5.3. Note that, for this source, the basic unit to be mapped into the GDM sample (or the GCM item) is the *aliquot*,[13] which is derived from one input

[12] A Clinical Supplement is a collection of information about demographics, medical history (i.e., diagnosis, treatments, follow-ups, and molecular tests), and family relationships (i.e., exposure and history) of a particular patient. A Biospecimen Supplement instead includes information associated with the physical sample taken from a patient and its processing.

[13] An aliquot is the smallest data unit of GDC, referring to the biological material analyzed from a single patient of the study `https://docs.gdc.cancer.gov/Encyclopedia/pages/Aliquot/`. We consider the aliquot as the *urelement* for GDC. We map it on the GCM item and

GDC file for most data types, from three files for the Gene Expression Quantification data type, and from four files for Masked Somatic Mutation data type.

On the top left corner, we consider Biospecimen and Clinical Supplements; they are organized by *patient* (identified by the `bcr_patient_uuid` attribute), with a patient typically related to many aliquots. Multiple OpenGDC metadata files are created, one for each aliquot reported in the patient biospecimen file. We replicate the full content of the Clinical Supplement of a patient's overall metadata files regarding the aliquots of the patient. The resulting metadata attribute keys start with the `clinical__` prefix. A Biospecimen Supplement, instead, contains a unique section on the patient, but also distinct sections on multiple samples, their portions, and the resulting aliquots. In each aliquot metadata file, we replicate the common parts about the patient (and, in case, about related samples/portions), while the remaining content of the biospecimen file is divided among the different metadata files according to the specific aliquot each of them refers to. The resulting metadata attribute keys start with the `biospecimen__` prefix.

On the bottom left corner of Figure 5.3, we query the GDC Data Model[14] content using the GDC RESTful API links. We call the API services once for each aliquot listed in a Biospecimen Supplement and each data type of interest, by specifying the `aliquot_uuid` and the `data_type`, and then associate with each OpenGDC data file all the information retrieved in the obtained response. The extracted attributes describe a data file along different GDC Data Model conceptual areas (i.e., administrative, biological, clinical and analysis). Relevant administrative entities include the Program (i.e., the broad framework of goals to be achieved by multiple experiments, such as TCGA), the Project (i.e., the specifically defined piece of work that is undertaken or attempted to meet a single requirement, such as TCGA-LAML – which refers to Acute Myeloid Leukemia), the Case (i.e., the collection of all data related to a specific subject in the context of a specific project, such as a patient). Among biological entities, there is the Sample (i.e., any material sample taken from a biological entity for testing, diagnostic, propagation, treatment, or research purposes) and the Aliquot (i.e., pertaining to a portion of the whole; any one of two or more samples of something, of the same volume or weight). With the aim of importing GDC data into the META-BASE repository, all such elements have been mapped to the GCM schema according to the mapping rules defined in Section 4.6.

Clinical entities include Treatment (i.e., therapeutic agents provided, or to be provided, to a patient to alter the course of a pathologic process) and Diagnosis (i.e., data from the investigation, analysis, and recognition of the presence and nature of disease, condition, or injury from expressed signs and symptoms). Analysis entities include harmonization pipelines such as "Copy Number Variation" and "Methylation Liftover", each related to one data type. As the GCM does not include a specific space for clinical metadata – because they are not common to all genomic data sources but source-specific – this information is retained in the form of pairs of keys and values.

GDM sample, our genomic basic data unit. We refer the reader to our discussion on granularity in Section 4.1.

[14] The GDC Data Model is available at `https://gdc.cancer.gov/developers/gdc-data-model`

In case an OpenGDC data file corresponds to n original GDC files (e.g., OpenGDC gene expression data files that derive from 4 input files), the JSON response to the corresponding API call is divided into n partitions, each containing information on one single GDC original file and on the related aliquot (the information of the latter one is replicated in each partition). Then, in the final OpenGDC metadata file, we group the information from the original files (by concatenating multiple values in a single key-value pair), while we consider the aliquot information only once. All these metadata attribute names are prefixed with `gdc__` and obtained by flattening the hierarchical structure of the JSON responses, i.e., through concatenation of JSON keys at each traversed level of the response structure.

As an addition to GDC inputs, we generate a set of manually curated key-value pairs (gathered in the group of metadata keys prefixed with *manually_curated__*). These contain information that is missing in GDC and derived from other sources or specified by our system. We add the data format (e.g., BED file textual format), URLs of the data and metadata files on the FTP server publicly offered by OpenGDC, the genome built (i.e., reference assembly), the *id, checksum, size* and *download date* of the data file, and the status of the tissue, which indicates if it is of a normal or control sample.

Combining Clinical/Biospecimen Supplement information with GDC Data Model information leads to value redundancy, which is due to the fact that there does not exist a specific data model for the Supplement data and it is impossible to determine *a priori* which information are non-overlapping. We ascertained the presence of attributes holding different names but the same semantics and associated values. We profiled all input data, obtaining sets of different keys that present the same values within the same metadata file. Example groups of key-value pairs with different keys and same value, along with the corresponding chosen candidate key preserved in each group, are shown in Table 5.3.

Preserved	Different attributes	Values
	biospecimen__bio__analyte_type	RNA
×	gdc__cases__samples__portions__analytes__analyte_type	RNA
×	biospecimen__admin__day_of_dcc_upload	31
	clinical__admin__day_of_dcc_upload	31
×	gdc__cases__primary_site	Ovary
	gdc__cases__project__primary_site	Ovary
×	gdc__cases__samples__portions__analytes__aliquots__concentration	0.17
	gdc__cases__samples__portions__analytes__concentration	0.17

Table 5.3 Example of choices produced by the data redundancy solver.

The preliminary profiling activity was used to provide guidance to create a list of data redundancy heuristics — with the aim to remove the redundant metadata attributes and their values — applied by the *Data redundancy solver* (at the center of Figure 5.3).

The heuristics have been primarily devised as a result of a long email exchange with the GDC Support team (`support@nci-gdc.datacommons.io`) that helped us to understand how their ingestion process works: a restricted number of attributes from the supplements are already provided with a defined mapping to the data model attributes, while for others the relation is still uncertain (i.e. not curated yet by the GDC) — for these, we reconstructed common semantics through a semi-automated approach. Moreover, clinical and biospecimen supplements cover partially overlapping semantics spaces. Thus we make the deliberate decision of extracting only one of them. Finally, the new data model entities are non-overlapping but the APIs provide their content in a nested fashion (a problem similar to the one of ENCODE, addressed in the *Cleaner* module explanation – Section 4.5). For example, a project is related to a case with a functional dependency, therefore the project information can be uniquely reached through the case entity. As a consequence, any information related to the `case__project` group is redundant w.r.t. the one given by a dual attribute with the same suffix. Analogously, aliquots are comprised of analytes (N aliquots are in 1 analyte), therefore we keep the information that is most specific, pertaining to the aliquot.

Note that this is a classical problem of data fusion. Among well-known strategies [129], the OpenGDC problem of intra-source redundancy is resolved with a simple rule-based method. We have summarized our approach to solving redundancy in four rules, derived from the above-mentioned heuristics, which are executed in a specific order. At the time of writing, these cover the whole space of possibilities; however, this set will be updated as the need for new rules will arise, in conjunction with updates of OpenGDC scheduled releases:

1. verify mappings on the official GDC GitHub repository,[15] specifying which fields from the BCR Supplements correspond to the GDC API fields: when redundant, keep the second ones;
2. when a field from the BCR Biospecimen Supplement is redundant w.r.t. a field of the BCR Clinical Supplement, keep the first one;
3. when a field belonging to the `case` group is redundant w.r.t. a `case__project` group field, keep the first one;
4. when a field belonging to the `analytes` group is redundant w.r.t. a `analytes__aliquots` group field, keep the second one.

To facilitate the use of metadata key-value pairs, in case keys are very long and cumbersome, we simplify them through the *Cleaner* (Section 4.5).

The GDC data source was very complex, therefore required to be handled separately from the META-BASE architecture. This was the result of a collaboration with the Roma Tre University group of researchers who realized the practical implementation of the system. We stress the point that all the operations performed within this pipeline are coherent with META-BASE steps (download, transform, clean). The datasets produced by OpenGDC contain GDM-compliant samples, that

[15] The GDC GitHub repository is available at `https://github.com/NCI-GDC/gdcdatamodel/tree/develop/gdcdatamodel/xml_mappings`

are integrated within the META-BASE repository and undergo the mapping process as all other sources.

5.3 Towards Automated Integration of Unstructured Metadata

The data of the Gene Expression Omnibus is of fundamental importance to the scientific community for understanding various biological processes, including species divergence, protein evolution, and complex disease. The number of samples in the database is growing exponentially (see Figure 5.4), and while tools for retrieving information from GEO datasets exist [217], large-scale analysis is complicated due to heterogeneity in the data processing across studies and most importantly in the metadata describing each experiment. When submitting data to the GEO repository, scientists enter experiment descriptions in a spreadsheet (see Figure 5.5) where they can provide unstructured information and create arbitrary fields that need not adhere to any predefined dictionary.[16] The validity of the metadata is not checked at any point during the upload process,[17] thus the metadata associated with gene expression data, usually does not match with standard class/relation identifiers from specialized biomedical ontologies. The resulting free-text experiment descriptions suffer from redundancy, inconsistency, and incompleteness [171, 411].

In the work hereby described, we developed automated machine learning methods for extracting structured information from the heterogeneous GEO metadata, with the aim to populate the META-BASE repository, using the attributes represented by the Genomic Conceptual Model (Section 3.2), which recognizes a limited set of concepts supported by most genomic data sources. In particular, from the text we extract instances that can be mapped into GCM concepts, assuming they are interconnected by GCM's relationships. Note that in the specific application described in this section, we focused on extracting information that is already present in the unstructured text. However, the used learning technique, allowed us to infer also null values that are existing but are unknown [34].[18]

Two different approaches were evaluated for the metadata extraction problem, both leveraging recent advances in Deep Learning for text analysis:

(1) The first approach builds a *multi-label classifier* to predict metadata attribute values using a deep embedding, and will serve as our baseline for later experiments. We used RoBERTa [252], a variation on the BERT [126] language model, a self-attention based transformer model [387].

---

[16] Gonçalves and Musen report in [171] that, in the description field, submitters have typed the concept "age" in at least 30 ways, using different choices of upper/lower case letters (age, AGE, Age), adding special characters (Àge), specifying measure units in many creative ways.

[17] Information regarding the NGS sequences submission is at `https://www.ncbi.nlm.nih.gov/geo/info/seq.html`.

[18] Note of the author, August 2023. The translation model described in this section has later been exploited for building an active-learning framework for improving the extraction of metadata from GEO experiment descriptions [362].

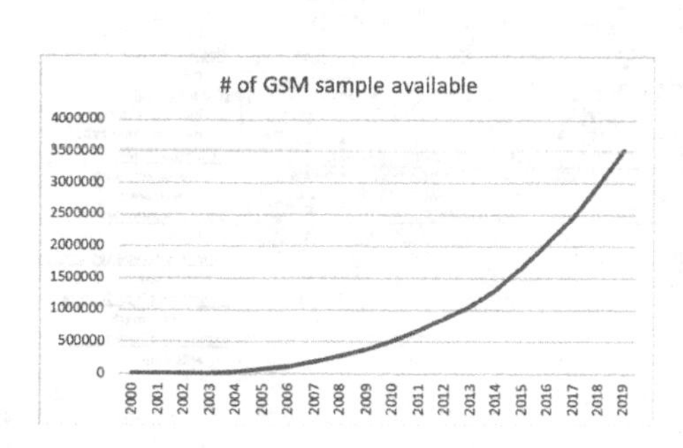

Fig. 5.4 Growth over time of samples available in the GEO database.

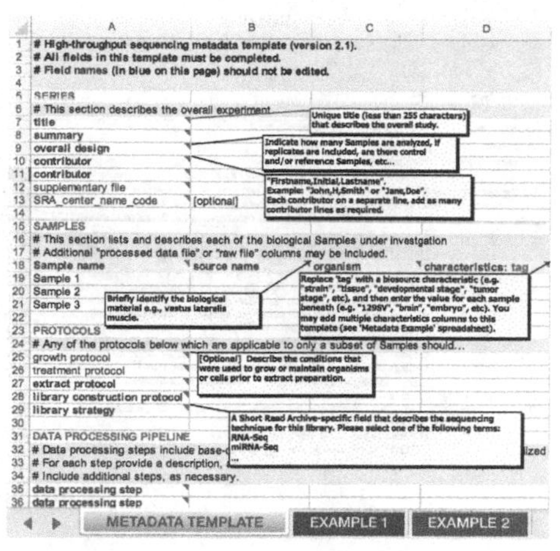

Fig. 5.5 Spreadsheet available to researchers to annotate their study and data submissions in GEO.

(2) The second makes use of a novel *translation-based approach* where powerful sequence-to-sequence models are leveraged to solve the metadata extraction problem in a more elegant and extensible fashion. We use the Encoder-Decoder LSTM with a Luong attention [262] mechanism and also OpenAI GPT-2, a more powerful sequence-to-sequence pre-trained language model [324], whose structure is based on Transformer Decoders [387].

To train our models we made use of data from GEO, Cistrome, and ENCODE. Each training sample is composed of *input-output* pairs, where *input* corresponds to the textual description of a biological sample and *output* is a list of attribute-value pairs. Figure 5.6 shows an example translation task: on the left, a metadata record from the GEO repository describing a human biological sample, in the middle the target schema, and on the right the resulting output pairs. The text output produced by the translation model should be human and machine-readable, so we used a dash-separated list of "*key: value*" pairs, `Cell Line: HeLa-s3 - Cell Type: Epithelium - Tissue: Cervix`.

5.3.1 Experiments

We designed three experiments to validate our proposal. Two experiments aimed to evaluate and compare the results of the two translation-based models mentioned above; a third experiment, instead, tested the performance of the best-proposed model

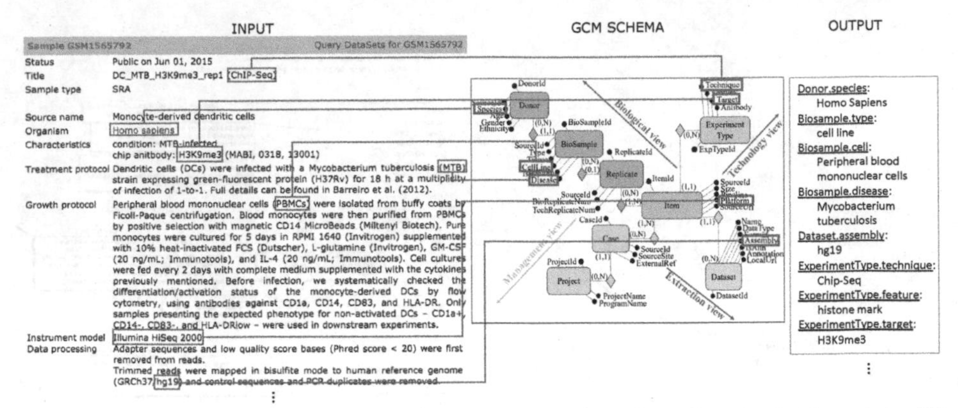

Fig. 5.6 Example mapping task: from GEO sample GSM1565792 input text, into GCM attributes, to finally produce output key-value pairs.

Attributes	%"None"	#distinct values
Cell line	52	519
Cell type	19	152
Tissue type	29	82
Factor	0	1252

Table 5.4 Cistrome attributes: percentage of "None" and count of distinct values.

Attributes	%"None"	#distinct values
Age	1	169
Age units	32	6
Assay Name	0	26
Assay type	0	9
Biosample term name	0	9
Classification	1	6
Ethnicity	74	15
Genome assembly	16	11
Health status	53	65
Investigated as	48	22
Life stage	1	17
Organism	1	5
Project	0	3
Sex	1	10
Target of assay	48	344

Table 5.5 ENCODE attributes: percentage of "None" and count of distinct values.

on randomly chosen instances from GEO. For experimental setup, we refer interested readers to Appendix B.

5.3.1.1 Datasets

We make use of data from GEO, Cistrome, and ENCODE for our experiments.

- **GEO**: Input text descriptions were taken from the GEOmetadb database [419]. We extracted the *Title*, *Characteristics_ch1*, and *Description* fields, which in-

Model	# Epochs	Accuracy	Precision	Recall
RoBERTa	69	0.90	0.89	0.91
LSTM + Attention	15	0.62	0.65	0.62
GPT-2	47	0.93	0.93	0.93

Table 5.6 Experiment 1: overall accuracy, precision, and recall. Precision and recall are weighted by the number of occurrences of each attribute value.

clude information about the biological sample from the `gsm` table. We formatted the input by alternating a field name with its content and separating each pair with the dash "-" character, e.g., `Title: [...] - Characteristics: [...] - Description: [...]`, thus allowing the model to learn possible information patterns, for example, information regarding "Cell Line" is often included in the "Title" section. Syntactical pre-processing was also performed.

- **Cistrome**: We selected this source as its samples have been manually curated and annotated with the *cell line*, *cell type*, *tissue type*, and *factor name*. We downloaded in total 44,843 metadata entries from Cistrome Data Browser[19] with the four mentioned attributes. As indicated in Table 5.4, three of the fields contain many "None" values, but these should not be interpreted as missing, since they actually indicate that the specific sample does not carry that kind of information.
- **ENCODE**: we downloaded 16,732 metadata entries from ENCODE web portal,[20] by requesting the fields listed in Table 5.5 for each experiment sample. The free text input related to each sample was retrieved by either: (i) exploiting a reference to the GEO GSM (only available for 6,233 entries) or (ii) concatenating the additional ENCODE fields *Summary*, *Description* and *Biosample Description*.

5.3.1.2 Setup and Results

Experiments 1 and 2 allowed us to compare the performances of the three analyzed models on two different datasets: Cistrome (with input from GEO) and ENCODE (with input both from GEO and ENCODE itself). We evaluated the performances of LSTM (with attention mechanism) and GPT-2 seq2seq models against RoBERTa, using samples from Cistrome (Experiment 1) and samples from ENCODE (Experiment 2).

In both experiments, overall GPT-2 outperforms both Encoder-Decoder LSTM and RoBERTa, as it can be observed in Tables 5.6-5.7. Results divided by class are shown in Figure 5.7 for Experiment 1 and in Figure 5.8 for Experiment 2.

[19] `http://cistrome.org/db/#/bdown`

[20] `https://www.encodeproject.org/`

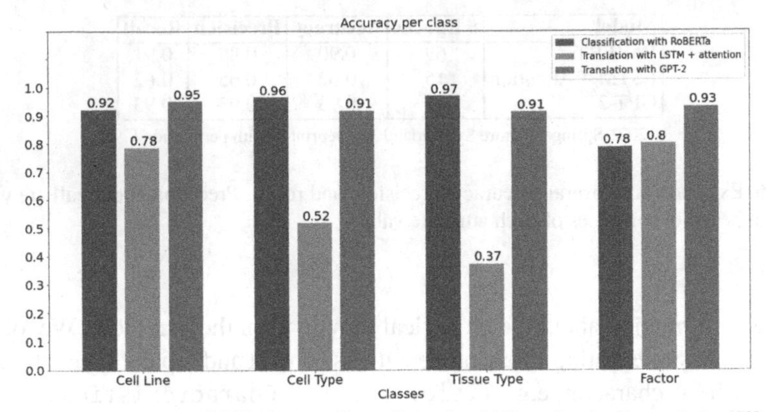

Fig. 5.7 Experiment 1: per-class accuracy for the three models on Cistrome data.

Experiment 1 considerations. From Figure 5.7, RoBERTa seems to perform better for classes that contain a low number of distinct values, i.e. *cell type* and *tissue type* (which contain 380 and 249 possible values). Instead, for *cell line* and *factor* (both with more than a thousand possible values) GPT-2 outperforms RoBERTa. The number of "None" values is taken into consideration (Table 5.4), the classes *cell line*, *cell type* and *tissue type* present a relevant percentage of "None", the weighted precision and recall analysis, however, shows high scores, despite the unbalance of values count; this implies that the models were able to correctly classify samples which lack of labels for certain classes.

Experiment 2 considerations. From Figure 5.8, we appreciate a similar behavior as in Experiment 1, i.e., translation models perform better for attributes with a larger amount of distinct values. The attributes *target of assay* and *biosample term name* present the highest number of distinct values and GPT-2 far exceeded RoBERTa in terms of accuracy. Instead, this experiment highlights how the LSTM model with attention does not perform well for a larger amount of target attributes, at least with the tested model size. The labels *health status* and *ethnicity* presented several "None" values (74% and 53%), but both RoBERTa and GPT-2 were able to predict correctly almost the totality of samples, producing results with high weighted precision and weighted recall.

Previous works aimed to extract a restricted set of labels (such as *age* and *sex*) with unsatisfactory results; they often limited the target *age unit* to "years" or "months" and the target *sex* to only "Male" and "Female". A lot of different scenarios for the input text made it impossible – for previous work – to extract correctly the target attributes (for example cases for which the information needs to be inferred, or when the experiment presents multiple cells, consequently multiple ages and multiple sex). This experiment shows that our proposed translation approach can outperform state-of-the-art approaches, additionally handling a different number of non-standard cases.

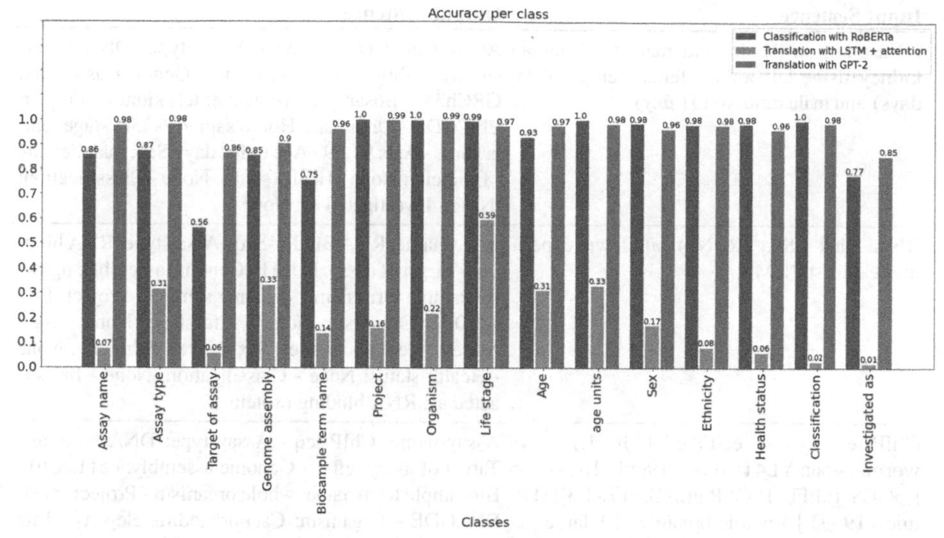

Fig. 5.8 Experiment 2: per-class accuracy of the three models on ENCODE data.

Model	# Epochs	Accuracy	Precision	Recall
RoBERTa	71	0.90	0.89	0.90
LSTM + Attention	22	0.19	0.19	0.19
GPT-2	48	0.96	0.96	0.96

Table 5.7 Experiment 2: overall accuracy, precision, and recall. Precision and recall are weighted by the number of occurrences of each attribute value.

Condition	Accuracy	Precision	Recall
Label **present in** the input	0.83	0.70	0.68
Label **absent from** the input	0.062	0.038	0.038

Table 5.8 Experiment 3: Results of prediction of 200 manually labeled samples for ENCODE class *biosample term name*.

Experiment 3. With the last experiment, we studied the behavior of GPT-2 in a realistic scenario involving randomly chosen samples from GEO. These samples were not selected based on presence in a database (Cistrome or ENCODE) and thus provide a realistic test scenario for the proposed use-case of the system. No reference labels were available for the randomly selected set of 200 input descriptions, so each instance was manually checked to provide ground-truth labels. The system was trained using both the Cistrome and ENCODE datasets We note the heterogeneity between the training and test examples for this experiment.

Input Sentence	Output Sentence
DNase-seq on 91 day old male fetal human kidney tissue left kidney female embryo (59 days) and male embryo (91 day)	Assay name: DNase-seq - Assay type: DNA accessibility - Target of assay: None - Genome assembly: GRCh38 - Biosample term name: left kidney - Project: ENCODE - Organism: Homo sapiens - Life stage: embryonic - Age: 91,59 - Age units: day - Sex: male,female - Ethnicity: None - Health status: None - Classification: tissue - Investigated as: None
RNA Bind-n-Seq (RBNS) pulldown experiment against PUM1	Assay name: RNA Bind-n-Seq - Assay type: RNA binding - Target of assay: PUM1 - Genome assembly: hg19 - Biosample term name: cell-free sample - Project: ENCODE - Organism: None - Life stage: None - Age: None - Age units: None - Sex: None - Ethnicity: None - Health status: None - Classification: None - Investigated as: RNA binding protein
ChIP-seq against eGFP-efl-1 in L1 larva worms. strain YL418 (unc-119(ed3)III; vrIs65 [pGES-1::EFL-1::GFP FLAG:EFL-1 3'UTR, unc-119 (+)]) whole organism L1 larva genetically modified using bombardment	Assay name: ChIP-seq - Assay type: DNA binding - Target of assay: efl-1 - Genome assembly: ce11,ce10 - Biosample term name: whole organism - Project: modENCODE - Organism: Caenorhabditis elegans - Life stage: L1 larva - Age: unknown - Age units: None - Sex: mixed - Ethnicity: None - Health status: None - Classification: whole organisms - Investigated as: cofactor,transcription factor
datatype: DnaseSeq; datatype description: DNaseI HS Sequencing; labversion: baseAlignCounts.pl v 1,fseq v 1.84, iff_generic_male,p-value cutoff: 0.05; replicate: 1,2; cell: Colo829; cell organism: human; cell description: malignant melanoma; cell karyotype: cancer; cell sex: M; labversion: baseAlignCounts.pl v 1 Duke human Colo829 DNase-seq COLO829 malignant melanoma	Assay name: DNase-seq - Assay type: DNA accessibility - Target of assay: None - Genome assembly: GRCh38,hg19 - Biosample term name: COLO829 - Project: ENCODE - Organism: Homo sapiens - Life stage: adult - Age: 45 - Age units: year - Sex: male - Ethnicity: Caucasian - Health status: metastatic melanoma - Classification: cell line - Investigated as: None

Table 5.9 Examples of GPT-2 translations of randomly selected GEO descriptions

Table 5.9 shows examples of the input and output of our procedure. Overall, the results showed that the OpenAI GPT-2 performed considerably better than others. In addition to surprising performances, we noted interesting findings regarding particular cases. In particular, we found cases in which the output contains a label that is:

1. *unseen* in training data, e.g., no sample contained *target of assay*: *MYC-1*.
2. *absent* from input description, e.g., for the input *"HNRNPK ChIP-seq in K562 K562 HNRNPK ChIP-seq in K562"* the output correctly contained: *Organism: Homo sapiens - Age: 53 - Age units: year - Sex: female - Health status: chronic myelogenous leukemia (CML)*, etc.

3. *multi-valued*: e.g., a particular GEO record contained samples from both male and female donors[21], and the output correctly noted both genders: *"Sex: male,female ..."*.
4. *reordered* with respect to the input, e.g., an input containing *"Tfh2_3 cell type: Tfh2 CD4+ T cell; ..."* correctly produced the output *"Biosample term name: CD4-positive Tfh2"*.

This preliminary experimentation showed that this is a very promising approach to address structured metadata extraction from data sources such as GEO where no data model is imposed a priori. We are working towards embedding this procedure in the META-BASE pipeline, which will require also a transformation of region data files.

5.3.2 Related Works

There is a compelling need to structure information in large biological datasets so that metadata describing experiments is available in a standard format and is ready for use in large-scale analysis [201]. In recent years, several strategies for annotating and curating GEO database metadata have been developed (see Wang *et al.* [395] for a survey). We group the approaches into five non-exclusive categories: 1) manual curation, 2) regular expressions, 3) text classification, 4) named-entity recognition, and 5) imputation from gene expression. Their limits have been overcome using the method described in this Section, where translation models are used in a completely novel manner to translate free text into a well-structured schema.
Manual curation. Structured methods for authoring and curating metadata have been promoted by numerous authors [186, 294, 249]. Moreover, a number of biological metadata repositories (e.g. RNASeqMetaDB [181], SFMetaDB [247] and CREEDS [396]) manually annotate their datasets, guaranteeing high accuracy. This option is, however, highly time-consuming and hardly practicable as the volume and diversity of biological data grows.
Regular expressions. The use of regular expressions for extracting structured metadata fields from unstructured text is common [165]. This simple technique is limited, however, to matching patterns that are:

- *expressible*: yet identifiers for biological entities often do not follow any particular pattern (e.g., IMR90, HeLa-S3, GM19130 all represent cell lines);
- *explicit*: therefore matching, e.g., the cell line K562 cannot possibly produce the implied (non-appearing) knowledge on sex, age, or disease information;[22] and

[21] The input in this case was: *microRNA profile of case NPC362656 survival status (1-death,0-survival): 0; gender (1-male,2-female): 1; age (years): 56; ...*

[22] K562 is a widely known cell line originally extracted from tissue belonging to a 53-year-old woman affected by myeloid leukemia

- *unique*: i.e., the technique cannot discern between multiple string matches in the same documentwith possibly different semantics (breast tissue vs. breast cancer).

Text classification. Machine learning techniques can be used to predict the value of metadata fields based on unstructured input text. Posch *et al.* [318] proposed a framework for predicting structured metadata from unstructured text using Term Frequency - Inverse Document Frequency (TF-IDF) and topics modeling-based features. The limitations of the classification approach include that a separate model needs to be trained for each attribute to extract and that values of the attribute need to be known in advance.
Named-entity recognition. NER models are often used to extract knowledge from free text data including medical literature. They work by identifying spans within an input text that mention named entities and then assigning these spans into predefined categories (such as "Cell Line", "Tissue", "Sex", etc.). By learning their parameters and making use of the entire input sentence as context, these systems overcome the limitations of simple regular expression-based approaches. In particular, certain works [156, 201, 345] have employed NER to map free textual description into existing concepts from curated specialized ontologies that are well-accepted by the biomedical community to improve the integrated use of heterogeneous datasets. In practice, training NER models can be difficult, since the training sequences must be labeled on an individual word level. This is especially time-consuming in the genomics domain, where biomedical fields require specific and technical labels.
Imputation from gene expression. The automated label extraction (ALE) platform [165], trains ML models based on high-quality annotated gene expression profiles (leveraging on text-extraction approaches based on regular expression and string matching). However, the information is limited to a small set of patient characteristics (i.e., gender, age, tissues). Authors in [139] also predict sample labels using gene expression data; a model is built and evaluated for both biological phenotypes (sex, tissue, sample source) and experimental conditions (sequencing strategy). The approach is applied to repositories alternative to GEO (i.e., training from TCGA samples, testing on GTEx [255] and SRA).

Chapter 6
Searching Genomic Data

"When someone seeks – said Siddhartha – then it easily happens that his eyes see only the thing that he seeks, and he is able to find nothing, to take in nothing. [...] Seeking means: having a goal. But finding means: being free, being open, having no goal."

— Herman Hesse

In our ongoing effort to provide the genomics community with useful concepts and tools, our next challenge is to make the content of the META-BASE repository searchable. We act in a context that shows a growing interest in exploiting the richness of ontological knowledge and facilitating the exploration of scientific datasets and their features; therefore we put a strong effort into allowing *semantic search* over metadata.

As described in Section 4.7, along with the Genomic Conceptual Model we implemented the multi-ontology semantic knowledge base of genomic terms and concepts. The union of GCM and the related ontological Knowledge Base has given birth to the Genomic Knowledge Graph (GKG). We selected ten attributes from GCM; values associated with each of the ten attributes were semantically enriched by using the respective best ontologies, after a careful domain-specific selection process. We assigned an ontological term, we described synonyms and other syntactic or semantic variants and provided a hierarchy of hyperonyms and hyponyms.

Semantic search technology, which is fueling the main search engines developed by Google, Microsoft, Facebook, and Amazon, is empowered by the use of large knowledge graphs, supporting search at the semantic level. In these systems, when the query string can be reliably associated with a given entity, other similar instances which are associated with that entity are also retrieved and displayed together with the entity properties. For instance, if the query string is associated with a book, then the semantic interpretation allows search engines to retrieve similar books and display the book's properties.

Inspired by the successful exploitation of knowledge graphs in search engines, we envisioned a semantic search approach empowered by our Genomic Knowledge Graph. However, our approach to semantic search differs from the above paradigm; we focus only on *domain specific* outputs and take into account that the target

A. Bernasconi: *Model, Integrate, Search... Repeat*, LNBIP 496, pp. 119–155, 2023.
https://doi.org/10.1007/978-3-031-44907-9_6

scientific community requires that users have *control on inference*, by choosing the appropriate data description level: from original values to normalized values, synonyms, and hyponyms. With this goal, we propose two different search interfaces that target different audiences:

- **GeKnowGraph**, a Neo4j graph-based database with a visual interface, equipped with explanation and exploration mechanisms for supporting semantic queries (specific for the scientific domain, where users are experts and need full control of inference);
- **GenoSurf**, a simpler interface where the graph is not explicit but multiple semantic levels can be set to tune the search needs of non-expert users.

Chapter organization. In Section 6.1, we hint at the power of knowledge expression in the genomic domain. Section 6.2 describes how we explain inferences that are performed in order to extract matching items for a given query. Section 6.3 briefly discusses the Neo4j data conversion from tabular format into the GKG and overviews how a user can drive more inferences through explicit navigation of the GeKnow-Graph; we show significant patterns of interaction enabled by the use of the graph. Section 6.4 presents the GenoSurf interface, overviewing the functionality and interplay of all its sections, presenting example use cases, and providing a validation of the system, executed on a group of 40 users. Sections 6.5 mentions related work on genomic and biological search systems.

For ease of explanation, in the following, we employ many examples using counts of items that match specific conditions. Please note that these counts refer to the version of the META-BASE content of July 2019, which in the meantime has been updated several times; readers may notice some mismatch with counts retrieved from the system, as of today.

6.1 Issues in Exploiting Semantic Knowledge in Genomics

We discuss the intricacies of exploiting biological enrichment by focusing on three attributes that represent the biological aspects of the analyzed samples, namely *Disease*, *Tissue*, and *Cell*. Indeed, although semantic enrichment search applies to ten attributes, most of the complexity has to do with these three attributes, as many search values match all of them; these values are substantially enriched, hence providing both opportunities (finding connected concepts) and risks (finding inappropriate concepts). We provide two typical examples of search, the former shows that terms may match distinct entities, and the latter shows that term hierarchies enable widening the search.

Search example 1 (based on GCM). As shown in the examples of Table 6.1, by using the same search keyword the user can reach various subsets of items, according to the different interpretations of the input keyword "brain":

- When it is understood as the tissue of provenance of a biological sample, 4 items originating from the "midbrain" tissue are extracted.
- When it is understood as the particular location of an organism's cells, 5 items from pericytes (i.e., multi-functional cells sustaining formation and functionality of the blood-brain barrier) are extracted.
- When it is understood as a pathology, 9,188 items are extracted, related to a brain cancer that has its material basis in glial cells.

Similar alternative interpretations are associated with the terms "lung" or "cervical". Different interpretations must be explained, as they may or may not be matching the user's intended query.

Keyword	*Tissue*: #items	*Cell*: #items	*Disease*: #items
"brain"	mid**brain**: 4	**brain** pericyte: 5	**Brain** Lower Grade Glioma: 9188
"lung"	Embryonic **lung**: 13	fibroblast of **lung**: 159	large cell **lung** cancer: 6
"cervical"	**cervical**: 1	HeLa-S3 **Cervical** Carcinoma Cell Line: 41	**cervical** cancer: 10

Table 6.1 Keywords spanning multiple attribute domains and possible interpretations.

Search example 2 (based on GKG). The capability of extracting data is amplified by normalization (using ontological terms and exact/broad/related synonyms) and enrichment (using ontological hierarchies). Thus, we can support a much more powerful search. To clarify this, we briefly discuss the case of search keywords related to the broad "uterus" concept. Figure 6.1 is an excerpt of the Uberon ontology [291], useful to grasp the ontological structure that contains the concepts interesting for this example. The "uterus" concept (ID:0000995) includes, among others, three parts: "body of uterus" (ID:0009853), "uterine cervix" (ID:0000002), and "uterine wall" (ID:0000459). The last one has the "endometrium" part (ID:0001295). Each concept can be related to exact or broad synonyms, related adjectives, and alternative syntax.

Table 6.2 reports which values (Search keyword) in our system have been clustered using an ontological Term ID and how many data items they allow to retrieve when the search uses only the original values (`Orig.`), the normalized values (`Syn.`), or the hierarchical hyponyms deriving from "uterus" (`Exp.`). Depending on the interpretation, the number of matches changes significantly.

By concentrating on the first entry, the search of "uterus" using plain term interpretation of GCM retrieves 57 data items; when the `Syn.` search is enabled, it returns 1708 items (as it matches also "uterus nos", which stands for "uterus, not otherwise specified"); when the `Exp.` search is enabled, it returns 16851 items (as it matches also all other terms in Table 6.2). Table 6.2 shows the matches of eleven terms originating from the "uterus" root in Uberon, at the three levels of enabled search.

The user thus reaches documents relevant to her search by setting the levels, without knowing the specific nomenclature used in the various integrated data sources.

However, knowing the actual matches is critical, as the user may not wish to extract all matched items; it depends on the specific needs.

Term ID	Search keyword	Orig.	Syn.	Exp.
0000995	uterus	57	1708	16851
	uterus nos	1651	1708	16851
0009853	body of uterus	0	9535	9535
	corpus uteri	9535	9535	9535
0000002	uterine cervix	0	5585	5585
	cervix uteri	5417	5585	5585
	cervix	167	5585	5585
	cervical	1	5585	5585
0000459	uterine wall	0	0	23
0001295	endometrium	21	23	23
	endometrial	2	23	23

Table 6.2 Number of retrieved items using different keywords from the "uterus" concept area. `Orig.` stands for search at the basic level, `Syn.` for search enriched by equivalent terms and synonyms, `Exp.` for search on the ontological descendant hierarchy.

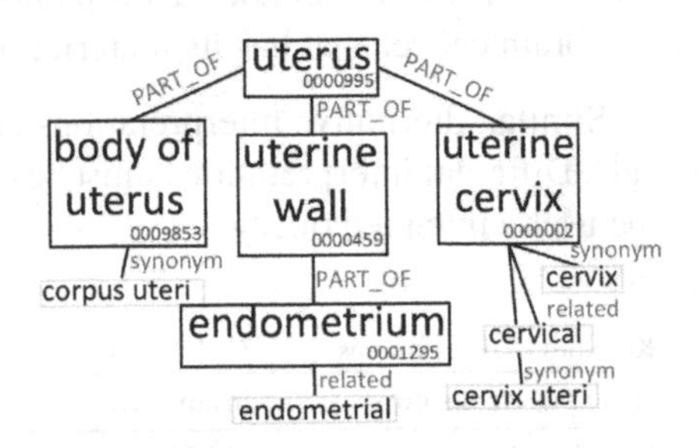

Fig. 6.1 Excerpt of the Uberon sub-tree originating from the "uterus" root. For space reasons, we only report the elements that are relevant to our example.

6.2 Inference Explanation

Users should be provided with *system's explanations*, giving evidence of the cognitive process involved in the inference, so that they can verify it (they may actually provide corrections when the inference is wrong). Therefore, for any inferred results, we must show the chain of deductions followed to infer the information.

Behind the scenes, the implemented keyword-based search is driven by a precise inference mechanism, which is tuned according to the user's choice of query option (`Original`, `Synonym` and `Expanded`) and is based on semantic enrichment. To illustrate the relational links that are traversed in the different kinds of searches over our system, we introduce the concept of *deduction chain*, which describes the internal path in the database that links the retrieved Item to the table where the match with the search keyword is found. The deduction chain shows the steps of the inference process that are activated according to the requested search level. A search may be performed considering:

1. the source original metadata key-value pairs;
2. the GCM attributes;

3. additionally, the ontological synonym annotations;
4. additionally, the ontological hierarchical expansions.

Level	#Items	Deduction chain
1	789	⟨Item⟩-⟨Key: `biosample__organ_slims`, Value: **brain**⟩
	4,670	⟨Item⟩-⟨Key: `gdc__project__disease_type`, Value: **Brain** Lower Grade Glioma⟩
	126	⟨Item⟩-⟨Key: `clinical__lgg__family_history_of_primary_`**`brain`**`_tumor`, Value: NO⟩
	2,463	⟨Item⟩-⟨Key: `clinical_patient__history_lgg_dx_of_`**`brain`**`_tissue`, Value: yes⟩
2	15,714	⟨Item⟩-⟨Replicate⟩-⟨BioSample.*Tissue*: **Brain**⟩
	9,188	⟨Item⟩-⟨Replicate⟩-⟨BioSample.*Disease*: **Brain** Lower Grade Glioma⟩
	10	⟨Item⟩-⟨Replicate⟩-⟨BioSample.*Cell*: smooth muscle cell of the **brain** vascula⟩
3	13	⟨Item⟩-⟨Replicate⟩-⟨BioSample.*Cell*: Fetal brain⟩ -⟨Vocabulary: **brain**, UBERON_0000955⟩
4	10	⟨Item⟩-⟨Replicate⟩-⟨BioSample.*Tissue*: Pons⟩ -⟨Vocabulary: pons, UBERON_0000988⟩ -⟨Vocabulary: regional part of **brain**, UBERON_0002616⟩
	8	⟨Item⟩-⟨Replicate⟩-⟨BioSample.*Tissue*: globus pallidus⟩ -⟨Vocabulary: globus pallidus, UBERON_0001875⟩ -[PART_OF]-⟨Vocabulary: pallidum, UBERON_0006514⟩ -[IS_A]-⟨Vocabulary: **brain** gray matter, UBERON_0003528⟩ -[PART_OF]-⟨Vocabulary: **brain**, UBERON_0000955⟩

Table 6.3 Available search levels and examples of their results for the "brain" search keyword.

The best way to illustrate deductive chains is by showing an example of use. A biologist or bioinformatician may be interested in the keyword "brain", intending to request all data items related to this concept (i.e. those that contain this string in their metadata). Table 6.3 shows how quantitative results in our system (i.e. numbers of items) can be explained: according to the different search levels (first column), we indicate the number of found data files (second column), and we show the deduction chain (third column). At the first level, the search produces key-value pairs corresponding to unchanged original metadata directly linked to the ⟨Item⟩. Thus, the first four rows of Table 6.3 link an ⟨Item⟩ directly to a ⟨Key,Value⟩; at this level, term matching can be performed on either keys or values. For instance, the first row indicates that 789 found items are associated with the pair ⟨`biosample__organ_slims, brain`⟩.

At the second level, the search is performed on the attributes of the core schema; results in Table 6.3 match values contained either in the *Tissue*, *Disease* or *Cell* attributes. ⟨Replicate⟩, further connected to a ⟨BioSample⟩, which contains the "Brain" value for the *Tissue* attribute. Instead, by looking at the last row, the user learns that there exist ten additional items connected to a ⟨Replicate⟩, further connected to a ⟨BioSample⟩, whose *Cell* field is "smooth muscle cell of the brain vascula".

At the third level, the search is based on ontological vocabularies plus their synonyms. The example in Table 6.3 shows that we found 13 items whose original

Cell value is "fetal brain", a synonym of "brain" annotated with the Uberon ontology term 0000955.

At the fourth level, the search is based on ontological vocabularies, their synonyms, plus their hyponyms, with more complex chains, where ontological terms are linked by relations expressing containment or generalization (i.e., *IS_A* and *PART_OF* relationships). For instance, the first row in Table 6.3 for the fourth search level indicates that 10 items are associated with the term "pons", which is a "regional part of brain" according to the Uberon ontology terms 0000988 and 0002616.

6.3 GeKnowGraph: Exploration-Based Interface

The structure of the knowledge graph should be exposed to knowledgeable users; these should be provided with *exploration capabilities* for accessing entities and relationships, e.g., by navigating from given experiments to the cell lines or tissues of provenance to the donors with their demography and phenotypes, and to the extraction process with the used technology and device, explicitly indicating the enrichment level of the terms of GKG. GeKnowGraph is a Neo4j-based query-answering system that responds to this need.

Among many available graph databases (e.g., Neptune,[1] Cosmos[2] or Titan[3]) we have chosen Neo4j,[4] currently the leading open source graph database, used by several companies also in the bioinformatics domain (e.g., EBI,[5] Intermine[6] and Reactome[7]). Neo4j implements the property graph model at the logical and storage level. *Nodes* are entities in the graph: they can hold any number of attributes, called *properties*, and can be tagged with *labels*, which represent different roles. *Relationships* provide connections between two node entities that are directed, named, and semantically relevant.

We employ a pipeline that extracts data from the relational database represented in Figure 4.11 and builds a Neo4j property graph instance. Specifically, we map GCM entities onto Neo4j node labels (with a 1:1 correspondence); GCM entity attributes onto node properties; GCM foreign keys onto directed relationships. We associate just one label to nodes and edges.

With a bash script, we input the information useful to define nodes' labels, edges' labels, properties, and how the results of the relational queries should be converted as elements of the graph. The specific transformation is applied using a batch loader tool into the property graph.

---

[1] https://aws.amazon.com/neptune/

[2] https://docs.microsoft.com/en-us/azure/cosmos-db/

[3] http://titan.thinkaurelius.com/

[4] https://neo4j.com/

[5] https://www.ebi.ac.uk/ols/docs/neo4j-schema

[6] https://github.com/intermine/neo4j

[7] https://reactome.org/dev/graph-database

The graph schema is represented in Figure 6.2, covering four layers of semantic detail, with a growing level of abstraction from the bottom towards the top. These four layers correspond also to the semantic search levels explained in Section 6.2. Original metadata, GCM schema, and ontology terms have been thoroughly described in the previous chapters. Here we highlight that terms are linked through relationships that represent subsumption (*IS_A*) – thus including hyperonyms and hyponyms of the stored terms – and containment (*PART_OF*) – thus including their holonyms and meronyms.

The current instance presents: 250K item nodes in Genomic Conceptual Model, 320K nodes for Genomic Conceptual Model (including Items, Biosamples, Donors, Datasets, ExperimentTypes. . .), 1.6M relationships among GCM nodes, 100K nodes for Ontological Terms (including annotation to ten properties on GCM nodes, up to 3 levels on the hierarchy), and 45M original metadata key-value pairs. The most updated version of the Neo4j database is made available on our servers at `http://geco.deib.polimi.it/dump-gkg/`.

For visualization purposes, we exploit the Neo4j native browser, which allows us to query the underlying graph database through Neo4j's query language (i.e., Cypher [152]) and gives the possibility to browse query results in tabular form and in a graphical, customizable, format. In the current instance, we exploit the graph layout automatically obtained in the native Neo4j console, which is a force-directed algorithm [35], with the aim to position the nodes in the two-dimensional screen so that the edges are of similar length and the number of crossing edges is minimized. However, as the layout of intermediate and final graphs may considerably impact the *comprehensibility* of the graph, in the future, we will explore more sophisticated layout algorithms and libraries.[8] Additionally, there is a problem of *scalability* of the graph, as it can become quickly very big, considering the size of source-specific attribute pairs and of ontological layers. At this moment, this is handled by a default limit imposed by the Neo4j browser. However, in the following works this issue may be addressed by clustering the nodes and their relationships at different levels of abstraction (as it was proposed in [378]), or by enforcing fish-eye views on the full graph (see [157]). Such extension may be investigated on both syntactic and semantic levels, by proposing several types of abstraction [6].

6.3.1 Exploration Interaction

The most interesting use of the GeKnowGraph occurs with visual exploration, which however requires a user to understand the entities and relationships of GCM, as well as their linking to the vocabulary and then to navigate the generalization *IS_A* and the containment *PART_OF* relationships. The user exploration may start from

[8] Many libraries of front-end programming languages – such as HTML and JavaScript – are available to employ sophisticated layout algorithms. See, for example, yFiles (`https://www.yworks.com/pages/visualizing-a-neo4j-graph-database.html`), Linkurious (`https://linkurio.us/`), or d3.js (`https://github.com/eisman/neo4jd3`).

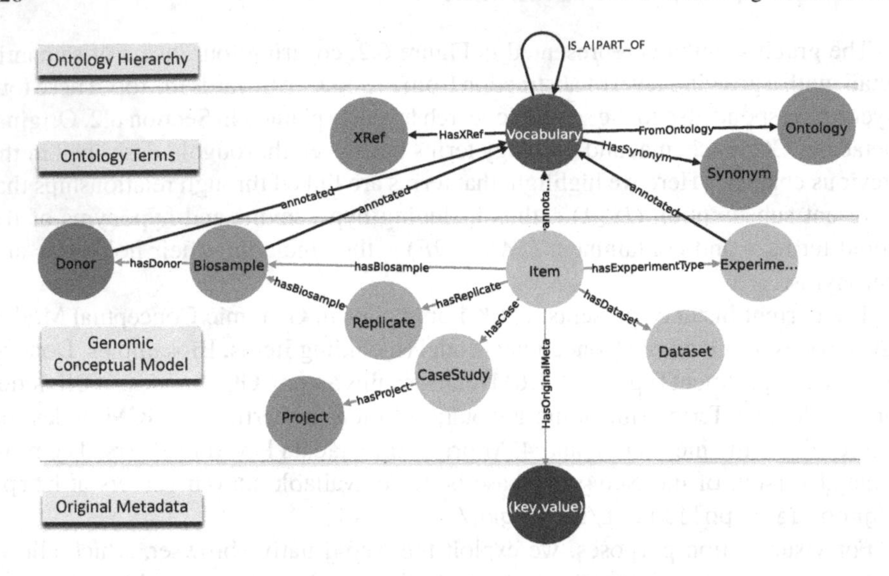

Fig. 6.2 Visual representation of the GKG components in a Neo4j graph schema.

GCM entities or from the vocabulary terms. We next explain 4 typical patterns of exploration: finding items of a given dataset, of a given patient, of a given case study, and associated with a given term.

Finding other items from the same datasets. A typical three-step exploratory interaction from an Item to a different ITEM of the same DATASET is shown in Figure 6.3. Entity instances are represented as circles that include the value of entity identifiers or some relevant properties; directed edges, carrying the relationship names, connect entity instances. At all times, one of the entity instances is the *navigation handler*, and its attributes can be (on request) extensively represented in a box presented below the diagram. The end of the navigation is shown in Figure 6.3 (C), where the navigation handler points to entity ITEM ENCFF42, but the navigation starts from ITEM ENCFF58 in Figure 6.3 (A).

We use Figure 6.3 (A) to illustrate the typical organization of a GCM instance, centered of the ITEM ENCFF58 (gray color, in the center), connected to the other entities REPLICATE, BIOSAMPLE, DONOR (colors from pink to dark red, along the biological view), to CASESTUDY and PROJECT (yellow colors, along the management view) and to EXPERIMENTTYPE (green color, along the technology view). In Figure 6.3 (B) we show that the user navigates to the DATASET entity (blue color, along the extraction view), where several other ITEM instances of the same DATASET are illustrated; then, Figure 6.3 (C) shows the end of the navigation. Navigation occurs by double-clicking on entity instances, while attributes of a given entity instance (in this case, of ITEM) are displayed by single-clicking.

Finding all the datasets of a given patient. Another typical search query asks for all data types pertaining to a specific cancer patient; associating the same patient

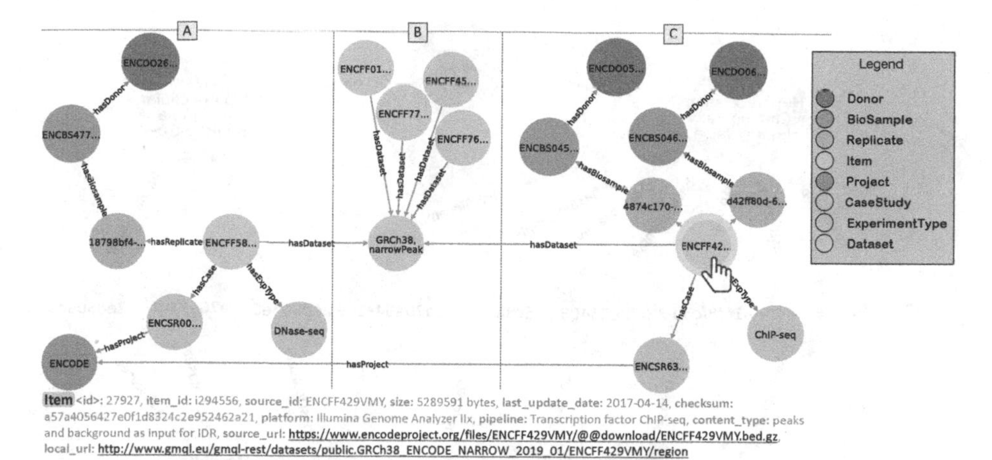

Fig. 6.3 Sequential interaction, from panel (A)—centered on ITEM ENCFF58—to panel (B)—centered on GRCh38 narrowPeak DATASET—to panel (C)—centered on ITEM ENCFF42. Note that the items in (A) and (C) share the same PROJECT, ENCODE.

with heterogeneous data types is highly valuable in order to understand the possible research questions that can be asked to the underlying data repository. However, this query must be explored patient by patient, as each patient may be associated with a highly variable number of data types.

As shown in Figure 6.4, we represent DONORS through their ethnicity, gender, and age (in this specific case through values [asian, male, 49y]). The database stores two biological samples extracted from this patient, who is affected by "Liver Hepatocellular Carcinoma". One sample is tumoral and the other one is healthy (i.e., a control). By further expanding the nodes, the user reaches the Item level, thereby extracting 9 data Items which belong to 7 different DATASETS, each showing the type of data described in the region files (e.g., mutations, methylation levels, copy number variations, and RNA or miRNA gene expression).

Exploring the organization of a given case study. Figure 6.5 shows another typical exploration. Assume that a user is not aware of what constitutes a case of study in the ENCODE data source and wants to discover it. Thus, she starts with a given CASESTUDY entity ENCSR63, shown at the bottom of the figure. This entity represents a set of ITEMS that are gathered together because they contribute to the same research objective. The interaction first allows us to visualize the group of eight ITEMS associated with this case study, belonging to the hg19_narrowPeak and GRCh38_narrowPeak DATASETS (respectively having cardinality five and three). Then, the underlying biological views are revealed, by showing that all the ITEMS are associated with chains originating from two distinct DONORS.

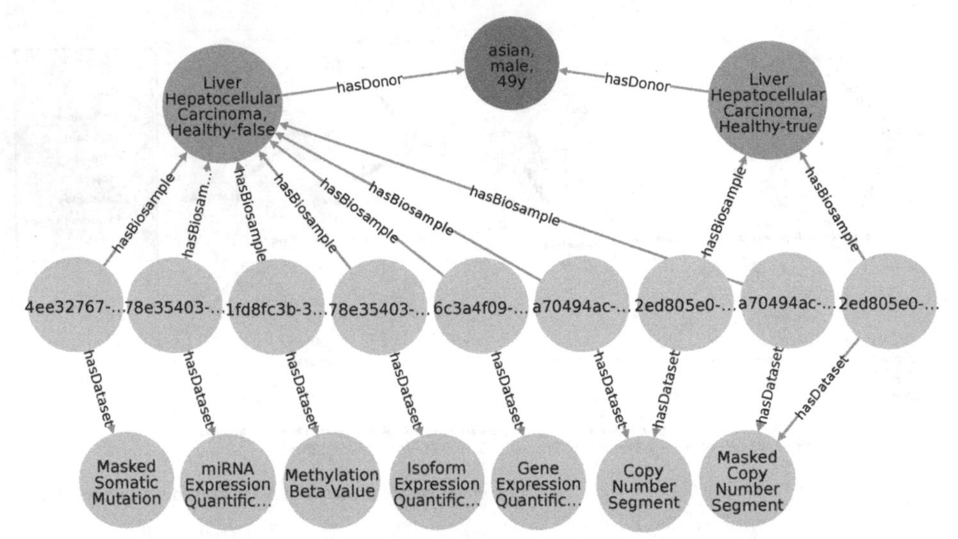

Fig. 6.4 Exploration starting from a DONOR, providing tumoral and normal tissues, which are used to provide *Items* belonging to different DATASETS. Note that here we omit REPLICATE nodes for space reasons; they have 1:1 correspondence with BIOSAMPLES.

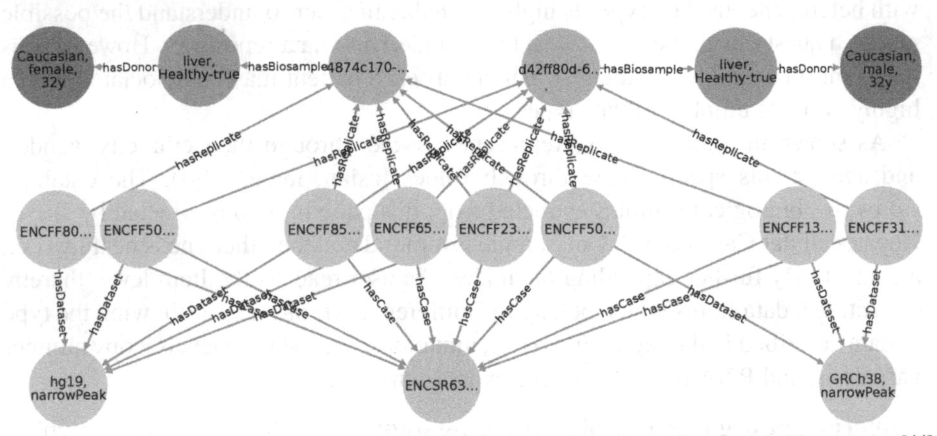

Fig. 6.5 From bottom to top: a CASESTUDY contains multiple ITEMS, which derive from two different REPLICATES/BIOSAMPLES/DONORS and are contained in two DATASETS based on the reference assembly of the genome.

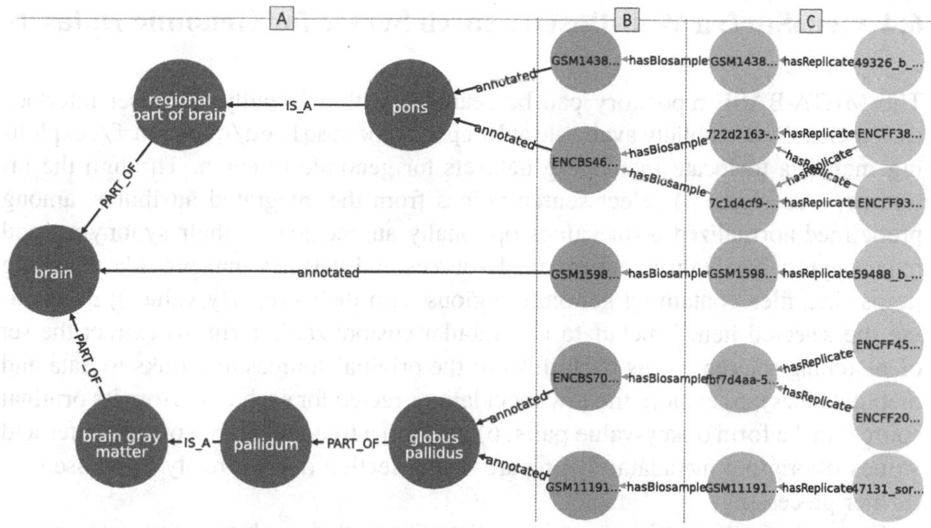

Fig. 6.6 Search starting from ontological terms. Essentially, (A) contains the ontological terms, (B) contains annotated BioSamples, and (C) the Replicates (pink) and derived Items (grey). This example shows the intricacy of how BioSamples, Replicates and Items connect to each other.

Ontological exploration. Another powerful use of the GeKnowGraph starts from ontological terms; this search allows a user to take advantage of the standardization and enrichment efforts performed on the different integrated data values. By starting from terms, the user may see how each term is connected to different entities, thereby typically exploring the hierarchical structure of ontological terms.

The visual representation is of great help in understanding the complex relationships among data, otherwise hard to describe. Figure 6.6 shows how multiple Items (grey nodes on the right) can be retrieved by using different deduction chains starting from the same hierarchical ancestor, ⟨brain⟩. A typical search may start from this entity, which already has a number of connected BioSamples (i.e., samples which have been *annotated*, as related to brain concept) and progressively discover all its sub-concepts up to the level where terms annotate other BioSamples. Then, the exploration connects BioSamples to their Replicates and eventually to Items.

Note that, in the figure, ⟨brain⟩ directly annotates a BioSample and is an indirect hyperonym of ⟨pons⟩ and ⟨globus pallidus⟩, each connected to two BioSamples. Note also that five BioSamples give rise to six Replicates and then to seven Items, and also note that some Items are associated with two Replicates. Once Items are reached, the user may be interested in understanding from which datasets or experiment types they derive; this is possible by further exploring from the Item nodes, using the first pattern of exploration discussed in this Section.

6.4 GenoSurf: a Web-Based Search Server for Genomic Datasets

The META-BASE repository can be searched with a friendly web user interface called GenoSurf, publicly available at `http://www.gmql.eu/genosurf/`, exploiting metadata to locate interesting datasets for genomic research. Through the interface, a user can: 1) select search values from the integrated attributes, among predefined normalized term values optionally augmented by their synonyms, and hyperonyms; 2) obtain a summary of sources and datasets that provide matching items (i.e., files containing genomic regions with their property values); 3) examine the selected items' metadata in a tabular customizable form; 4) extract the set of matching references (as back-links to the original sources and links to data and metadata files); 5) explore the raw metadata extracted for each item from its original source, in the form of key-value pairs; 6) perform a free-text search on attributes and values of original metadata; 7) prepare data selection queries ready to be used for further processing.

Search is facilitated by drop-down lists of matching values; aggregate counts, describing resulting files, are updated in real-time. The metadata content is fueled by the META-BASE automatized pipeline thoroughly described in Chapter 4 and is stored in a PostgreSQL database whose logical schema was previously shown in Section 4.9 – containing the GCM (hereon called "core schema"), the knowledge base, as well as original metadata – including for each item a link to the original source storing the referenced data.

GenoSurf data items are in one-to-one mapping with the most recent version of the data files in the GMQL repository and share the same identifiers. Hence, the result of a GenoSurf search can be immediately used within the GMQL engine to extract and directly process comprehensively relevant genomic region data files and their metadata.

Users can search the integrated content by exploiting its describing metadata (enhanced with synonyms and hyperonyms) and retrieving a corresponding list of matching genomic data files. The interface is composed of five sections, described in Figure 6.7:

1) the *Menu Bar* to navigate the different services and their documentation;
2) intuitive *Query Utilities*;
3) the *Data search*: a search interface over the core database (i.e., whose content can be set on three levels: original metadata, synonyms/alternative syntax, and hierarchical ontological expansion;
4) the *Key-value search*: a search interface over key-value pairs, for searching over original metadata from the imported sources;
5) a *Result Visualization* section, showing the resulting items in three different aggregation sections.

The interface also enables an interplay between the core Data search and Key-value search, thereby allowing to build complex queries given as the logical conjunction of a sequence of core metadata and key-value search steps of arbitrary length; results are updated at each step to reflect the additional search conditions,

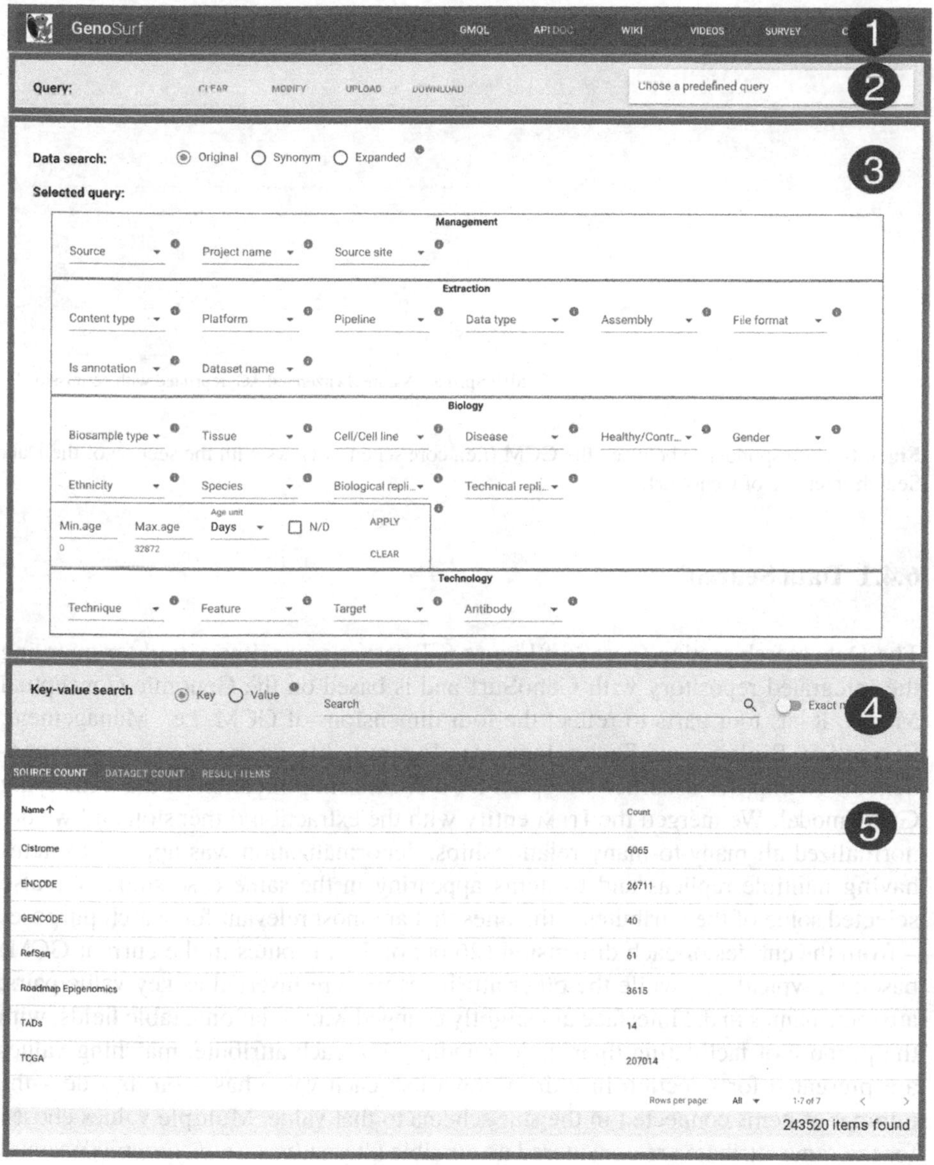

GenoSurf GMQL API DOC WIKI VIDEOS SURVEY 1

Query: CLEAR MODIFY UPLOAD DOWNLOAD Chose a predefined query 2

Data search: Original Synonym Expanded 3

Selected query:

Management

Source Project name Source site

Extraction

Content type Platform Pipeline Data type Assembly File format

Is annotation Dataset name

Biology

Biosample type Tissue Cell/Cell line Disease Healthy/Contr... Gender

Ethnicity Species Biological repli... Technical repli...

Min.age 0 Max.age 32872 Age unit Days N/D APPLY CLEAR

Technology

Technique Feature Target Antibody

Key-value search Key Value Search Exact 4

SOURCE COUNT DATASET COUNT RESULT ITEMS 5

Name ↑	Count
Cistrome	6065
ENCODE	26711
GENCODE	40
RefSeq	61
Roadmap Epigenomics	3615
TADs	14
TCGA	207014

Rows per page: All 1-7 of 7

243520 items found

Fig. 6.7 Sections of GenoSurf web interface: 1) Top menu bar; 2) Query utilities; 3) Data search; 4) Key-value search; 5) Results visualization.

and the counts are dynamically displayed to help users in assessing if query results match their intents. In the following, we describe the main GenoSurf sections more in detail.

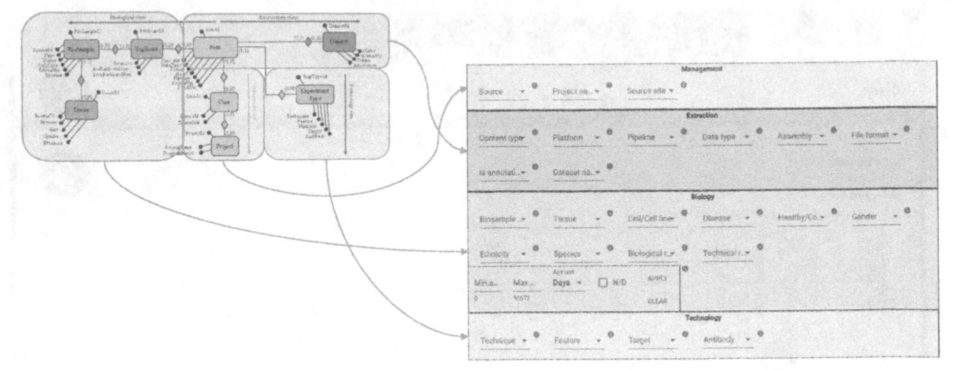

Fig. 6.8 Correspondence between the GCM (i.e., core schema) views with the section of the Data Search interface of GenoSurf.

6.4.1 Data Search

The Data search section (part 3 of Figure 6.7) serves as a primary tool for querying the integrated repository with GenoSurf and is based on the Genomic Conceptual Model. It has four parts to reflect the four dimensions of GCM, i.e., Management, Extraction, Biology, and Technology (see Figure 6.8 to appreciate the correspondence). To improve usability we opted for a drastic simplification of the underlying GCM model. We merged the Item entity with the extraction dimension and we denormalized all many-to-many relationships; denormalization was applied to items having multiple replicas and to items appearing in the same case study. We also selected some of the attributes – the ones that are most relevant for search purposes – from the entities of each dimension (26 out of 38 attributes in the current GCM) based on typical use, while the other attributes were re-inserted as key-value pairs; attribute names in the interface are slightly changed w.r.t. relational table fields, with the purpose of facilitating their understanding. For each attribute, matching values are presented for selection in a drop-down list; each value has – on its side – the number of items connected in the star schema to that value. Multiple values chosen for the same attribute are considered as possible alternatives (in *disjunction*); values chosen over different attributes are considered as conditions that should all be satisfied (in *conjunction*) by the resulting items. The special value "N/D" indicates `null` values and allows to select items for which a particular attribute is undefined. After each selection, a running query is progressively built and shown in the interface field "Selected query"; the current query is evaluated, and the number of matching items is displayed. The interface allows setting different levels of semantic enrichment: the `Original` option (search using metadata values provided by original data sources), the `Synonym` option (adding synonyms), and the `Expanded` option (adding hyperonyms and hyponyms). On the basis of our experience, too many levels of ontological enrichment would bring unnecessary sophistication, which would not be appreciated by users. Thus, we included ontology terms' hyperonyms and hyponyms up to three levels of depth, as a reasonable trade-off that also guarantees acceptable query performances.

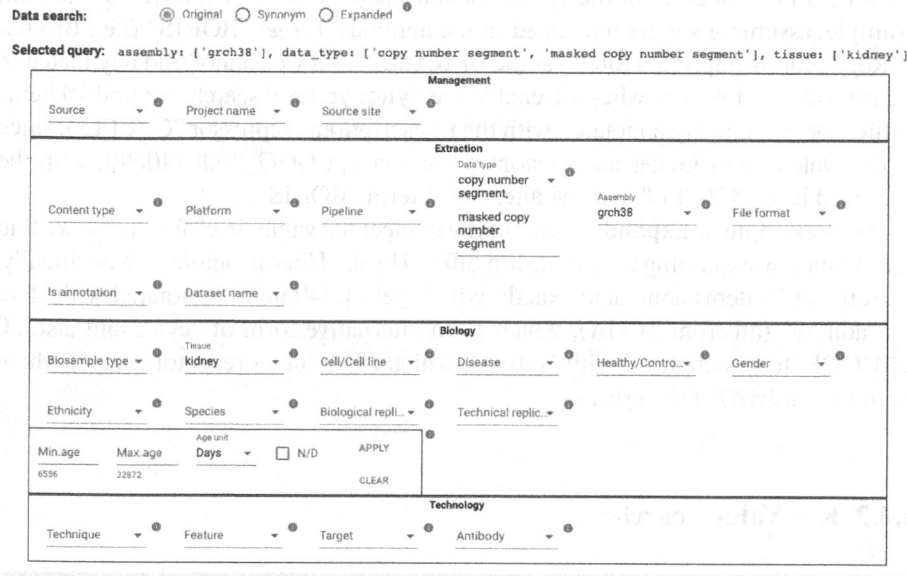

Fig. 6.9 Data search section of the GenoSurf web interface, highlighting attributes within the four dimensions of the repository core schema; values are entered by users and appear in drop-down menus for easing their selection.

In the example shown in Figure 6.9, the user is searching for all items that have *Data type* either "copy number segment" or "masked copy number segment", and that have *Assembly* "grch38" and *Tissue* "kidney"; the query option is set to `Original`. As a consequence of the attribute value selection, the field "Selected query" is compiled as follows.

```
assembly: ['grch38'], data_type: ['copy number segment', '
    masked copy number segment'], tissue: ['kidney']
```

Counts of items associated with attribute values are changed dynamically. For example, if initially, the *Data type* drop-down menu shows 22,371 items for "copy number segment" and 22,374 for "exon quantification", when the user selects the *Tissue* "kidney" (with count 19,357), the *Data type* drop-down menu shows 1,862 items for each of the two mentioned *Data types*, reflecting the reduction of matching items.

As an example of query with Data search option set to `Synonym`, when we only select the value "k562" for the *Cell/Cell line* attribute (which at the `Original` level had a 5,942 count), we obtain a count of 5,986 items, which is the same for all equivalent syntactic variants "k-562", "k562 cell", "k-562 cell" of the attribute. Indeed, the additional 44 items derive from a small set of items labeled with "k562 leukemia cells", which have been annotated with such synonym concept corresponding to the term EFO_0002067 in the Experimental Factor Ontology [268]. As a second example, assuming we are interested in the antibody *Target* "BORIS" (i.e., Brother of Regulator of Imprinted Sites), at the `Original` level we cannot find any match in the repository. However, when we enable the `Synonym` level search, we find 10 items (which were originally annotated with the transcriptional repressor "CTCFL"), since in the Ontology of Genes and Genomes the concept OGG_3000140690, with the preferred label "CTCFL", has the alternative term "BORIS".

As an example of `Expanded` search, if we select the value "eye" for *Tissue* we find 1,473 items by exploiting the expansion offered by the Uberon ontology. Specifically, we retrieve 13 items annotated exactly with "eye", 1,440 items annotated with "Eye and adnexa" (all from TCGA), which is an alternative form of "eye", and also 20 ENCODE items annotated with "retina", which *IS_A* "photoreceptor array", which is, in turn, a *PART_OF* "eye".

6.4.2 Key-Value Search

The Key-value search section (part 4 of Figure 6.7) allows searching metadata without having previous knowledge of the original metadata attribute names and values, or of the attribute names and data content of the GCM core schema, which stands behind the integration effort. In the Key-value search, the user can perform a case-insensitive search either over metadata attributes (using the Key option) or over metadata values (using the Value option). Users can search both keys and values that either exactly match or only contain the input string.

When input strings are searched within keys, in case a match is found among the core attributes of the Genomic Conceptual Model (which can also be considered as 'keys'), we provide an informative result: example values of each of the matched attributes and the number of distinct values available for that attribute. Conversely, when showing the results of a match on original attributes (i.e., keys), a list of all matching keys is provided in the output, equipped with the number of distinct values available for each of such keys; the user can then explore these values and select any of them. Figure 6.10 shows a search with the Key option and input string "disease".

Value search has a simpler interface, showing all possible matches in values, both for core attributes and original key-value pairs. Users can directly select desired key-value pairs among the ones shown in the result.

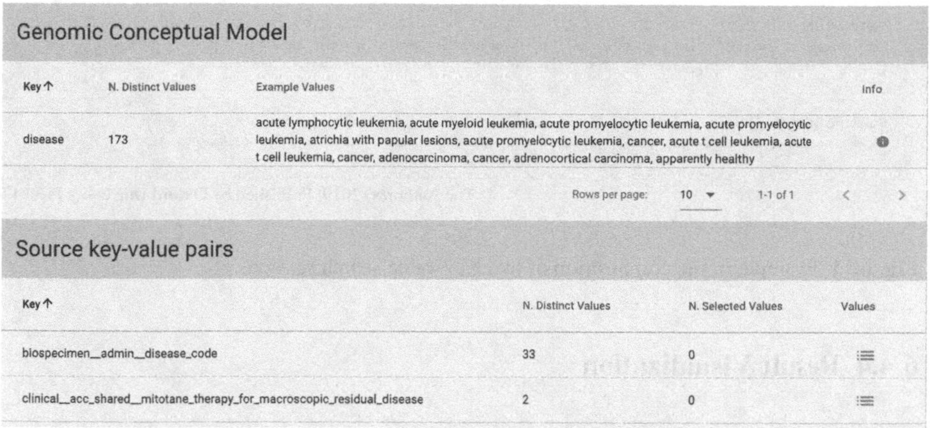

Genomic Conceptual Model

Key ↑	N. Distinct Values	Example Values	Info
disease	173	acute lymphocytic leukemia, acute myeloid leukemia, acute promyelocytic leukemia, acute promyelocytic leukemia, atrichia with papular lesions, acute promyelocytic leukemia, cancer, acute t cell leukemia, acute t cell leukemia, cancer, adenocarcinoma, cancer, adrenocortical carcinoma, apparently healthy	

Rows per page: 10 1-1 of 1

Source key-value pairs

Key ↑	N. Distinct Values	N. Selected Values	Values
biospecimen__admin__disease_code	33	0	
clinical__acc_shared__mitotane_therapy_for_macroscopic_residual_disease	2	0	

Fig. 6.10 Key-value search result using input string "disease" as a Key. The keyword is matched both in the GCM attributes (for each matching attribute we present the number of available distinct values and some example values) and in the original source attributes (each matching attribute enables exploration and selection of any corresponding values).

6.4.3 Query Sessions

A query consists of a sequence of search sessions, performed by alternating simple Data search and Key-value search; a sequence of searches produces items resulting from the conjunction of search conditions. Within each search session, multiple options for values (either for core attributes or as keys/values in Key-value search) are considered in disjunction. Figure 6.11 shows how a query can be composed using a sequence of two Key-value search sessions; steps can be deleted by rolling them back in any order. The query of Figure 6.11 corresponds to the predicate:

```
("biospecimen__admin__disease_code" = "chol" OR "
   biospecimen__admin__disease_code" = "kich" OR "
   clinical_patient__history_immunological_disease" = "
   hashimoto's thyroiditis") AND "
   biospecimen_sample__sample_type" = "primary tumor"
```

Every choice in the Data search and Key-value search sections impacts the results and their visualization at the bottom of the web interface page (part 5 of Figure 6.7). Acting on the Data search section, the result table is updated whenever the user either adds or removes a value from a drop-down menu or types/deletes text directly in a text field. In the Key-value search section, filters are instead applied/deleted by selecting the corresponding options – due to their greater complexity they are typically applied one-by-one, hence a dynamic update is not useful.

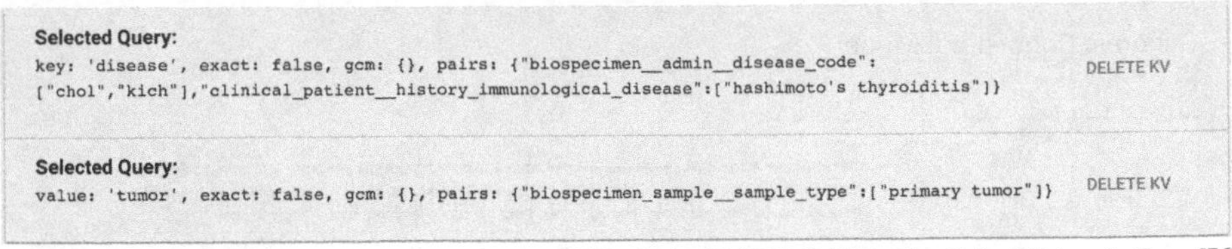

Selected Query:
key: 'disease', exact: false, gcm: {}, pairs: {"biospecimen__admin__disease_code": ["chol","kich"],"clinical_patient__history_immunological_disease":["hashimoto's thyroiditis"]}
DELETE KV

Selected Query:
value: 'tumor', exact: false, gcm: {}, pairs: {"biospecimen_sample__sample_type":["primary tumor"]}
DELETE KV

Fig. 6.11 Example of the composition of two Key-value search sessions.

6.4.4 Result Visualization

As shown in Figure 6.12, the result visualization includes three sections (see the blue bar at the top): 1) *Source count*, containing the number of found items aggregated by origin data source; 2) *Dataset count*, containing the number of found items aggregated by dataset name; 3) *Result items*, reporting the core metadata values of resulting items (to be navigated in batches of chosen cardinality, with suitable scroll options). Within the last section, a table is presented with one row for each item; for all found items, we also provide links to the data description page at the source location, and to data files at their source location ('Source URI') and at our repository location ('Local URI'). The user can visualize the original metadata key-value pairs of a data file by clicking on the row's 'Extra' button. In the bottom part of the table, the user can select how many rows should be visible on the page, up to a 1,000 limit; other pages can be scrolled using the left/right arrows. Fields can be arbitrarily sorted, included, or excluded from the tabular visualization ('Sort fields' button).

The user can change the one-item-per-row default view by using the 'Replicated/Aggregated' switch button; when items match many Replicates/Biosamples/-Donors, with the 'Aggregated' option, related information is aggregated by concatenating the possible distinct values through the pipe symbol "|".

6.4.5 Additional Functionalities

To provide a complete and useful environment to users, we allow them to modify, save, and load queries, as well as search results, in a customizable way; other functionalities allow them to use results produced by GenoSurf within the GMQL engine.

Interaction with Queries. To support the re-use of queries, we provide the possibility to download and upload text files in JSON format containing the query, or directly copy, paste, and modify JSON queries on the web interface. Furthermore, ten predefined queries are available to demonstrate practical uses of the interface.

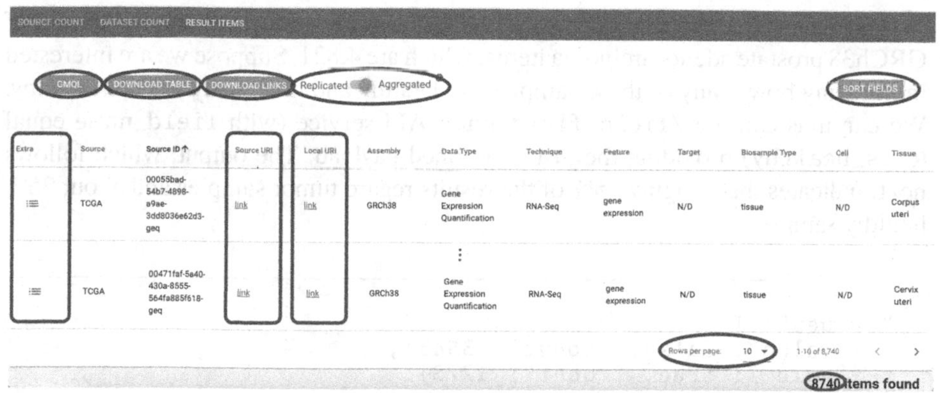

Fig. 6.12 Excerpt of the Result items table resulting from a search session. Red ellipses highlight relevant features. Top left: GMQL button to generate queries to further process related data files; DOWNLOAD buttons for Result items table and data file links; Replicated/Aggregated switch. Top right: SORT FIELDS button to customize the columns visualized in the table. Center: Extra, Source URI, and Local URI columns with clickable links. Bottom right: component to set the number of rows visible at a time; an indication of the total items corresponding to the performed query.

Use of Results. The genomic region data files retrieved by the performed search can be downloaded individually from the GenoSurf web interface using the 'Source URI', a clickable link to download the region data file from the origin source, or the 'Local URI', to download the region data file corresponding to the selected item from the GMQL system, when available (see Figure 6.12). Additionally, for each search query, we provide (through the buttons 'DOWNLOAD TABLE' and 'DOWNLOAD LINKS' in Figure 6.12): a text file containing all the URLs to download all the genomic region data files from our system, and a comma-separated file to download the entire results table.

Finally, the user can generate a GMQL query (button 'GMQL' in Figure 6.12) that can be used directly in our GMQL engine in order to select specifically the items found with a GenoSurf search for further processing.

RESTful API. All services used in the GenoSurf web interface are implemented using our GenoSurf API available at `http://geco.deib.polimi.it/genosurf/api/`. All POST services are based on the principle of setting a JSON payload that establishes the context for the next query. As an example, consider the JSON payload that follows.

```
{
  "gcm": {
    "disease": ["prostate adenocarcinoma"],
    "assembly": ["grch38"]
  },
  "type": "original",
  "kv": {}
}
```

This means that the next query (i.e., API request) is performed only on the set of GRCh38 prostate adenocarcinoma items, which are 4,821. Suppose we are interested in knowing how many of these samples are healthy and how many are non-healthy. We can thus call the `/field/field_name` API service (with `field_name` equal to `is_healthy`) providing the just mentioned payload. The output, which follows next, indicates that roughly 75% of the results regard tumor samples and about 25% healthy samples.


```
{
  "values": [
    {"value": false, "count": 3543},
    {"value": true, "count": 1278}
  ],
  "info": {
    "shown_count": 2,
    "total_count": 2,
    "item_count": 4821
  }
}
```


6.4.6 Use Cases

In this section, we show typical data retrieval queries performed by a hypothetical user of GenoSurf to select interesting subsets of the integrated repository.[9]

Extracting Cancer Patient Data. Suppose we are interested in extracting data of different types divided by patient for a specific cancer type. Evaluating genomic, epigenomic, and transcriptomic data of cancer patients in a comprehensive way provides a general view of their biomolecular system, possibly leading to novel findings. Let us consider as an example GRCh38 TCGA data for the disease "Cholangiocarcinoma". In total, the repository contains 401 related items divided into seven datasets, each of which contains between 45 and 85 different items, as can be observed in Figure 6.13.

In the 'RESULT ITEMS' table, the order of columns can be customized. For this particular query, it is useful to arrange *Source ID*, *Donor ID*, *Data Type*, and *Healthy* as the first columns. The resulting table can be sorted by *Donor ID* and downloaded as a CSV file. Groups of rows with the same *Donor ID* represent all available genomic region data files for each specific donor, with different data types and normal/tumor characterization. Table 6.4 shows an excerpt of the result, relative to two different patients having 9 and 14 items each. The first patient has normal/tumor data pairs for the Copy Number Segment and Masked Copy Number Segment data types, while the second patient has normal/tumor data for all available data types except for Copy Number Segment and Masked Somatic Mutation (in some cases

[9] More examples of interest can be found on the GenoSurf WIKI page at `http://www.gmql.eu/genosurf/`.

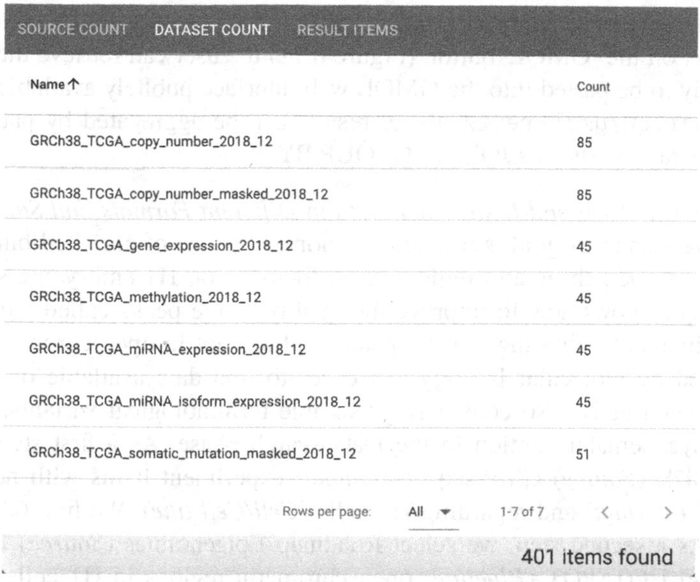

SOURCE COUNT DATASET COUNT RESULT ITEMS

Name ↑	Count
GRCh38_TCGA_copy_number_2018_12	85
GRCh38_TCGA_copy_number_masked_2018_12	85
GRCh38_TCGA_gene_expression_2018_12	45
GRCh38_TCGA_methylation_2018_12	45
GRCh38_TCGA_miRNA_expression_2018_12	45
GRCh38_TCGA_miRNA_isoform_expression_2018_12	45
GRCh38_TCGA_somatic_mutation_masked_2018_12	51

Rows per page: All 1-7 of 7

401 items found

Fig. 6.13 Available datasets for the performed GRCh38 TCGA Cholangiocarcinoma data search.

normal data is even repeated). This kind of quick data extraction can be conveniently used to understand how many same-patient data items are available for performing differential data analysis (i.e., comparison between certain characteristics of normal vs. tumor patients' signals and sequences).

Source ID	Donor ID	Data Type	Healthy	Technique	Source Site
3787a...	0775583e-c0a0-4f18-9ca2-8f89cedce3d6	Copy Number Segment	FALSE	Genotyping Array	Sapienza University of Rome
e6443...	0775583e-c0a0-4f18-9ca2-8f89cedce3d6	Copy Number Segment	TRUE	Genotyping Array	Sapienza University of Rome
f36ef...	0775583e-c0a0-4f18-9ca2-8f89cedce3d6	Gene Expression Quantification	FALSE	RNA-Seq	Sapienza University of Rome
f7b6d...	0775583e-c0a0-4f18-9ca2-8f89cedce3d6	Isoform Expression Quantification	FALSE	miRNA-Seq	Sapienza University of Rome
3787a...	0775583e-c0a0-4f18-9ca2-8f89cedce3d6	Masked Copy Number Segment	FALSE	Genotyping Array	Sapienza University of Rome
e6443...	0775583e-c0a0-4f18-9ca2-8f89cedce3d6	Masked Copy Number Segment	TRUE	Genotyping Array	Sapienza University of Rome
9aa16...	0775583e-c0a0-4f18-9ca2-8f89cedce3d6	Masked Somatic Mutation	FALSE	WXS	Sapienza University of Rome
1ba92...	0775583e-c0a0-4f18-9ca2-8f89cedce3d6	Methylation Beta Value	FALSE	Methylation Array	Sapienza University of Rome
f7b6d...	0775583e-c0a0-4f18-9ca2-8f89cedce3d6	miRNA Expression Quantification	FALSE	miRNA-Seq	Sapienza University of Rome
2cbc6...	20bf79af-3b0f-477d-b619-5597d42f5d5e	Copy Number Segment	TRUE	Genotyping Array	Mayo Clinic Rochester
c2d57...	20bf79af-3b0f-477d-b619-5597d42f5d5e	Copy Number Segment	TRUE	Genotyping Array	Mayo Clinic Rochester
d3b1d...	20bf79af-3b0f-477d-b619-5597d42f5d5e	Gene Expression Quantification	FALSE	RNA-Seq	Mayo Clinic Rochester
2649a...	20bf79af-3b0f-477d-b619-5597d42f5d5e	Gene Expression Quantification	TRUE	RNA-Seq	Mayo Clinic Rochester
016fd...	20bf79af-3b0f-477d-b619-5597d42f5d5e	Isoform Expression Quantification	FALSE	miRNA-Seq	Mayo Clinic Rochester
f002e...	20bf79af-3b0f-477d-b619-5597d42f5d5e	Isoform Expression Quantification	TRUE	miRNA-Seq	Mayo Clinic Rochester
5150...	20bf79af-3b0f-477d-b619-5597d42f5d5e	Masked Copy Number Segment	FALSE	Genotyping Array	Mayo Clinic Rochester
2cbc6...	20bf79af-3b0f-477d-b619-5597d42f5d5e	Masked Copy Number Segment	TRUE	Genotyping Array	Mayo Clinic Rochester
c2d57...	20bf79af-3b0f-477d-b619-5597d42f5d5e	Masked Copy Number Segment	TRUE	Genotyping Array	Mayo Clinic Rochester
80052...	20bf79af-3b0f-477d-b619-5597d42f5d5e	Masked Somatic Mutation	FALSE	WXS	Mayo Clinic Rochester
33585...	20bf79af-3b0f-477d-b619-5597d42f5d5e	Methylation Beta Value	FALSE	Methylation Array	Mayo Clinic Rochester
d8106...	20bf79af-3b0f-477d-b619-5597d42f5d5e	Methylation Beta Value	TRUE	Methylation Array	Mayo Clinic Rochester
016fd...	20bf79af-3b0f-477d-b619-5597d42f5d5e	miRNA Expression Quantification	FALSE	miRNA-Seq	Mayo Clinic Rochester
f002e...	20bf79af-3b0f-477d-b619-5597d42f5d5e	miRNA Expression Quantification	TRUE	miRNA-Seq	Mayo Clinic Rochester

Table 6.4 Excerpt of result table from the extraction of GRCh38 TCGA Cholangiocarcinoma data, grouped by patient (i.e., Donor ID).

The datasets can be analyzed using a genomic data analysis tool such as GMQL. By clicking on the 'GMQL' button (Figure 6.12) the user can retrieve the selection query ready to be pasted into the GMQL web interface publicly available at `http://www.gmql.eu/gmql-rest/`; there, results can be aggregated by patient using specific operations such as JOIN or GROUP BY.[10]

Combining ChIP-seq and DNase-seq Data in Different Formats and Sources. Suppose the data analysis goal is to extract genomic regions of enriched binding sites that occur in open chromatin regions, e.g., focusing on H1 embryonic stem cells. This example shows how to improve the quality of the peaks called within ChIP-seq experiments by filtering out the peaks that are not in open chromatin regions (as required by molecular biology). In order to find data available on GenoSurf related to such cells, also considering possible terminological variants, we select the `Synonym` semantic option in the Data search phase. As a first step, we look for ENCODE (*Source*) ChIP-seq (*Technique*) experiment items with narrowPeak format (*File format*) and regarding H1 cells (*Cell/Cell line*). We find 601 items as a result. As a second step, we select Roadmap Epigenomics (*Source*) DNase-seq (*Technique*) HOTSPOT (*Pipeline*) open chromatin regions in H1 cells (*Cell/Cell line*). Such selection produces as a result 4 items. For this set, we decide to further restrict the selection to items with a false discovery rate (FDR) threshold of at least 0.01 (note that the HOTSPOT peak caller was used to call domains of chromatin accessibility both with an FDR of 1% and without applying any threshold). Since this is source-specific metadata information, we apply this filter by using the Key-value search interface: we first search metadata keys that contain the "FDR" string, obtaining the `manually_curated__fdr_threshold` key with values "0.01" and "none"; we then chose to apply the `manually_curated__fdr_threshold = 0.01` filter, which reduces our results to only one item, with the desired content. The obtained JSON query corresponding to this second step looks as follows.

```
{
  "gcm":{
    "source":["roadmap epigenomics"], "technique":["dnase-seq
      "],
    "pipeline":["hotspot"], "cell":["h1 cells"] }
  "type":"synonym",
  "kv":{
    "fdr_0":{
      "type_query":"key",
      "exact":false,
      "query":{
        "gcm":{},
        "pairs":{"manually_curated__fdr_threshold":["0.01"]}
      }
    }
  }
}
```

[10] For more details please refer to [275] and to the "GMQL introduction to the language" document at `http://www.bioinformatics.deib.polimi.it/genomic_computing/GMQLsystem/documentation.html`.

Such JSON document can be retrieved by selecting ‘MODIFY’ or ‘DOWNLOAD’ (at the top of the GenoSurf web interface) and can also be used as a payload in the RESTful API services.

The located data files can be either downloaded to be further processed or directly selected in GMQL. Indeed, the objective of this use case corresponds to performing a JOIN operation in GMQL between the regions in the data items found in the first step and those in the item from the second step.

Extracting Triple-Negative Breast Cancer Cases. Suppose we are working on a comparative Triple-Negative Breast Cancer analysis. This means we need to select breast tissue data from the TCGA-BRCA project characterized by the absence of all the three types of receptors known to fuel most breast cancer growth: estrogen, progesterone, and HER2. Such absence can be encoded in the data as a negative status of the receptors. To do so, first, in the GenoSurf Data search section we select: `project_name: [‘tcga-brca’] and tissue: [‘breast’]`, which reduces the result to 23, 581 items. Then, in the Key-value search section, we need to set the following conditions in conjunction.

```
"clinical__brca_shared__breast_carcinoma_estrogen_receptor_status":
["negative"]
AND
"clinical__brca_shared__breast_carcinoma_progesterone_receptor_status":
["negative"]
AND
"clinical__brca_shared__lab_proc_her2_neu_immunohistochemistry_receptor_status"
    :
["negative"]
```

Note that the exact name of the keys/attributes to query can be identified by previously performing a Key search for estrogen, progesterone, or HER2, respectively.

Figure 6.14 shows such a search on the GenoSurf Key-value interface, leading to the desired result. Note that for building conjunctive conditions each one must be in a separate panel; filters selected in the same panel result in a disjunction.

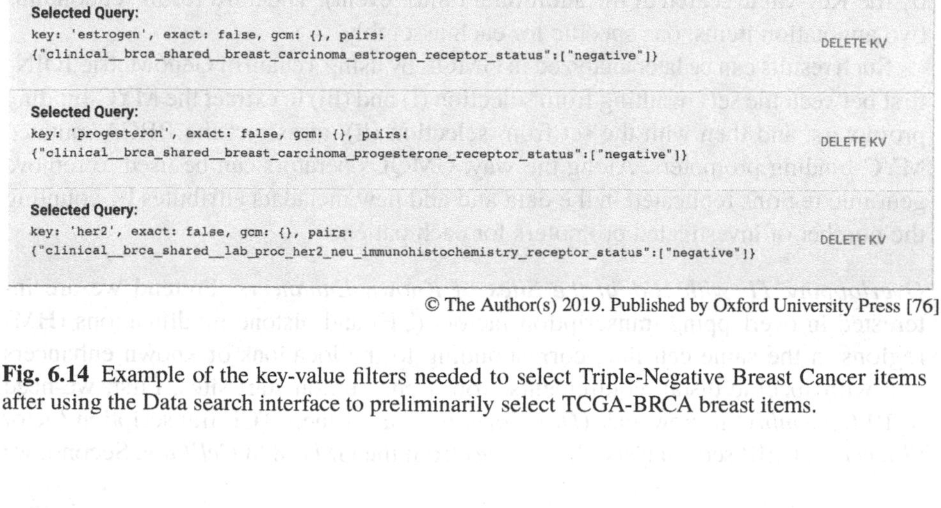

Selected Query:
key: 'estrogen', exact: false, gcm: {}, pairs:
{"clinical__brca_shared__breast_carcinoma_estrogen_receptor_status":["negative"]}
DELETE KV

Selected Query:
key: 'progesteron', exact: false, gcm: {}, pairs:
{"clinical__brca_shared__breast_carcinoma_progesterone_receptor_status":["negative"]}
DELETE KV

Selected Query:
key: 'her2', exact: false, gcm: {}, pairs:
{"clinical__brca_shared__lab_proc_her2_neu_immunohistochemistry_receptor_status":["negative"]}
DELETE KV

Fig. 6.14 Example of the key-value filters needed to select Triple-Negative Breast Cancer items after using the Data search interface to preliminarily select TCGA-BRCA breast items.

Extracting from Multiple Sources at a Time. Suppose we need to retrieve items of hg19 assembly from healthy brain tissue (and possibly its subparts) of male individuals up to 30 years old. In total, hg19 items in the repository are 123,965. Healthy tissue corresponds to choosing "true" in the *Healhty/Control/Normal* filter, which reduces the result to 18,090 items. Since *Tissue* is an attribute that benefits from ontological expansion, we select the Expanded semantic option, to be able to find items connected also to hyponyms of "brain". This filter selects 1,046 items (annotated with "brain" or "cerebellum"). *Gender* "male" gets 604 items, and finally the condition *Max.age* = 30 years (corresponding to 10,950 days in the API calls performed by the system) finds 56 items. As it can be observed in the 'SOURCE COUNT' tab, such output derives from the ENCODE (2 items) and TCGA (54 items) sources.

Combining Mutation and ChIP-seq Data. Suppose we are interested in identifying DNA promotorial regions bound by the MYC transcription factor and that present somatic mutation in breast cancer patients with tumor recurrence. To answer such a typical biological question a user can concentrate on hg19 assembly and perform three separate search sessions:

(i) selection of ENCODE (*Source*), hg19 (*Assembly*), ChIP-seq (*Technique*), narrowPeak (*File format*), MCF-7 (*Cell/Cell line*) – a breast cancer cell line, and MYC binding sites (*Target*);
(ii) selection of TCGA (*Source*), hg19 (*Assembly*), BRCA (*Project name*), DNA-seq data (*Technique*) of patients who encountered a new tumor occurrence – such latter information can be selected from the Key-value search part, for example using the value search string "new tumor";
(iii) selection of hg19 genomic region annotations describing promoter locations from RefSeq.

The first result set amounts to 16 items; these can be retrieved by using the filters in the GenoSurf Data search section. The second result set contains 3 items (first, 993 items are extracted in the Data search section; then, they are reduced to 3 items by the Key-value search of the additional tumor event). The third result set contains two annotation items, one specific for each assembly.

Such results can be later analyzed in GMQL by using a chain of GenoMetric JOINs first between the sets resulting from selection (i) and (iii) to extract the MYC-binding promoters, and then with the set from selection (ii), to extract the BRCA-mutated MYC-binding promoters. Along the way, GMQL operators can be used to remove genomic regions replicated in the data and add new metadata attributes by counting the number of investigated promoters for each patient.

Overlapping TF with HM in the Sites of Known Enhancers. Pretend we are interested in overlapping transcription factors (TF) and histone modifications (HM) regions in the same cell line, corresponding to the locations of known enhancers and with specific distal requirements from transcription start sites. First, we need hg19 (*Assembly*) narrowpeak (*DataFormat*) data of the CTCF transcription factor (*Target*) in ChIP-seq analysis (*Technique*) from the GM12878 *CellLine*. Second, we need hg19 narrow peak data from the H3K4me1 histone modification (*Target*) in ChIP-seq analysis from the GM12878 cell line (using Synonym semantic option in the Data search phase). Third, we look for annotation data about gene and enhancer locations (two possible values in *ContentType*). This use case is performed on the hg19 reference genome. The three search actions retrieve respectively 14, 5, and 5 items each. These datasets can then be used for a query in GMQL, with specific genomic distal operators.

6.4.7 Validation of GenoSurf

We here report our evaluation of the GenoSurf interface and henceforth of the previous process of data integration that resulted in the META-BASE repository. Specifically, we target the following research questions:

[RQ1] *Usability and Understandability*. How easy is it to understand and use the interface for an interdisciplinary user base?
 [RQ1.BIO] for users with a biology background?
 [RQ1.CS] for users with a computer science background?
[RQ2] *Effectiveness*. How effective is the interface in providing data integration results without errors?
[RQ3] *Efficiency*. How much does the interface lower the effort required to perform the same search queries on original systems?

6.4.7.1 RQ1: Usability and Understandability Experiment

We here report the results of a compendious empirical study whose target is a population composed of biologists, bioinformaticians, and computer scientists. The same setting was employed first to answer RQ1, addressing the general mixed population, second to answer RQ1.BIO (only involving users with knowledge of Biology) third to answer RQ1.CS (only with users in the Computer Science field). Most users were totally unaware of conceptual modeling guidelines and, as such, well represented the typical GenoSurf target user. Their feedback was particularly useful to plan future improvements to the interface.

Study Rationale. For evaluating the usability and understandability of our interface, we planned the empirical study process that is shown in Figure 6.15; it consists of presenting a questionnaire to a group of biologists, bioinformaticians, and computer scientists/software developers with an interest in Genomics. Before being engaged with the search system, users were provided with WIKI documentation and video tutorials. We planned questions of progressive levels of difficulty; each question presents a specific research scenario and participants are asked to use our interface for extracting items, thereby simulating the typical search task. After the submission of answers, we show the right answers to users, and provide explanations of each answer; we expect that during the process users can develop a better understanding and progressively master the search system. After such training, we ask the users to evaluate the overall experience and specify the degree of expertise in the domain.

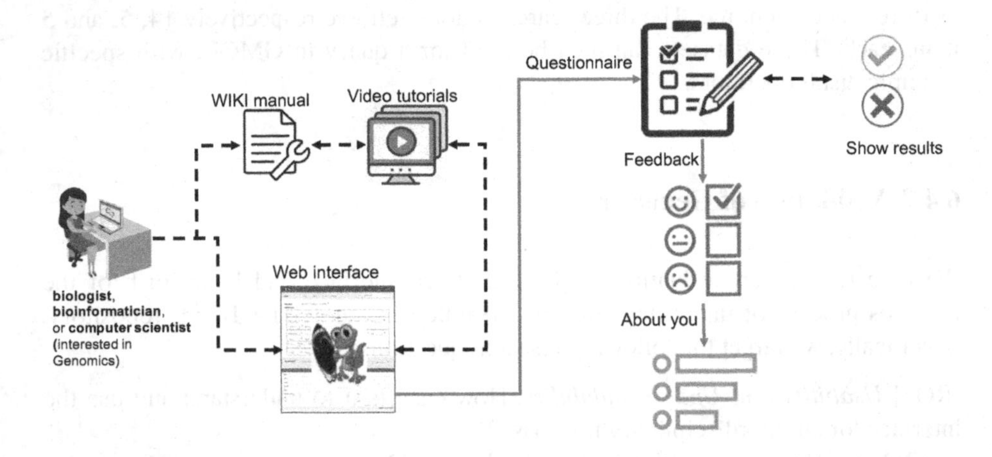

Fig. 6.15 Overview of empirical validation process for GenoSurf.

Q1. How many datasets do we provide from the source TCGA with assembly GRCh38?
Q2. How many items do we provide for TCGA, assembly GRCh38, in the normal (**a**) / tumoral (**b**) cases?
Q3. Which TCGA GRCh38 project among COAD (Colon adenocarcinoma), LUAD (Lung adenocarcinoma), and STAD (Stomach adenocarcinoma) has more gene expression data?

Q4. How many sources contain data annotated with the human fetal lung cell line IMR-90 (both using original spelling (**a**) and alternative syntaxes (**b**))?
Q5. How many sources contain data annotated with the tissue uterus (both using original spelling (**a**) and the broadest possible intepretation (**b**))?
Q6. In ENCODE, how many items of ChIP-Seq can you find for the histone modifications H3K4me1, H3K4me2, and H3K4me3?
Q7. Assume you want to retrieve items from the TADs source that correspond to combined replicates (i.e., they belong to at least 2 biological replicates). How many items can you find?

Q8. We would like to retrieve items of hg19 assembly from healthy brain tissue (and possibly its subparts) of male gender, up to 30 years old. How many items can you find with these characteristics in the sources ENCODE (**a**) and TCGA (**b**)?
Q9. We are interested in ovarian cancer patients at clinical Stage III and IV. Select TCGA-OV project data. Then, select pairs with the key 'clinical_patient__clinical_stage' corresponding to the stage iii and iv (e.g., stage iiia, stage iiib, ...). How many items can you retrieve?
Q10. Suppose you need to identify DNA promotorial regions bound by the MYC transcription factor that present somatic mutations in breast cancer patients. For each of the following steps, provide the number of retrieved items. First, get from ENCODE source, ChIP-seq narrowpeak data from the cell line MCF-7, regarding MYC binding sites (**a**). Second, DNA-seq data is needed from TCGA BRCA patients which encountered a new tumor occurrence (**b**). Third, genomic region annotations describing promoters locations should be retrieved from RefSeq (**c**).

Table 6.5 Proposed survey questions.

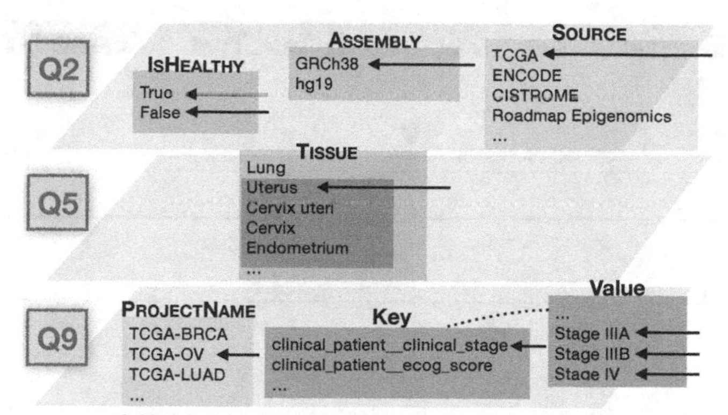

Fig. 6.16 Q2 describes a case in which the user selects from the *IsHealthy* attribute list first the value "True" and then the value "False", corresponding to two sub-questions. Then, she selects "GRCh38" among the possible values in the *Assembly* attribute list and "TCGA" as a *Source*. Q5 presents an enriched list of values for the attribute *Tissue* — note that "Cervix uteri" and "Cervix" are synonyms and, together with "Endometrium", they are hyponyms of uterus. For Q9, after selecting the *ProjectName*, the user explores keys and values through a specific interface.

Group	Question	Sub-questions	Desired output	Cross-dimension attributes	Logical disjunction	Semantic enrichment	Combination original/integr.	Complete study
1	Q1	1	#D	×				
	Q2	2	#I	×				
	Q3	1	#I	×				
2	Q4	2	#S			×		
	Q5	2	#S			×		
	Q6	1	#I	×	×			
	Q7	1	#I	×				
3	Q8	2	#I	×		×		
	Q9	1	#I		×		×	
	Q10	3	#I	×			×	×

Table 6.6 Input features tested in the survey. Desired output column contains numbers of items (#I), datasets (#D), or sources (#S).

Experiment Design. During the conception of the survey, we followed a number of study design principles. We attempted to lower the ambiguity of the questions and to provide guidance to the users; when asking users to resolve a search problem, we used questions that could have exact answers (i.e., numbers), to lower the possible interpretation biases; we stratified questions by complexity, to capture different levels of understanding of the interface and its structure; we diversified the challenges addressed in the questions, to overview all search possibilities encompassed by our system.

In Table 6.5 we show the complete list of 10 proposed questions (some of which contain two or three sub-questions). We divided the questionnaire according to three

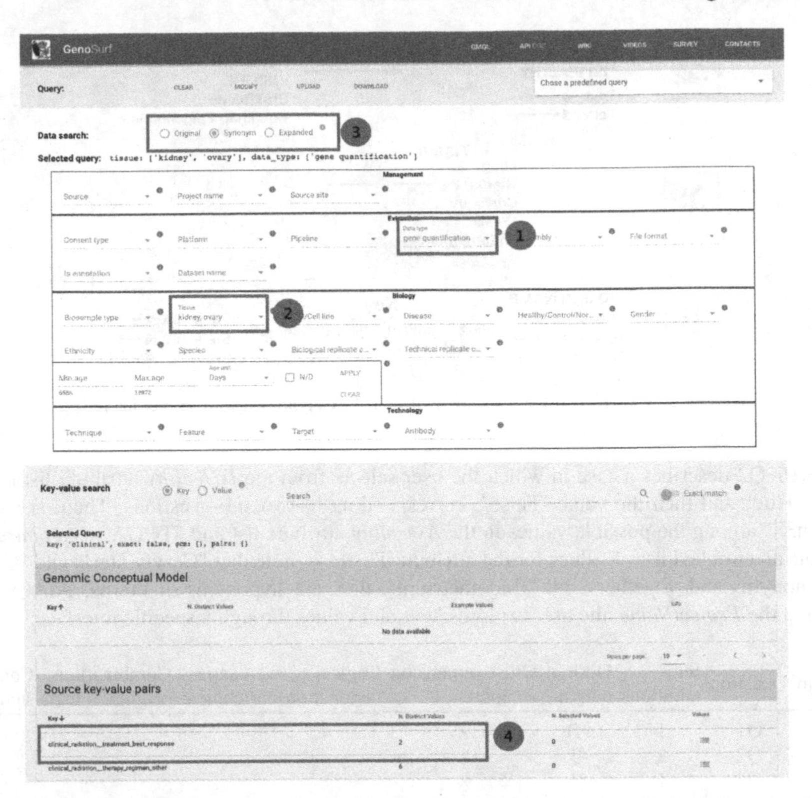

Fig. 6.17 Graphic representation of tested user operations: the red stamp (1) and (2) represent the combination of attribute filters coming from different dimensions; (2) uses value filters in disjunction; (3) indicates the semantic enrichment options; (1), (2), and (4) allow to combine selection on the key-value-based interface with metadata from the GCM.

groups of questions, in order of complexity: the first provides a simple scenario with incremental addition of filters: first a source with the assembly (Q1), then selection of normal/tumor patients (Q2) and of specific disease projects (Q3); the second explores peculiar (i.e., less standard) features of the search, e.g., semantic enrichment with synonyms (Q4), ontological hierarchies (Q5), disjunction of attribute values (Q6), and aggregate attributes (Q7); the third one builds three more complex cases: the combination of many filters (Q8), joined use of original metadata (in key-value format) and structured metadata (Q9), and the composition of three selections from data sources to simulate a complete study (Q10). Figure 6.16 visually explains the process of attribute selection and value provisioning required by questions Q2, Q5, and Q9.

As depicted in Figure 6.17 we tested: the ability to compose queries by combining attribute filters coming from different dimensions, the use of value filters in disjunction one with the other, the understanding of semantic enrichment options, the combined used of original metadata filters – using a key-value-based interface – with structured integrated metadata—based on the GCM, the execution of a complete study. In Table 6.6 we map these challenges into the different questions that allowed us to test them. With respect to the interplay between original and structured metadata: the query interface must enable interaction with both (in the key-value pairs it is important that people can ask separately what are the key — typically defining the property associated with the item — and what are the values — associated to the specific property). In different questions, we alternatively asked to report the number of items, datasets, or sources.

Study Execution. The experiment target users were sourced from within our research group (GeCo) and from several collaborating institutions (such as Politecnico di Torino, Istituto Nazionale dei Tumori, Università di Torino, Università di Roma Tre, Istituto Italiano di Tecnologia, Radboud Universiteit Nijmegen, Freie Universität Berlin, Harvard University, Broad Institute, National University of Singapore, University of Toronto), including researchers with different backgrounds (computational and molecular biology, bioinformatics, and computer science) but also students and pure software developers with interest in Genomics. Out of about 60 invitations, we received 40 completed responses.

Results. First of all, we consider the self-assessment of users about their experience in the field of genomic data analysis (see histograms in Figure 6.18), highlighting that our sample was roughly divided into two balanced sets with different expertise. Overall, users present expertise scores that range from "None" to "Expert". When asked about their use of the platform to find data for analysis and about their need to combine inter-sources data, answers ranged from "Never" to "Daily", confirming that our users' test set was well-assorted.

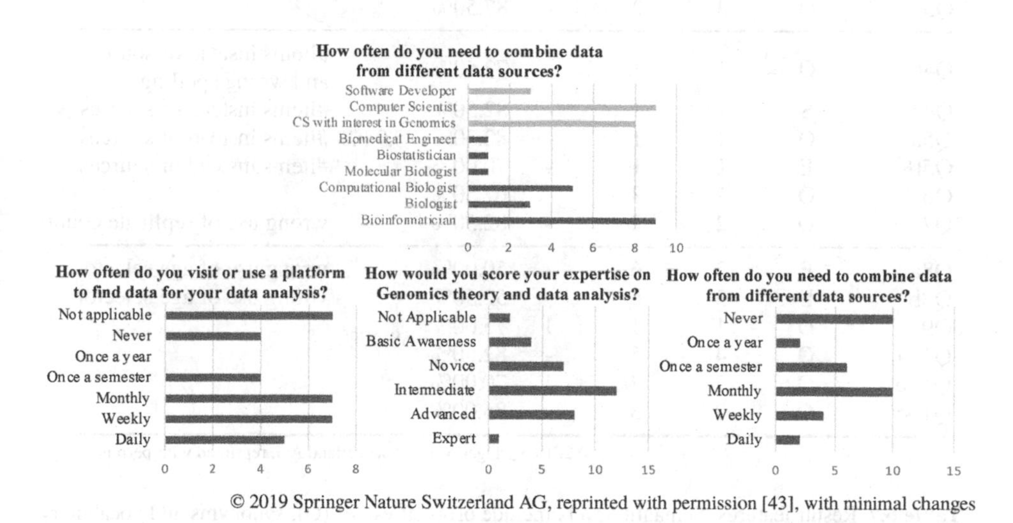

Fig. 6.18 Histograms showing the user's expertise on genomic data analysis.

Based on these results, we formed one BIO group, with users who answered: "Bioinformatician", "Biologist", "Computational Biologist", "Molecular Biologist", "Biostatistician", or "Biomedical Engineer" and one CS group, with users who answered: "Computer Scientist with interest in Genomics", "Pure Computer Scientist", or "Software Developer".

To answer RQ1 we first considered the general population of 40 individuals and then repeated the same analysis by considering only users with a computer science background (20) or with a biology background (20). As a threat to the validity of this partition of the user base, we point to the fact that self-assessment is not always accurate. We did include some queries to understand the user profile, however, we did not consider this information sufficient to build a strictly balanced number of partitions of the user base.

Correct Answers. In Table 6.7 we report: the required semantic level to set at the beginning of the query, the numbers of dimensions, integrated attributes and original keys involved in the query. Then we show percentages of correct answers (scores) of each specific sub-question and aggregated by group of questions. The correctness of answers is established on an exact-match base. In this assessment, we chose not to consider a finer concept of distance between the correct number and other numbers, possibly accounting for typos or wrong choices of the users.

Question	Semantic level	#dim.	#integr. attributes	#orig. keys	Scores	Group score	Typical errors
Q1	O	2	2	-	97.50%		#items instead of datasets
Q2a	O	3	3	-	97.50%	93.33%	
Q2b	O	3	3	-	92.50%		
Q3	O	3	2	-	87.50%		
Q4a	O	1	1	-	72.50%		#items instead of sources and wrong spelling
Q4b	S	1	1	-	82.50%		#items instead of sources
Q5a	O	1	1	-	82.50%	75.94%	#items instead of sources
Q5b	E	1	1	-	70.00%		#items instead of sources
Q6	O	2	3	-	67.50%		
Q7	O	2	2	-	82.50%		wrong use of replicate count
Q8a	E	3	6	-	50.00%		wrong use of age selector
Q8b	E	3	5	-	52.50%		wrong use of age selector
Q9	O	1	1	1	75.00%	68.47%	
Q10a	O	4	5	-	82.50%		
Q10b	O	2	2	1	70.00%		
Q10c	O	2	3	-	85.00%		

Table 6.7 Result features. Semantic levels include original values (O), synonyms and vocabulary terms (S), or the expanded option, with also hierarchical hyponyms (E).

Note that, if we consider together the performances of each group, as expected, group 1 reached a high percentage of correct answers (93.33%), group 2 a little less (75.94%), while group 3 had the worse score (68.47%). Some typical errors spotted in many answers are also reported. Question 8 had a low rate of correct answers

(50% and 52.63%); we asked to retrieve the number of items in two sources for a specific assembly from a healthy tissue (using the semantic option that includes ontological hierarchy) of one gender in a restricted age range. Such questions combined many elements (six data search filters, use of semantic expansion, age feature). Overall, users replied correctly to 78.92% of the questions (grouping together the sub-questions of the same entry). Five users answered correctly to all questions. On average, it took them less than 44 minutes to answer all 10 questions.

Question	Scores BIO	Group score BIO	Scores CS	Group score CS
Q1	100.00%		95.00%	
Q2a	100.00%	95.00%	95.00%	91.67%
Q2b	90.00%		95.00%	
Q3	90.00%		85.00%	
Q4a	85.00%		60.00%	
Q4b	90.00%		75.00%	
Q5a	90.00%	86.88%	75.00%	65.00%
Q5b	80.00%		60.00%	
Q6	80.00%		55.00%	
Q7	95.00%		70.00%	
Q8a	55.00%		45.00%	
Q8b	60.00%		45.00%	
Q9	75.00%	74.17%	75.00%	62.78%
Q10a	100.00%		65.00%	
Q10b	75.00%		65.00%	
Q10c	95.00%		75.00%	
Total	85.00%		70.94%	

Table 6.8 Results divided by user profile: 20 users with a Biology background (BIO), 20 users with a Computer Science background (CS).

In Figure 6.8, instead, we appreciate the results divided by user groups. Overall, all three groups of questions, as well as the total calculated as an average of all questions, highlighted that users with a Biology background performed better than the ones with a Computer Science background.

Lessons Learned. In retrospect, we made mistakes in the formulation of some of our queries. Users were confused when we asked them to count the containers of data (e.g. sources or datasets) instead of the data items, probably because they did not understand the notions of sources and of datasets: distinguishing the dataset and data source storing the items probably requires a computer science background that was not present in many users. As in these cases users made the exact choices of attributes and values and just provided a wrong numerical answer, we considered their answers as valid. In two questions (Q6 and Q8) users did not reach a satisfying percentage: in one case, this was probably due to the misunderstanding between the use of "comma" in natural language and the correspondence with a logical AND; in the other case, to the misinterpretation of some filters. These observations provided us with important insights into the low usability of some parts of the interface. We made an effort to understand what caused the errors in the answers and reported the most obvious reasons in the last column of Table 6.7. In the future, we shall produce a more advanced evaluation framework that allows us to infer or exactly link each wrong answer to its causative action by the user.

In spite of these mistakes, our user study provided us with important feedback. We were forced to de-normalize and simplify the conceptual schema, but the logical organization of our simplified schema, centered on the item with selected attributes and organized along four dimensions, still proved to be effective; it facilitated both the training and the search interface organization. Clustering attributes along the four dimensions allowed us to explain them first collectively and then individually; users understood well their meaning and in most cases were able to translate narrative questions into the correct choice of attributes embedding the questions' semantics.

6.4.7.2 RQ2: Effectiveness

Effectiveness referred to the data integration process was answered in Section 4.10, where we addressed the aspect of lossless integration (both on the intensional and extensional perspective) as well as the semantic enrichment correctness. Here we propose to measure the effectiveness of the GenoSurf search system based on the users' satisfaction after its use.

After filling out the first part of the questionnaire, we asked users if they learned from the system and if they liked it, and to give us hints on how to proceed in our work (possibly with open suggestions on how to improve it). Answers to this part of the questionnaire are shown in Figure 6.19.

Two-thirds of users declared that answering to the proposed questions was "Moderately easy" or "Neither easy nor difficult". Most users either did not use the documentation or found it moderately/very useful, while users who watched the video tutorials were generally satisfied with them. When asked to perform a query to reach items useful to their own research, most users declared it was moderately easy. The majority were satisfied with the data sources available in the interfaces and were quite likely to recommend the platform to colleagues and researchers in the field.

These few results may further support the positive evidence of the usability and readability of the system shown in RQ1. For a more precise assessment, obviously, users' responses may have been measured by means of efforts and time spent in learning GenoSurf usage; however, in this evaluation, we did not collect this kind of data.

Some specific feedback and the observation of users' mistakes allowed us to improve the instructions for using the search interface, as we eliminated some sources of ambiguities and misunderstanding. We also received important indications about missing data sources according to users' experience; this information has driven us in selecting the next sources to be integrated into the META-BASE repository. A

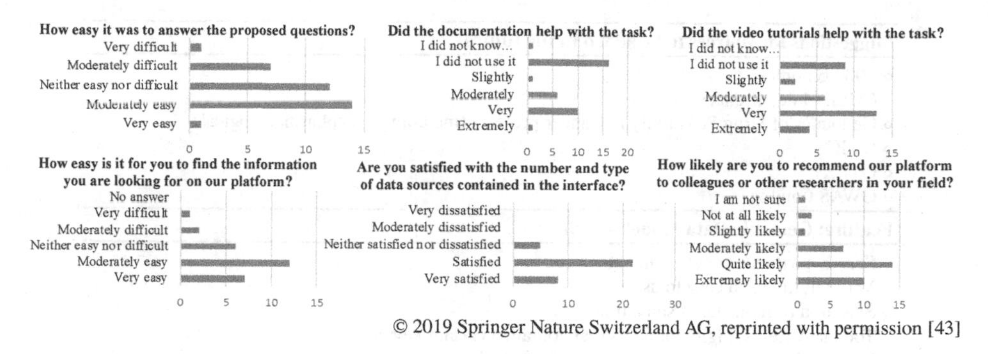

Fig. 6.19 Histograms showing the user's evaluations of the search system.

representative taxonomy of received comments is provided in Table 6.9, including user suggestions on the addition of sources, comments on the usability of specific features, hints for future work, and general criticisms.

6.4.7.3 RQ3: Efficiency

To evaluate the efficiency of our integration process supported by the search interface, we compared the effort made by a user to query original sources as opposed to our integrated system. As benchmarks, we employed six complex queries described in the Use Cases (see Section 6.4.6).

For these examples, we assume that the goal of a user of our system is to find data for her data analysis. She measures the value of her search by considering it successful when a number of items useful to her data analysis are returned by the system, as opposed to a search that does not produce results.

Table 6.10 shows our findings on these six example queries: we indicate the number of data sources involved in the results of the query, the number of filters set on the GenoSurf Data Search (filter) and Key-Value Search (kv), and our assessment of the possibility of performing the same query on original data source systems. Note that the shown examples demonstrate the added value provided by our integration effort, which is materialized in META-BASE and provided through GenoSurf. The first four listed examples could not be performed on any of the original sources, because: i) the source system did not allow sorting or search by a given attribute; ii) there was no interface for choosing filters on metadata; iii) the results of a source query are not files/items ready to be processed (instead, lists of experiments).

In addition to these, we point to other reasons why GenoSurf provides users with a more efficient environment with respect to having to query all sources separately:

i) querying multiple sources together, with the same filters saves time;
ii) original sources do not include (or in a very limited way) ontological support, resulting in additional effort to navigate parts of ontologies sub-trees and to perform queries to match hyponyms;

Suggestions for sources to be added in the future
⋄ 1000 genomes
⋄ CCLE (3 occurrences)
⋄ Genomics of Drug Sensitivity in Cancer projects (including also pharmacological)
⋄ ICGC
⋄ GTEX
⋄ GWAS Catalog
Feature: Genomic Data Model search
⋄ Search attributes are clear and powerful
⋄ Very intuitive and easy to use
⋄ Selection of items takes some time
⋄ Too many fields; highlight fields that contain a certain keyword
⋄ Request for allowing to copy content of single fields
Feature: Original-Synonym-Expanded search options
⋄ Good (2 occurrences)
⋄ Very useful
⋄ "Expanded" option not completely clear
⋄ May be confusing, but once understood they are very useful
⋄ Why is original the default?
⋄ Synonyms could be somehow shown grouped
Feature: Key-Value search
⋄ Consider adding a search feature on both keys and values simultaneously (3 occurrences)
⋄ Good (2 occurrences)
⋄ Useful
⋄ Sorting of value items seems strange
⋄ Results appear in a bit complicated way
Feature: GMQL query builder
⋄ Very good
⋄ Not used
Feature: Download of data files links
⋄ Good (2 occurrences)
⋄ Very good
⋄ Useful and easy
⋄ Downloading sometimes is very slow
Hints for future work
⋄ Focusing more on the metadata search through text
⋄ Enabling the search directly through known ontologies
⋄ Implementing queries history or undo/redo
⋄ Providing more practical examples of the combined use of the various parts
⋄ Providing a complete use case to follow
Criticisms
⋄ Tool useful for people with bioinformatic background, difficult for doctors or biologists
⋄ User experience quite overwhelming: a user should see a window only when needed
⋄ Interface is not intuitive, especially the possibility to use it jointly with GMQL

Table 6.9 Taxonomy of suggestions provided by study participants of the questionnaire.

iii) original sources are not directly integrated with the GMQL analysis environment—this is an indirect advantage provided by GenoSurf, which outputs files ready to be processed in a Bioinformatic Tertiary Analysis computational environment.

Benchmark	DS	#Actions GenoSurf	Expressible outside GenoSurf
Extracting Cancer Patient Data	1	3 filters, reordering table	✗: TCGA does not allow to order by aliquot
Combining ChIP-seq and DNase-seq Data	2	4 filters (ENCODE), 4 filters + 1 kv (Roadmap)	✗: Roadmap does not have an interface with metadata filters
Extracting Triple-Negative Breast Cancer Cases	1	1 filter + 3 kv	✗: TCGA does not allow filters on XML clinical information
Extracting from Multiple Sources at a Time	2	5 filters (1 with Exp.)	✗: Filters available in both sources, but no ontological expansion possible, health status filter not available in ENCODE, user extracts experiments, not files
Combining Mutation and ChIP-seq Data	3	5 (ENCODE), 4 + 1 kv (TCGA), 3 (annotations)	✓: Filters available on ENCODE and TCGA, files to be parsed for annotations (no interface)
Overlapping TF with HM in the sites of known enhancers	3	5 (ENCODE), 5 (1 with Syn.) (ENCODE), 2 (annotations)	✓: Filters available on ENCODE and TCGA, files to be parsed for annotations (no interface)

Table 6.10 Results of efficiency assessment.

Overall, we can ascertain that GenoSurf offers users a wide set of instruments to perform easier search sessions than from the original sources. Starting from the original sources users would have to apply many more transformations and perform a much greater effort to achieve similar results. Future work may address more specifically the added value to users' interests, trying to assign a cost to actions, as it was done in more business-oriented contexts, e.g., by Even and Shankaranarayanan [140, 141], even if our domain concerns research and does not seem directly monetizable.

6.5 Related Works

Graphs for Genomics. As described in a survey by Angles and Gutierrez [16], Graphs have been investigated in the past as a means to store genome maps. Such approaches [172, 189] were mostly theoretical and did not use the metadata information as a driver for genomic data selection. Among various integrated databases in the bioinformatics domain that employ graph-based paradigms, we cite: BioGraphDB [283], a resource to query, visualize and analyze biological data belonging to several online available sources (focused on genes, proteins, miRNAs, pathways); Bio4j [308], a platform integrating semantically rich biological data (focused on proteins, and functional annotations); ncRNA-DB [68], integrating associations among non-coding RNAs and other functional elements. In [408] the authors demonstrate that graph databases efficiently store complex biological relationships (such as protein-protein interaction, drug-target, etc.) and have the potential to reveal novel links among heterogeneous biological data.

Semi-automatic generation of knowledge graphs is discussed extensively in the Semantic Web community [160, 128]. For the generation of GeKnowGraph, we did not use these techniques because genomics data sources are very heterogeneous and for many of them we were not able to learn a schema (sometimes the metadata is not even exposed with APIs or structured HTML pages). Since a priori we cannot assume regularity in the metadata structure of sources, we strongly believe that a curated approach, driven by a sound conceptual schema such as GCM, leads to better and more useful results, for the biology community.

Browsing interfaces for genomics. DNADigest [231] investigates the problem of locating genomic data to download for research purposes. The study is also documented more informally in a blog.[11] This work differs from ours since, while allowing the dynamical and collaborative curation of metadata, they only provide means to locate raw data, while we provide data to be used by our genomic data management system (described in Section 4.9).

UCSC Xena [168] provides a strong browsing interface, with more powerful tools than GenoSurf, as it also includes multiple visualization features and the possibility to store private datasets. However, it only encompasses data sources relevant to cancer genomics.

Terra[12] of the Broad Institute is a new platform that aggregates genomic data from different sources, also including cloud computational environments. Metadata are curated but datasets can only be browsed source by source, therefore without benefiting from an integrative view. Their integration pipeline is not a general framework but a set of different ETL scripts that are written *ad hoc* for each new imported source.

qcGenomics [62] is a platform for retrieving, transforming (using quality assessments), and visualizing genomic datasets in comparative views; the focus is on allowing users to perform multidimensional data integration online.

We consider DeepBlue [9] as the most similar easy-to-use platform with respect to GenoSurf. This data server was developed to search, filter, and also process epigenomic data produced within a multi-center research consortium. Some of its modeling choices are similar to ours (e.g., the distinction between region data and metadata, management of both experimental and annotation datasets, and a set of mandatory attributes and key-value pairs to store additional metadata). However, DeepBlue focuses on epigenomics (i.e., the study of epigenetic modifications on the cell), a limited branch of genomics. Instead, GenoSurf allows us to browse a repository resulting from a much broader integration effort, as we consider a larger spectrum of different data/experiment types. The DeepBlue database identifies five mandatory metadata attributes (three of them are standardized to externally controlled vocabularies and equipped with synonyms and hierarchies), while GenoSurf accounts for eight entities with thirty-nine attributes (ten of which are normalized, also including synonyms, hierarchies, and external references). The look-and-feel of the platform is similar to ours, with the possibility to select specific values for each attribute and a table of results; however, we give users the possibility to select many

[11] https://blog.repositive.io/

[12] https://terra.bio/

more aspects and provide dynamically updated counts of items available for each specific value.

As opposed to DeepBlue, GenoSurf does not provide functionalities to process region data, but it is well-coupled with GMQL, which is a powerful tool for genomic region manipulation. GenoSurf builds extraction queries to be used directly in GMQL.

Empirical Studies. In the last few years, the focus of empirical studies dedicated to conceptual modeling has ranged from works on tools based on CM [414], to process mining [24] and to artifact sampling [261]. A broad experiment has compared traditional conceptual modeling with ontology-driven conceptual modeling [388]. Empirical studies on conversational interfaces-based systems for biology or data science have been conducted in [211, 143, 113].

Chapter 7
Future Directions of the Data Repository

"You got to have a problem that you want to solve, a wrong that you want to right, it's got to be something that you're passionate about because otherwise, you won't have the perseverance to see it through."

— Steven Paul Jobs

In the previous chapters, we have presented our approach to modeling genomic data (Chapter 3), building a sound repository of genomic datasets using data integration techniques (Chapters 4-5), and exposing its content through a user interface that is rich in functionalities and data complexity (Chapter 6). Excitingly, the realized work opens up many interesting challenges, which are briefly described in the following, pointing at the next research directions:

- We aim to include other data sources and data types (among both well-established sources and just-born initiatives). As expected, while up to now we only targeted open sources, some of the new ones raise problems of privacy protection in genomic data (Section 7.1).
- We are working towards the proposal of solutions that help achieve better data quality during the process of integration, i.e., while building the integrated repository (Section 7.2).
- We are aware that there exist many interoperability issues that could be addressed by using tools made available from the active community of semantic web technologies for life sciences; this aspect will be further investigated (Section 7.3).
- In the broader context of the GeCo project, which includes the repository described in this thesis, but also a query language and engine for answering complex biological queries (i.e., GMQL), our next purpose is to simplify the use of these tools and connect them within an integrated environment that can be used by biologists and clinicians, even with a limited technological background; this will be realized in the new GeCoAgent project (Section 7.4).
- As a specific addition to GeCoAgent, we envision the creation of a repository of best practices that can be selected by users within our environment or also offered as an external service in the form of a marketplace of bioinformatic tertiary analysis procedures (Section 7.5).

A. Bernasconi: *Model, Integrate, Search... Repeat*, LNBIP 496, pp. 157–164, 2023.
https://doi.org/10.1007/978-3-031-44907-9_7

7.1 Including New Data Sources

In Chapter 2 we discussed the main genomic data players such as repositories, international consortia, and integrators focused on given diseases (e.g., cancer for TCGA), or on scientific aspects (e.g., epigenetics for Roadmap Epigenomics). The next obvious target is other sources that are most commonly accessed in bioinformatics/biologists everyday work; we will start from the ones suggested by GenoSurf evaluation participants described in Section 6.4.7 (i.e., ICGC, GTEx, GEO, COSMIC, CCLE).

Then, we would like to investigate the possibility of determining new suitable additions in an automatic way, possibly setting up a service that is able to suggest and orient a stream of data sources acquisitions. Such a service requires, as a prerequisite, the provision of a more accurate methodology to measure the value achieved by adding new datasets incrementally, impacting the results of both pure search activities and downstream analysis not detailed in this thesis.

Such measure could be based on several factors, such as the dimension of the source, the geographical area (or population ethnicity/race) covered by the sample collection, the type of samples (e.g., from healthy or non-health tissues, case or control populations), or the type of data (i.e., we have many epigenomics sources but only one – even if it is considered the most important in the community – regarding cancer genomics). Notably, it would be important to measure the value of adding a new source with respect to given search goals provided by different classes of users, either driven by their typical profiles or by shared sets of queries of interest.

This service would greatly improve the variety of datasets offered in META-BASE and GenoSurf but could be of great interest also for external systems, if offered as an API, actively orienting collaborative streams of acquisitions.

Considering new trends, in recent years we are assisting the worldwide emergence of a new generation of large-scale genomic *national* initiatives [375]. Some employ population-based sequencing, such as *All of US* [107] from NIH in the United States (aiming at sequencing 1 million American volunteers' genomes), the *Million Veteran Program* (>1 million participants) [161], the *China's Precision Medicine Initiative* [116], *GenomeDenmark* [125], the *Estonian Genome Project* [243], the *Qatar Genome Programme* [322], and the *Korean Reference Genome Project* [101]. Others are testing large numbers of cancer or rare disease patients, for example, the *100,000 Genomes Project* [85] (a UK Government project that is sequencing whole genomes from UK National Health Service patients), the *Saudi Human Genome Program* [3], and the *Turkish Genome Project* [14]. Still, other nations are focused on developing infrastructure to later achieve similar results (for instance *FinnGen* [149] and *GenomeCanada* [75]). Also private companies are participating in this race: the Icelandic project (now deCODE Genetics [188]) that has been bought by the U.S. biotechnology firm Amgen, 23andMe,[1] and Human Longevity Inc.[2]

[1] `https://www.23andme.com/`

[2] `https://www.humanlongevity.com/` to mention a few

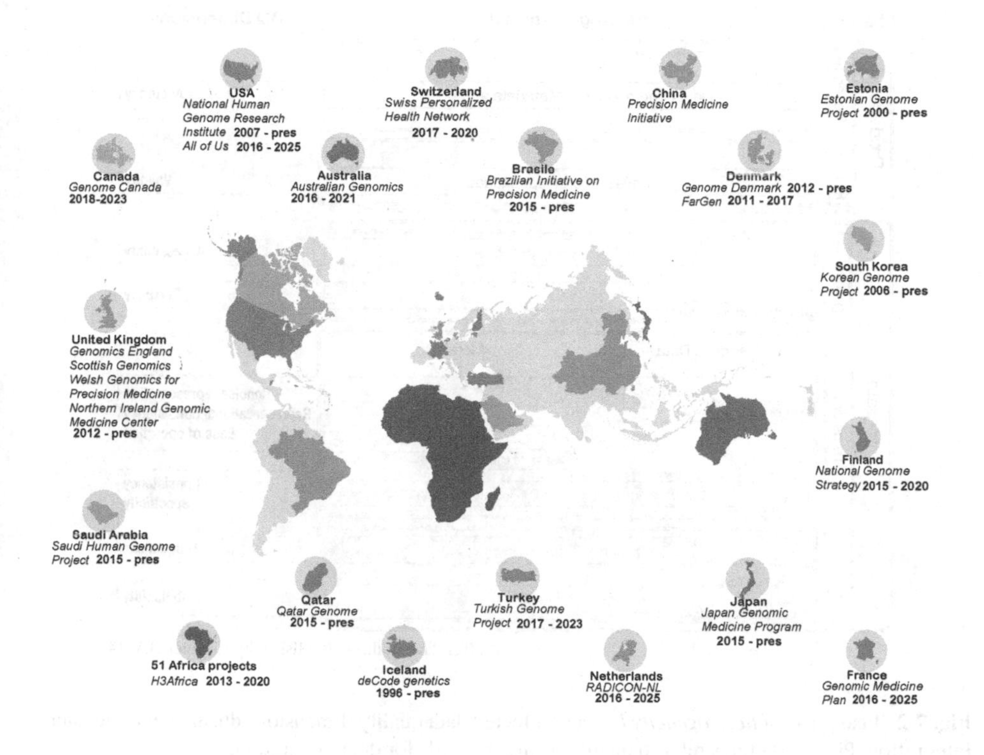

Fig. 7.1 Map of currently active government-funded national genomic-medicine initiatives (information extracted from [60]).

Figure 7.1 describes recently born initiatives in different countries worldwide. Their data access models are still unarguably not open for research (sharing issues of data governance and privacy protection [117]), therefore we do not discuss them in this thesis; no integrators include them yet either. However, all these projects certainly represent a wealth of information, which will need to be considered within the scope of future data integration efforts, giving a new and substantial boost to the potential of genomic data analysis.

7.2 Improving Genomic Data Quality Dimensions

The integrated use of data coming from different data sources is very challenging, as heterogeneity is met at multiple stages of data extraction (e.g., download protocols, update policies), integration (e.g., conceptual arrangement, values, and terminologies), and interlinking (e.g., references and annotation). While integrating genomic datasets, either for *ad hoc* use in a research study, or for building long-

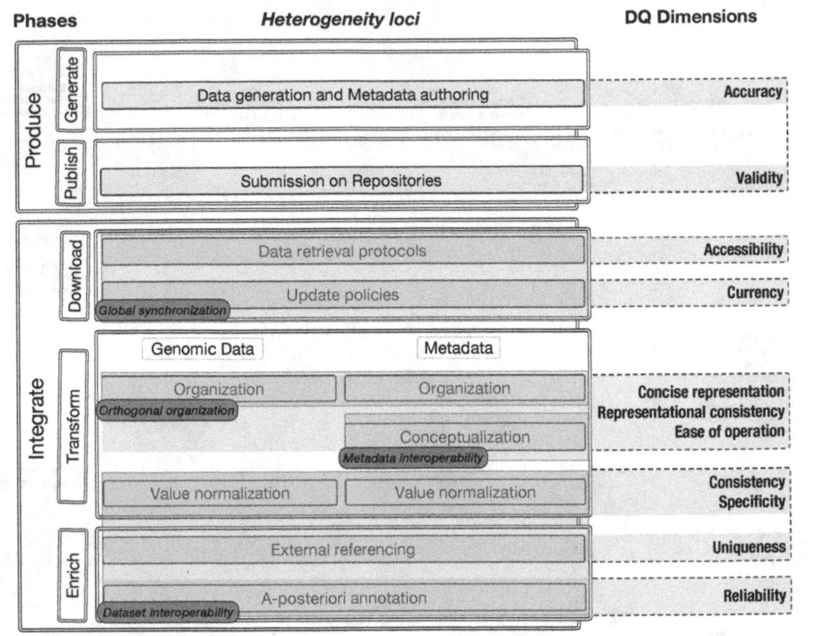

Fig. 7.2 Taxonomy of *heterogeneity loci* and affected data quality dimensions during genomic data integration. Pink rectangles refer to quality-aware methods for data integration.

lasting integrated data warehouses, we deal with various complexities that arise during three phases: i) download and retrieval of data from the (potentially multiple) sources; ii) transformation and manipulation, providing fully or partially structured data in machine-readable formats; iii) enrichment, improving the interoperability of datasets.

With *heterogeneity locus* we refer to an activity or phase within the genomic data production/integration process of multiple sources that exhibits heterogeneity issues, thus undermining the quality of resulting resources and hindering the seamless reconciliation of records and information. The taxonomy in Figure 7.2 keeps track of all the phases in which a genomic data user may need to resolve problems related to non-standardized ways of producing data, making it accessible, organizing it, or enhancing its interoperability; the *heterogeneity loci* (listed in the central column) are grouped by production and integration phases (on the left) and are related to data quality dimensions on the right (such as accuracy, consistency, currency, and reliability) that are critical in the represented heterogeneity aspects.[3] We have here focused on the specific case of semi-structured data extraction. It has to be noted that, in cases in which the downloaded information is unstructured text, the dimensions become more complex, possibly including lexical and syntactic accuracy, "readabil-

[3] We refer to widely used state-of-the-art definitions of data quality dimensions [394, 328] as well as to more recent ones [20, 34].

ity" and text comprehension, and local/global coherence [34]. Instead of considering the quality of genomic signals extracted from raw data (as this problem is extensively studied for each signal extraction technology), we are planning to approach a novel angle: addressing data quality dimensions while diverse data sources are being integrated together to enable further applications. A starting point could be that of conceiving a measure for the degree of heterogeneity between two sources (or between one new source and the global META-BASE repository) so that the effort required to reconcile such different datasets could be defined as a function of this measure. Aside from being of great contribution to our framework, this service could be exploited by any company wishing to sell integration services, based on a systematic calculation of their price

7.3 Towards Better Interoperability

While in this thesis we have approached data integration as a *low-level* effort, based on our experience in building solid data warehouses, we are aware of the trend that is emerging in many applied informatics communities, including the bioinformatics one: *on-the-fly* data integration based on interoperability of systems. Many initiatives are targeting the production of Findable, Accessible, Interoperable, Reusable (FAIR) data. Fairness is a broad concept that has been addressed by different communities and several definitions have been provided. In particular, the bioinformatics community embraces the principles drafted in [402] by a big group of researchers that produced the first formal publication on this topic. Findability has to do with unique and persistent identifiers, rich metadata that explicitly references the described data and is indexed in searchable resources. Accessibility refers to the possibility of retrieving data through standard communications protocol, that is generally open or allows for authentication procedures, while metadata is always accessible even when data is not anymore. Interoperability represents the adoption of a broadly used knowledge representation language, vocabularies that are FAIR themselves, and references to other (meta)data. Reusability ensures the availability of rich documentation, data usage licenses, detailed provenance, and community standards relevant to the domain of data. Optimizing these four characteristics in genomics data and bio-repositories is becoming key to successful resources in the genomic research community.

Some initiatives are starting to promote FAIRness and open/sharable science: FAIRsharing [347]; CEDAR [293], a system for development, evaluation, use, and refinement of genomics (biomedical in general) metadata; BioSchemas.org [173], that applies schemata to online resources to make them easily findable; DNAdigest [231], promoting efficient sharing of human genomic datasets; DATS [346] to boost their discoverability.

From these baselines, we envision a data integration process that includes a seamless evaluation of quality parameters, relating them to the specific genomic analyses that are targeted time by time (see Batini and Scannapieco [34] for a comprehensive discussion on the dependency of quality dimension metrics from the addressed goal), preparing data that are more directly employable in biological discovery.

Predictably, future data integration approaches (such as the one initiated in [55, 56]) will include more and more a data quality-aware *modus operandi* with the following characteristics:

i) currency-driven synchronization with sources;[4]
ii) concise, orthogonal, and common data representations;
iii) light and interoperable data descriptions;
iv) reliability-tailored dataset linkage.

7.4 Simplifying Data and Tools for End Users

So far, GeCo's building blocks have been used in about 50 research studies with very positive feedback. However, its achievements have employed principally a model/system-driven approach, leading to significant limitations in the usability and intuitiveness of the interfaces. Above all, to fully exploit computational strategies as well as GeCo technology, a bioinformatics background and hands-on attitude toward computational resources are needed.

We have become aware that bioinformatics has to translate research objectives into "customized" search routines and identify suitable tools to get out actionable genomics information. We need to treat applications as first-class citizens and direct major efforts to produce a workflow-driven approach that makes data search and analysis processes more attractive for domain experts (with strong competence in the field, but low computer science and programming knowledge).

To achieve this goal, we have started the development of GeCoAgent, a fully integrated, user-centered web platform aimed at empowering end-user competencies for building cloud-based big data applications. To make GeCo resources accessible to a wide audience – including clinicians and biologists – GeCoAgent uses: dialogues to interact with computational tools, a grammar-driven conversational agent translated into a chatbot, and a dashboard where several data summarization/analysis visual objects – progressively built by the system – are shown to the user. A dialogic interface is the easiest form of interaction, both in terms of time required to accomplish the operations and minimizing the user's errors; a sort of soft virtual bioinformatics assistant that works like "Ok Google/Alexa".

The overall process supported by GeCoAgent can be appreciated in Figure 7.3, where has four phases are identified in a typical genomic data-extraction-and-analysis

[4] Note that in genomics it is essential that a centralized integrated repository is updated with respect to sources as the richest and best-quality data files must be captured (see our Downloader implementation in Section 4.3). Instead, addressing the temporal dimension of integration is typically less relevant, as data that is related to different timestamps from various sources, is usually not referring to pre-existing individuals/entities. Longitudinal studies are an exception, however, we do not discuss them in this thesis.

pipeline. The phases resulting from our high-level conceptualization are concerned with data extraction, exploration, analysis, and visualization.

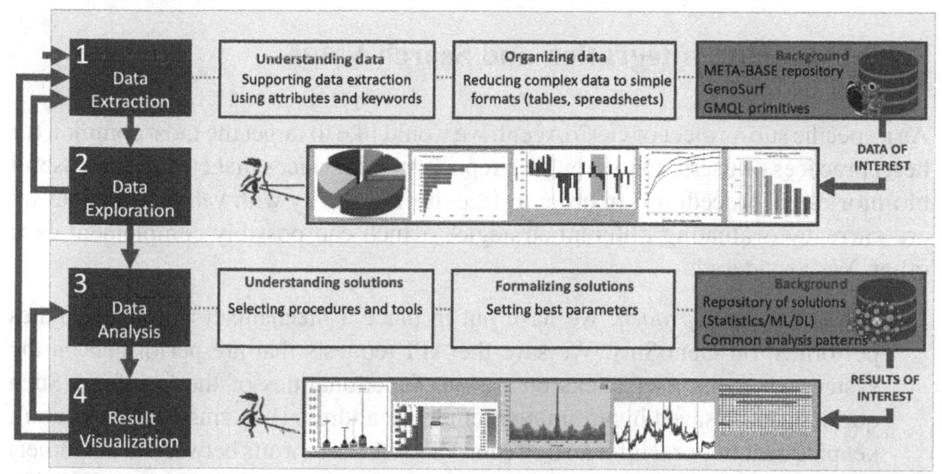

Fig. 7.3 Overall tertiary data analysis process adopted by GeCoAgent, which consists of data extraction followed by data analysis. Data extraction includes the two macro phases of defining objectives, for constructing the "universe of interest" out of a wider information basis, and of inspection of that universe by means of data visualization tools. Data analysis includes the two macro phases of defining the data analysis procedures out of a repository of techniques, and then inspecting the results produced by those techniques, typically through data visualizations.

For *extracting the data of interest*, the researcher should define a set of involved resources: which data, how organized, how retrieved. As a result, users define a universe of interest that can be further explored and evaluated. A second *data exploration* phase consists of inspecting the obtained universe, taking advantage of many possible statistical qualitative/quantitative visualization techniques. Once the data is extracted, it can be analyzed. The third phase involves *defining an analysis procedure*, i.e., understanding suitable procedures/tools, appropriate parameters to be set, taking advantage of an existing repository of methodologies (e.g., Statistics, Machine Learning, and Deep Learning libraries), and commonly employed solutions, ranked accordingly to their suitability for the problem at hand. When the result of interest has been generated, the fourth phase of *result inspection* can be performed with the support of other visualization tools.

As it can be observed in Figure 7.3, Various paths can combine the four phases in many ways; the user may first observe the whole universe – represented through appropriate abstractions that relieve the possible information overload – to get inspiration from all the available data and their interactions, then focus on a specific portion of it; or instead, select datasets one by one and evaluate their characteristics. During the analysis, algorithms, and parameters are progressively adapted until the user is confident with the results; at any stage, the user can also decide that the initial datasets do not fit the needs of the analysis anymore and thus go back to the data extraction phase to re-iterate the process.

7.5 Monitoring Integration and Search Value

As a specific sub-project of GeCoAgent, we would like to target the most common (or best) practices of users, so to provide a repository of services that encapsulate useful bioinformatics procedures. In order to feed this repository with valuable content we are currently evaluating different strategies, which can possibly complement each other. We considered:

- a *data-driven approach*: we have put in place a mechanism to trace searches performed on GenoSurf. We save the API requests that are performed on the system when the user clicks on various functionalities of the interface. Such queries can be saved by grouping them by IP address (i.e., single-user session), keeping the temporal information (how much the user waits between one another) and the order (the sequence is explicit, based on the same IP address). We plan to use this data to model the search process on our systems, building a meta-repository that traces all performed queries (plus the involved attributes/values), clustered by kind of research.
- a *model-driven approach*: we reviewed a number of publications, technical reports, and master students' theses completed within the GeCo group and outside (so far about 50 units), and categorized them according to the research objective, the single performed tasks – both data extraction and data analysis – and the sets of tasks combined together to achieve the research goal. This led us to specify "macros", consisting of different queries that are typically performed together (or in sequence) to achieve common extraction/analysis tasks.
- an *empirical-study-driven* approach, running an interview-based user study to elicit a hierarchical task tree of the tertiary bioinformatics research process. This aspect is currently being investigated by other members of the GeCo group [113].

We are considering using a formalism such as Business Process Model and Notation (BPMN) to better capture the concept of bioinformatic search and analysis as a process, which is different between different user profiles. This would allow us to look for interesting patterns and also enforce process mining (by, for example, employing *frequency-driven usage metadata* for recommendation [34]).

This kind of research will be of great interest both for internal use within GeCo-Agent – achieving better results of usability and effectiveness – but even more for the general public of researchers: a "queries marketplace service" that includes a library of best-practice BPMN processes, assisting users in selecting the appropriate process for their goal, based on maximizing the value of their queries.

Part II
Viral Sequence Data Integration

Chapter 8
Viral Sequences Data Management Resources

"The basic point is so important I'll repeat it: RNA viruses mutate profligately. [...] A mutation in that strain might have made it especially aggressive, efficient, transmissible, and fierce."
— David Quammen, Spillover: Animal Infections and the Next Human Pandemic

The outbreak of the COVID-19 disease has presented novel challenges to the research community, which is rushing towards the delivery of results, pushed by the intent of rapidly mitigating the pandemic effects. During these times, we observe the production of an exorbitant amount of data, often associated with a poor quality of describing information, sometimes generated by insufficiently tested or not peer-reviewed efforts. But we also observe contradictions in published literature, as it is typical of a disease that is still in its infancy, and thus only partially understood.

In this context, the collection of viral genome sequences is of paramount importance, in order to study the origin, wide-spreading and evolution of SARS-CoV-2 (the virus responsible for the COVID-19 disease) in terms of haplotypes (i.e., clusters of inherited variations at single positions genomic sequence), phylogenetic tree (i.e., a diagram for representing the evolutionary relationships among organisms) and new variants. Since the beginning of the pandemic, we have observed an almost exponential growth in the number of deposited sequences within large shared databases, from a few hundred in March 2020, up to thousands; indeed, it is the first time that Next Generation Sequencing technologies have been used for sequencing a massive amount of viral sequences. In August 2020 the total number of sequences of SARS-CoV-2 available worldwide reached about one hundred thousand. In several cases, also relevant associated data and metadata are provided, although their amount, coverage, and harmonization are still limited.

Several institutions provide databases and resources for depositing viral sequences. Some of them, such as NCBI's GenBank [356], preexist the COVID-19 pandemic, as they host thousands of viral species – including, e.g., Ebola, SARS and Dengue, which are also a threat to humanity. Other organizations have produced a new data collection specifically dedicated to the hosting of SARS-CoV-2 sequences, such as the Global Initiative on Sharing All Influenza Data (GISAID) [370, 137] – originally created for hosting virus sequences of influenza – which is soon becoming the predominant data source.

A. Bernasconi: *Model, Integrate, Search... Repeat*, LNBIP 496, pp. 167–182, 2023.
https://doi.org/10.1007/978-3-031-44907-9_8

Most data sources hereby reviewed, including GenBank, the COronavirus disease 2019 Genomics UK Consortium (COG-UK) [335] and some new data sources from China, have adopted a fully open-source model of data distribution and sharing. Instead, GISAID is protecting the deposited sequences by controlling users, who must log in from an institutional site and must observe a Database Access Agreement;[1] probably, such protected use of the deposited data contributes to the success of GISAID in attracting depositors from around the world.

Given that viral sequence data are distributed over many database sources, there is a need for data integration and harmonization, so as to support integrative search systems and analyses; many such search systems have been recently developed, motivated by the COVID-19 pandemic.

This Chapter provides readers with a complete background representation of the context where the contribution of this Part II of the thesis should be set.

Chapter organization. We start by describing the database sources hosting viral sequences and related data and metadata, distinguishing between fully open-source and GISAID (Section 8.1). We then discuss the data integration issues that are specific to viral sequences, by considering schema integration and value harmonization (Section 8.2). Then, we present the various search systems that are available for integrative data access to viral resources (Section 8.3); our own final product, ViruSurf – that will be described along Chapters 9, 10, and 11 – is already mentioned, positioned among other available integrated search systems.

Finally, we discuss the current obstacles to the goals of integration (Section 8.4).

Readers should note that, when describing viral sequences available in databases and search systems, this review chapter refers to a frozen snapshot of available resources, captured at the beginning of August 2020, corresponding to a stage of the COVID-19 pandemic that was less critical than the period of March-June 2020, and in which resources had started to be more data-rich and organized. For consistency and the benefit of comparison, we do not update the reported counts to later dates.

8.1 Landscape of Data Resources for Viral Sequences

The panorama of relevant initiatives dedicated to data collection, retrieval, and analysis of viral sequences is broad. Many resources previously available for viruses have responded to the general call to arms against the COVID-19 pandemic and started collecting data about SARS-CoV-2.

According to the WHO's code of conduct [400], alternative options are available to data providers of virus sequences. The providers who are not concerned about retaining ownership of the data may share it within the many databases that provide full open data access. Among them, GenBank assumes that its submitters have "received any necessary informed consent authorizations required prior to submitting sequences," which includes data redistribution. However, in many cases, data

[1] https://www.gisaid.org/registration/terms-of-use/

providers prefer data-sharing options in which they retain some level of data ownership. This attitude has been established since the influenza pandemic (around 2006) when the alternative model of GISAID EpiFlu^{TM} emerged as dominant.

A general view of relevant resources and initiatives dedicated to data collection, retrieval, and analysis of virus sequences is shown in Figure 8.1. Rectangles represent resources identified using their logo. We partitioned the space of contributors by considering: institutions that host data sequences (1), primary sequence deposition databases (2), tools provided for directly querying and searching them (3), secondary data analysis interfaces that also connect to viral sequence databases (4) portals directly exposing NCBI/GISAID databases (5). Below, we include the integrative search systems (6) – also our own ViruSurf and its GISAID-specific version – that are transversal to the above divisions. Satellite resources (7) are growing, linked externally by viral sequences databases.

We next focus on the four upper levels depicted in Figure 8.1, by starting with the sources that provide full open access and then presenting GISAID.

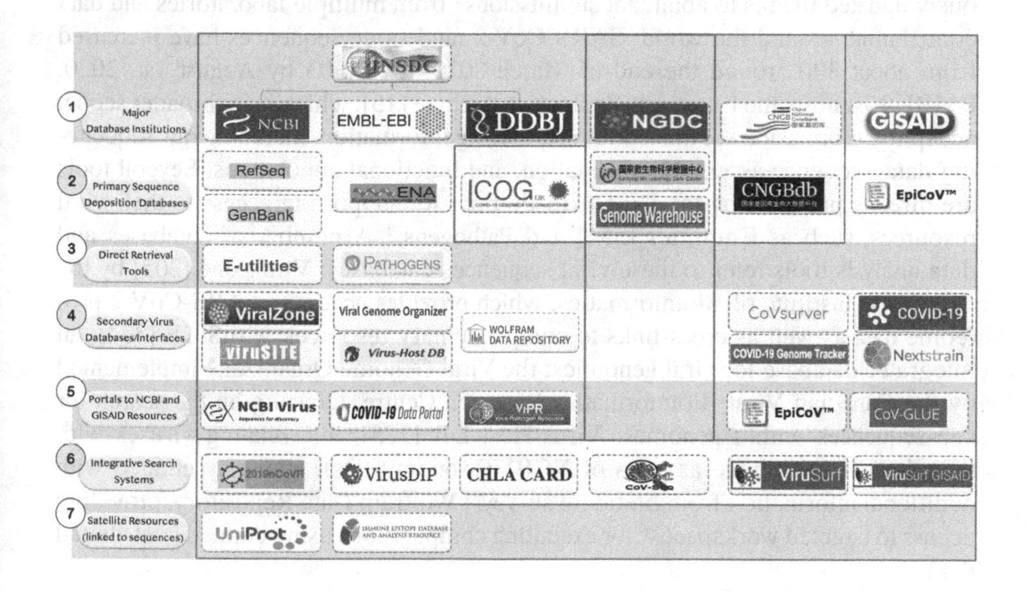

Fig. 8.1 Current relevant resources and initiatives dedicated to data collection, retrieval, and analysis of virus sequences, divided by open and registered access.

8.1.1 Fully Open-Source Resources

8.1.1.1 Resources Coordinated by INSDC

The three main organizations providing open-source viral sequences are NCBI (US), DDBJ (Japan), and EMBL-EBI (Europe); they operate within the broader context of the International Nucleotide Sequence Database Collaboration (INSDC[2]). INSDC provides what we call an *integration by design* of sequences (i.e., three institutions provide agreed submission pipelines, curation process, and points of access to the public, coupled by the use of the same identifiers and rich interoperability between their portals).

NCBI hosts the two most relevant sequence databases: GenBank [356] contains the annotated collection of publicly available DNA and RNA sequences; RefSeq [301] provides a stable reference for genome annotations, gene identification/characterization, and mutation/polymorphism analysis. GenBank is continuously updated thanks to abundant submissions[3] from multiple laboratories and data contributors around the world; SARS-CoV-2 nucleotide sequences have increased from about 300 around the end of March 2020, to 13,303 by August 1st, 2020. EMBL-EBI hosts the European Nucleotide Archive [15], which has a broader scope, accepting submissions of nucleotide sequencing information, including raw sequencing data, sequence assembly information, and functional annotations. Several tools are directly provided by the INSDC institutions for supporting access to their viral resources, such as E-utilities [354] and Pathogens.[4] A number of databases and data analysis tools refer to these viral sequences databases: ViralZone [205] by the SIB Swiss Institute of Bioinformatics, which provides access to SARS-CoV-2 proteome data as well as cross-links to complementary resources; viruSITE [374], an integrated database for viral genomics; the Viral Genome Organizer,[5] implemented by the Canadian Viral Bioinformatics Research Centre, focusing on the search for sub-sequences within genomes; Virus-Host DB [285], interrelating viruses with their hosts (represented as pairs of NCBI Taxonomy IDs), manually curated with additional information from literature surveys; Wolfram Data Repository, providing access to targeted workspaces[6] for executing computations using curated contributed data.

[2] http://www.insdc.org/

[3] Users and laboratories can submit their data to GenBank through https://submit.ncbi.nlm.nih.gov/.

[4] https://www.ebi.ac.uk/ena/pathogens/

[5] https://4virology.net/virology-ca-tools/vgo/

[6] The Genetic Sequences for the SARS-CoV-2 Coronavirus are provided at https://datarepository.wolframcloud.com/resources/Genetic-Sequences-for-the-SARS-CoV-2-Coronavirus/.

8.1.1.2 COG-UK

The COronavirus disease 2019 Genomics UK Consortium (COG-UK) [335] is a national-based initiative launched in March 2020 thanks to big financial support from three institutional partners: UK Research and Innovation, UK Department of Health and Social Care, and Wellcome Trust. The primary goal of COG-UK is to sequence about 230,000 SARS-CoV-2 patients (with priority to healthcare workers and other essential workers in the UK) to help track the virus transmission. They provide data directly on their webpage, open for use, as a single FASTA[7] file;[8] this is associated with a CSV file for metadata. As of August 1st, 2020, the most updated release is dated 2020-07-28, with 38,124 sequences (as declared on the Consortium's web page).

8.1.1.3 Chinese Sources

Since the early outbreak of COVID-19, several resources were made available in China:

- The Chinese National Genomic Data Center [332] provides some data resources relevant for COVID-19 related research, including the Genome Warehouse,[9] which contains genome assemblies with their detailed descriptive information: biological sample, assembly, sequence data, and genome annotation.
- The National Microbiology Data Center (NMDC[10]) provides the "Novel Cov National Science and Technology Resource Service System" to publish authoritative information on resources and data concerning 2019-nCoV (i.e., SARS-CoV-2 virus) to provide support for scientific studies and related prevention/-control actions. The resource is provided in Chinese language with only some headers and information translated into English. Its FTP provides a collection of sequences from various coronaviruses, including many from NCBI GenBank, together with a restricted number of NMDC original ones.
- The China National GeneBank DataBase [106] (CNGBdb[11]) is a platform for sharing biological data and application services to the research community, including internal data resources; it also imports large amounts of external data from INSDC databases.

Around the world there are many other sequence collections not yet included within international repositories, which are hardly accounted for; one of them is the CHLA-CPM dataset collected by the Center for Personalized Medicine (CPM[12]) at the

[7] A text-based format for representing either nucleotide sequences or amino acid (protein) sequences, in which nucleotides or amino acids are represented using single-letter codes.

[8] The COG-UK data is available at `https://www.cogconsortium.uk/data/`

[9] `https://bigd.big.ac.cn/gwh/`

[10] `http://nmdc.cn/`

[11] `https://db.cngb.org/`

[12] `https://www.chla.org/center-personalized-medicine`

Children's Hospital, Los Angeles (CHLA), resulting from an initiative launched in March 2020 to test a broad population within the Los Angeles metropolitan area.

8.1.2 GISAID and its Resources

During the COVID-19 pandemic, GISAID has proposed again its solution in the form of the new database EpiCoV™, associated with similar services as the ones provided for influenza. The GISAID-restricted open-source model has greatly facilitated the rapid sharing of virus sequence data, but it contemplates constraints on data integration and redistribution, which we later describe in Section 8.4. At the time of writing, GISAID has become the most used database for SARS-CoV-2 sequence deposition, preferred by the vast majority of data submitters and gathering 75,507 sequences by August 1st, 2020.

It is also the case that GISAID formatting/criteria for metadata are generally considered more complete and are thus suggested even outside of the direct submission to GISAID; the SARS-CoV-2 sequencing resource guide of the US Centers for Disease Control and Prevention (CDC[13]) reports, in the section regarding recommended formatting/criteria for metadata [87], that the user is invited to submit always using the submission formatting of GISAID EpiCoV™ "which tends to be more comprehensive and structured". However, in order to check such a format, the user is invited to create an account on GISAID, which probably leads to using GISAID directly instead of going back to GenBank.

Some interesting portals are "enabled by data from GISAID", as clearly stated on the top of their pages, with different focuses. NextStrain [185][14] overviews emergent viral outbreaks based on the visualization of sequence data integrated with geographic information, serology, and host species. A similar application for exploring and visualizing genomic analysis has been implemented by Microreact, which has a portal dedicated specifically to COVID-19.[15] CoVsurver,[16] which had a corresponding system for influenza virus called FluSurver,[17] enables rapid screening of sequences of coronaviruses for mutations of clinical/epidemiological importance. CoV Genome Tracker[18] [7] combines in a dashboard a series of visualizations based on the haplotype network, a map of collection sites and collection dates, and a companion tab with gene-by-gene and codon-by-codon evolutionary rates.

[13] `https://www.cdc.gov/`

[14] `https://nextstrain.org/ncov/`

[15] `https://microreact.org/project/COVID-19/`

[16] `https://corona.bii.a-star.edu.sg`

[17] `http://flusurver.bii.a-star.edu.sg`

[18] `http://cov.genometracker.org`

8.2 Integration of Sources of Viral Sequences

Next Generation Sequencing is successfully applied to infectious pathogens [183], with many sequencing technology companies developing their assays and workflows for SARS-CoV-2 (see Illumina [207] or Nanopore [296]). Some sources provide the corresponding raw data (see European Nucleotide Archive [15] of the INSDC network), but most sources present just the resulting sequences, typically in the form of FASTA, together with some associated metadata. In this section, we do not address the topic of sequence pipeline harmonization (we refer interested readers to forum threads [123] and recent literature contributions [224]). We focus instead on the data integration efforts required for their metadata and value integration.

8.2.1 Metadata Integration

Metadata integration is focused on provisioning a global, unified schema for all the data that describe sequences within the various data sources [33]. In the context of viral sequences, as the amount of data is easily manageable, it is common to import data at the integrator site; in this way, data curation/reprocessing can be performed in a homogeneous way. In such context, one possible solution is to apply conceptual modeling (i.e., the entity-relationship approach [32]) as a driver of the integration process. In the variety of resources dedicated to viruses [367], very few works relate to conceptual data modeling. Among them, [377] considers host information and normalized geographical location, while [259] focuses on influenza A viruses. CoV-GLUE [371] includes a basic conceptual model for SARS-CoV-2.[19] In comparison, the Viral Conceptual Model (VCM, [50]), which will be extensively described in Chatper 9, works seamlessly with any kind of virus, based on the molecule type, the species, and the taxonomic characteristics. VCM has many dimensions and attributes, which are very useful for supporting research queries on virus sequences; it uses the full power of conceptual modeling to structure metadata and organize data integration and curation.

8.2.2 Value Harmonization and Ontological Efforts

Besides schema unification, data values must be standardized and harmonized in order to fully support integrated query processing. The following value harmonization problems must be solved:

- Virus and host species should refer to dedicated controlled vocabularies (the NCBI Taxonomy [144] is widely recognized as the most trusted, even if some

[19] http://glue-tools.cvr.gla.ac.uk/images/projectModel.png

concerns apply to the ranking of SARS-CoV-2 as a species or just an isolate/-group of strains).[20]
- Sequence completeness should be calculated using standard algorithms, using the length and the percentage of certain indicative types of basis (e.g., unknown ones = N).
- The information on sequencing technology and assembly method should be harmonized, especially the coverage field, which is represented in many ways by each source.
- Dates – both collection and submission ones – must be standardized; unfortunately, they often miss the year or the day, and sometimes it is not clear if the submission date refers to the transmission of the sequence to the database or to a later article's publication.
- Geographical locations, including continent, country, region, and area name, are encoded differently by each source.
- In some rare cases, sequences come with gender and age information, hidden in the middle of descriptive fields.

A number of efforts have been directed at the design of ontologies for solving some of these problems:

- The Infectious Disease Ontology (IDO) has a focus on the virus aspects; its curators have proposed an extension of the ontology core to include terms relevant to COVID-19 [22, 59].
- CIDO [194] is a community-based ontology to integrate and share data on coronaviruses, more specifically on COVID-19. Its infrastructure aims to include information about the disease etiology, transmission, epidemiology, pathogenesis, host-coronavirus interactions, diagnosis, prevention, and treatment. Currently, CIDO contains more than 4,000 terms; in observance of OBO Foundry principles, it aggregates already existing well-established ontologies describing different domains (such as ChEBI [191] for chemical entities, Human Phenotype Ontology [228] for human host phenotypes, the Disease Ontology [359] for human diseases including COVID-19, the NCBI taxonomy, and the IDO itself) – so to not create unnecessary overlaps. New CIDO-specific terms have been developed to meet the special needs arising in the research of COVID-19 and other coronavirus diseases. The work on host-pathogen interactions is described in depth in [410], while the inclusion in CIDO of aspects related to drugs and their repurposing is described in [251].
- The COVID-19 Disease Map [304] is a visionary project by Elixir Luxembourg, that aims to build a platform for exploration and analyses of molecular processes involved in SARS-CoV-2 interactions and immune response.

For the sequence annotation process, there are two kinds of ontologies that are certainly relevant: the Sequence Ontology [135], used by tools such as SnpEff [104] to characterize the different subsequences of the virus, and the Gene Ontology [19],

[20] The problem of organizing viruses in a taxonomy is far from being solved, as reported by Koonin *et al.* [229], who propose a mega taxonomy of the virus world.

which has dedicated a page to COVID-19[21] that provides an overview of human proteins that are used by SARS-CoV-2 to enter human cells, divided by the 29 different virus' proteins.

8.2.3 Replicated Sequences in Multiple Sources

Record replication is a recurrent problem occurring when integrating different sources; it is solved by "Entity Resolution" tasks, i.e., identifying the records that correspond to the same real-world entity across and within datasets [163]. This issue arises for SARS-CoV-2 sequences, as many laboratories use to submit sequences to multiple sources; in particular, sequences submitted to NCBI GenBank and COG-UK are often also submitted to GISAID.

Such a problem is resolved in different manners by the various integrative systems. The main approaches – detailed for specific sources in the following – aim to either resolve the redundancy by eliminating from one source records that appear also in another one or by linking records that represent the same sequence, adding to both records an "external reference" pointing to the other source. Along this second solution, a template proposal for data linkage (defined in [102]) is provided by the CDC [86]: a simple lightweight line list of tab-separated values to hold the name of the sequence, as well as IDs from GISAID and GenBank. Note that no advanced methodologies of record linkage are requested in this case since, to find correspondence between two records, an exact match algorithm between sequences could be run at any time.

8.3 SARS-CoV-2 Search Systems

In this section, we compare the systems that provide search facilities for SARS-CoV-2 sequences and related metadata, possibly in addition to those of other viruses. With respect to Figure 8.1, we here discuss the initiatives related to levels (5) and (6). In Table 8.1 we summarize the content addressed by each system; in the first section we indicate the target virus species, which either includes the SARS-CoV-2 virus only, or also similar viruses (e.g., Coronavirus, other RNA single-stranded viruses, other pandemic-related viruses), or an extended set of viruses. In the second section, the table shows which sources are currently integrated by each system. The first five columns refer to portals to resources gathering either NCBI or GISAID data (level 5), while the following ones refer to integrative systems over multiple sources (level 6). These are described in the next two sections.

[21] http://geneontology.org/covid-19.html

8.3.1 Portals to NCBI and GISAID Resources

Native portals for accessing NCBI and GISAID resources are hereby described even if they do not provide integrative access to multiple sources, as they are recognized search facilities for SARS-CoV-2 sequences collected from laboratories all around the world:

- An interesting and rich resource (Virus Variation Resource [192]) is hosted by NCBI, targeting many viruses relevant to emerging outbreaks. At the time of writing, a version for coronaviruses – and SARS-CoV-2 in the specific – has not been released yet. Instead, for this virus users are forwarded to the NCBI Virus resource;[22] this portal provides a search interface to NCBI SARS-CoV-2 sequences, with several filter facets and a result table where identifiers are linked to NCBI GenBank database pages. It provides very quick and comprehensive access to SARS-CoV-2 data but is not well aligned with the API provided for external developers, making integration efforts harder, as differences need to be understood and synchronized.
- COVID-19 Data Portal[23] joins the efforts of ELIXIR and EMBL-EBI to provide an integrated view of resources spanning from raw reads/sequences, to expression data, proteins, and their structures, drug targets, literature and pointers to related resources. Coupling raw and sequence data in the same portal can be very useful for users who wish to resort to original data and recompute sequences or compare them with existing ones. We focus on their contribution to data search, which is given through a data table containing different data types (sequences, raw reads, samples, variants, etc.). The structure of the table changes based on the data types (the metadata provided for nucleotide sequence records are overviewed next).
- The Virus Pathogen Database and Analysis Resource (ViPR[24] [315]) is a rich repository of data and analysis tools for multiple virus families, supported by the Bioinformatics Resource Centers program. It provides GenBank strain sequences with UniProt proteins, 3D protein structures, and experimentally determined epitopes from the Immune Epitope Database (IEDB [389]). For SARS-CoV-2 many different views are provided for genome annotation, comparative genomics, ortholog groups, host factor experiments, and phylogenetic tree visualization. It provides the two functions "Remove Duplicate Genome Sequences" and "Remove Identical Protein Sequences" to resolve redundancy respectively of nucleotide and amino acid sequences.
- GISAID EpiCoV^{TM} portal provides a search interface upon GISAID metadata. Nine filters are available to design the user search, while the results table shows 11 metadata attributes. By clicking on single entries, the user accesses much richer information, consisting of 31 metadata attributes. The browsing power of the interface is limited by the use of a few attributes. While more information

[22] `https://www.ncbi.nlm.nih.gov/labs/virus/`

[23] `https://www.covid19dataportal.org/`

[24] `https://www.viprbrc.org/`

is given on single entries, the users are prevented to take advantage of these additional attributes to order results or see distinct values.

- CoV-GLUE[25] [371] has a database of replacements, insertions, and deletions observed in sequences sampled from the pandemic. It also provides a quite sophisticated metadata-based search system to help filter GISAID sequences with mutations. While other systems are sequence-based, meaning that the users can select filters to narrow down their search on sequences, CoV-GLUE is variant-based: the user is provided with a list of variants (on amino acids); by selecting given variants, the sequences that present such variants can be accessed.

		Portals to NCBI/GISAID resources					Integrative search systems				
		NCBI Virus	COVID-19DP	ViPR	EpiCoV™	CoV-GLUE	2019nCoVR	VirusDIP	CARD	CoV-Seq	ViruSurf
Content	SARS-CoV-2 specific		×		×	×		×	×	×	
	SARS-CoV-2 + similar						×				×
	Extended virus set	×		×							
Included sources	GenBank	×	×	×			×	×	×	×	×
	RefSeq	×	×	×			×	×	×		×
	GISAID				×	×	×	×	×	×	×
	COG-UK										×
	NMDC						×		×		×
	CNGBdb						×	×	×	×	
	Genome Warehouse						×		×		
	CHLA-CPM								×		

Table 8.1 Top part: characterization of each system based on its focus on general SARS-CoV-2 virus only, SARS-CoV-2 and similar viruses (e.g., other Coronavirus or pandemic-related viruses), or an extended set of viruses. Bottom part: integration of sequences by each portal (columns) from each origin source (rows).

8.3.2 Integrative Search Systems

The following systems provide integrative data access from multiple sources (see Figure 8.1, level 6), as indicated in Table 8.1:

- 2019nCoVR[26] [417] at the Chinese National Genomics Data Center (at the Beijing Institute of Genomics) is a rich data portal with several search facets, tables, and visual charts. This resource includes most sources publicly reachable, including GISAID; however, it is unclear if this is compliant with the GISAID data-sharing agreement (see Section 8.4 for a related discussion). 2019nCoVR handles sequence records redundancy by conveniently providing a "Related ID" field that allows one to map each sequence from its primary database to others that also contain it.

[25] http://cov-glue.cvr.gla.ac.uk/

[26] https://bigd.big.ac.cn/ncov/

- The Virus Data Integration Platform (VirusDIP[27] [391] is a system developed at CNGBdb to help researchers find, retrieve and analyze viruses quickly. It declares itself as a general resource for all kinds of viruses; however, to date includes only SARS-CoV-2 sequences.
- The COVID-19 Analysis Research Database (CARD [368]) is a rich and interesting system giving the possibility to rapidly identify SARS-CoV-2 genomes using various online tools. However, the data search engine seems to be still under development and to date does not allow building complex queries combining filters yet.
- CoV-Seq[28] [250] collects tools to aggregate, analyze, and annotate genomic sequences. It claims to integrate sequences from GISAID, NCBI, EMBL, and CNGB. It has to be noted that sequences from NCBI and EMBL are the same ones, as part of the INSDC. The search can only be done on very basic filters directly on the columns of the table, providing poor functionalities. It also provides a basic search system over a few metadata and computes the "Identical_Seq" field, where a sequence is mapped to many identical ones.
- ViruSurf[29] [77] is based on a conceptual model [50] that describes sequences and their metadata from their biological, technical, organizational, and analytical perspectives. It provides many options for building search queries, by combining – within rich Boolean expressions – metadata attributes about viral sequences and nucleotide and amino acid variants. Full search capabilities are used for open-source databases, while search over the GISAID database is suitably restricted to be compliant with the GISAID data-sharing agreement. ViruSurf also solves the problem of record redundancy among different databases by using different external reference IDs available and by exploiting in-house computations. While here it is presented among other systems for comparison, we will describe it thoroughly in Chapter 11.

8.3.3 Comparison

Table 8.2 shows a comprehensive view of which metadata information is included by each search system. Attributes are partitioned into macro-areas concerning the *biology* of the *virus* and of the *host organism sample*, the *technology* producing the sequence, the *sequence* details and the *organization* of sequence production. For each of them, we provide three kinds of columns: S = *Search filters* (attributes supporting search queries, typically using conjunctive queries), T = *columns in result Tables* (providing direct attribute comparisons), and E = *columns of single Entries*

[27] https://db.cngb.org/virus/ncov
[28] http://covseq.baidu.com/
[29] http://gmql.eu/virusurf/

	Attribute description	NCBI Virus		COVID-19DP		ViPR		EpiCoV™			CoV-GLUE			2019nCoVR			VirusDIP		CARD	CoV-Seq	ViruSurf	
		S	T	S	T	S	T	S	T	E	S	T	E	S	T	E	S	T	S	T	S	T
Biology: Virus	Accession	×	×		×	×	×	×	×	×	×	×	×		×	×	×	×	×	×	×	×
	Related ID														×	×			×	×		
	Strain				×	×	×	×	×	×	×	×			×	×	×	×	×	×	×	×
	Virus Taxonomy ID				×	×												×				×
	Virus Species	×	×			×	×											×			×	×
	Virus Genus		×			×	×			×												×
	Virus Subfamily					×																×
	Virus Family		×			×	×															×
	Lineage									×	×	×	×								×	×
	CoV-GLUE Lineage												×									
	Pangolin Lineage												×									
	Total LWR												×									
	MoleculeType				×		×															×
	SingleStranded																					×
	PositiveStranded																					×
	Passage detail								×	×												
Biology: Sample	Collection date	×	×	×	×	×	×	×	×	×	×	×	×	×	×	×	×	×	×	×	×	×
	Location	×	×	×		×	×	×	×	×	×	×	×	×	×	×	×	×	×	×	×	×
	Origin lab								×	×			×		×	×		×	×			×
	Host Taxonomy ID																				×	×
	Host organism	×	×	×		×	×	×	×	×				×	×	×	×	×	×		×	×
	Host gender									×									×		×	×
	Host age									×									×		×	×
	Host status							×		×												
	Environmental source	×																				
	Specimen source	×	×							×											×	×
	BiosampleId		×																			
Technology	Sequencing Technology									×								×			×	×
	Assembly method									×								×			×	×
	Coverage							×		×											×	×
	Quality Assessment								×	×				×	×	×						
	Sequence Quality													×	×							
	SRA Accession		×																			
Sequence	Complete	×	×			×		×						×	×	×			×		×	×
	Length	×	×				×		×					×	×			×	×	×	×	×
	IsReference		×																		×	×
	GC bases %																				×	×
	Unknown bases %													×							×	×
	Degenerate bases %													×								
Organization	Authors	×	×							×			×					×	×	×		
	Publications		×															×				
	Submission date							×	×	×					×		×	×	×		×	×
	Submission lab								×	×			×		×	×		×	×	×	×	×
	Submitter									×								×				
	Release date	×	×												×	×						
	Data source	×	×											×	×	×		×	×	×	×	×
	Last Update Time														×							
	Bioproject ID																				×	×

Table 8.2 Inspection of metadata fields in different search portals for SARS-CoV-2 sequences. × is used when the attribute is present in the Search filters (S), in the Table of results (T), and in single Entries (E).

or records (once a record is clicked, search systems enable reading rich metadata but only for each individual sequence entry).

The different systems provide the same attribute concepts in very many terminology forms. For example:

- "Virus name" in EpiCoV™ is called "Virus Strain Name" in 2019nCoVR, just "Virus" in CoV-Seq, "Strain" or "Title" CARD (as they provide two similar search filters), and "StrainName" in ViruSurf.

Data features	Portals to NCBI/GISAID resources				Integrative search systems					
	NCBI Virus	COVID-19DP	ViPR	EpiCoV™	CoV-GLUE	2019nCoVR	VirusDIP	CARD	CoV-Seq	ViruSurf
Haplotype network						×				
Pylogenetic tree	×		×		×	×	×	×		
Nucl. sequences	×	×	×							×
Aa sequences	×	×	×							×
Prec. nucl. variants						×		×	×	×
Prec. aa variants						×		×		×

Table 8.3 Additional features provided by search systems in addition to standard metadata.

- Geographic information is named "Geo Location" in NCBI Virus results table and "Geographic region" in the search interface; it is given using the pattern Continent/Country/Region/SpecificArea in GISAID EpiCoV™, while the four levels are kept separate in CARD and ViruSurf.
- The database source is referred to as "Sequence Type", "Data Source", "Data source platform", "Data_source", "Data_Source", or "DatabaseSource".

It must be mentioned that some systems include additional metadata that we did not add here, as they were not easily comparable. For example, GISAID has a series of additional location details and sample IDs that can be provided by submitters (but normally they are omitted), whereas NCBI Virus adds info about provirus, lab host, and vaccine strains (but these are omitted for SARS-CoV-2 related sequences).

Table 8.3 provides a quick report on which additional data features are provided in each search system, beyond classic metadata; they include haplotypes, phylogenetic tree, nucleotide (nuc.), and protein sequences (aa = amino acid) and their pre-calculated variants.

8.4 Discussion

We articulate the discussion along a number of directions: impact of GISAID's model, lack of metadata quality, (un)willingness of sequence sharing.

8.4.1 GISAID Restrictions

While most of the resources reviewed in this paper, and in particular all the open data sources and search engines, are available through public Web interfaces, the GISAID portal can be accessed just by using a login account; access is granted in response to an application, which must be presented by using institutional emails and requires agreeing to a Database Access Agreement.[30] Registered users are invited to "help GISAID protect the use of their identity and the integrity of its user base".

[30] `https://www.gisaid.org/registration/terms-of-use/`

Thanks to this controlled policy, GISAID has been able to gather the general appreciation of many scientists, who hesitate to share data within fully open-source repositories. Some concerns are actually legitimate, such as not being properly acknowledged; acknowledgment of data contributors is required of whoever uses specific sequences from the EpiCoV^{TM} database. However, GISAID policies impose limitations on data integrators. Sequences are not communicated to third parties, such as integration systems; interested users can only access and download them one by one from the GISAID portal. As the reference sequence can be reconstructed from the full knowledge of nucleotide variants, these are similarly not revealed.

8.4.2 Metadata Quality

Another long-standing problem is the low quality of input metadata. A commentary [358] from the Genomic Standards Consortium board (GSC[31]) alerts the scientific community of the pressing need for systematizing the metadata submissions and enforcing metadata sharing good practices. This need is even more evident with COVID-19, where the information on geo-localization and collection time of a sample easily becomes a "life and death issue"; they claim that the cost paid for poor descriptions about the pathogen-host and collection process could be greater than the cost paid for poor quality of nucleotide sequence record itself. To this end, EDGE COVID-19[32] [254] has made an attempt to facilitate the preparation of genomes of SARS-CoV-2 for submission to public databases providing help both with metadata and with processing pipelines.

As to what concerns variant information, it is important that they refer to the same reference sequence. We found that in some cases a different sequence was used as a reference for SARS-CoV-2 with respect to the one of NCBI GenBank, commonly accepted by the research community. If different reference sequences are used and original sequence data are not shared, it becomes very hard to provide significant statistics about variant impact.

The insufficient submission of enriched contextual metadata is generally imputed to the fact that individual researchers receive little recognition for data submission and that probably they prefer withholding information, being concerned that their data could be reused before they finalize their own publications. This is not only a problem of submission practice, but also of data sharing, as discussed next.

8.4.3 (Un)Willingness to Share Sequence Data

In times of pandemic, there is – as discussed next – a strong need for data sharing, creating big databases that can support research. Regardless of this necessity, many

[31] https://www.gensc.org/

[32] https://edge-covid19.edgebioinformatics.org/

researchers or research institutions do not join the data-sharing efforts. For example, it looks strange to us that searches for SARS-CoV-2 sequence data from Italy, witnessing one of the first big outbreaks of COVID-19 in the world, return only a few sequences (27 on GenBank and 239 on GISAID, of which only 117 with patient status information, as of August 1st, 2020); similar numbers apply to many other countries.

Successful provisioning of sequences is the result of a number of conditions: having funds for sequencing, high-quality technology to retrieve useful results, willingness to join the FAIR science principles [402]. Sampling activity in the hospitals is essential, as well as its timely processing and sequencing pipelines in laboratories; however, nowadays the most critical *impasse* is met at the stage of submitting sequences and associated metadata (if not even clinical information regarding the host), which has become almost a deliberate political act in the current times [319].

We also observed the opposite attitude: consortia such as the COVID-19 Host Genetics Initiative have been assembled around the objective and principles of open data sharing. As another significant case, the E-ellow Submarine [133] interdisciplinary initiative for exploiting data generated during the COVID-19 pandemic is fully committed to open data. Practical ecosystems for supporting open pathogen genomic analysis [61] will become more widespread if proactively encouraged by strong institutional support. We hope and trust that events such as the COVID-19 pandemic will move scientists towards open data sharing, as a community effort for mitigating the effects of this and future pandemic events.

Chapter 9
Modeling Viral Sequence Data

"Fear, left unchecked, can spread like a virus."
— Lish McBride, Necromancing the Stone

Despite the advances in drug and vaccine research, diseases caused by a viral infection pose serious threats to public health, both as emerging epidemics (e.g., Zika virus, Middle East Respiratory Syndrome Coronavirus, Measles virus, or Ebola virus) and as globally well-established epidemics (such as Human Immunodeficiency Virus, Dengue virus, Hepatitis C virus). The pandemic outbreak of the coronavirus disease COVID-19, caused by the "Severe acute respiratory syndrome coronavirus 2" virus species SARS-CoV-2 (according to the GenBank [356] acronym[1]), has brought unprecedented attention to the genetic mechanisms of coronaviruses. Thus, understanding viruses from a conceptual modeling perspective is very important. The sequence of the virus is the central information, along with its annotated parts (known genes, coding, and untranslated regions...) and the nucleotide/amino acids variants, computed with respect to the reference sequence chosen for the species. Each sequence is characterized by a *strain name*, which belongs to the virus that has been isolated (obtained) from an infected individual, rather than grown in a laboratory. Viruses have complex taxonomies (as discussed in [229]): a species belongs to a genus, to a sub-family, and finally to a family (e.g., Coronaviridae). Other important aspects include the host organisms and isolation sources from which viral materials are extracted, the sequencing project, the scientific and medical publications related to the discovery of sequences; virus strains may be searched and compared intra- and cross-species. Luckily, all these data are made available publicly by various resources, from which they can be downloaded and re-distributed.

Chapter organization. Section 9.1 proposes the Viral Conceptual Model (VCM), a general conceptual model for describing viral sequences (representing the central entity), organized along specific groups of dimensions that highlight a conceptual schema similar to our previously proposed model for human genomics datasets. In Section 9.2 we provide a list of interesting queries replicating newly released

[1] SARS-CoV-2 is generally identified by the NCBI taxonomy [144] ID 2697049.

A. Bernasconi: *Model, Integrate, Search... Repeat*, LNBIP 496, pp. 183–191, 2023.
https://doi.org/10.1007/978-3-031-44907-9_9

literature on infectious diseases; these can be easily answered by using VCM as reference conceptual schema. We review related works in Section 9.3.

9.1 Conceptual Modeling for Viral Genomics

In Section 3.2 we described the Genomic Conceptual Model (GCM, [54]), an Entity-Relationship diagram that recognizes a common organization for a limited set of concepts supported by most genomic data sources, although with different names and formats. In that case, the model is centered on the ITEM entity, representing an elementary experimental file of genomic regions and their attributes. Four views depart from the central entity; they respectively describe the *biological* elements involved in the experiment, the *technology* used in the experiment, the *management* aspects and the *extraction* parameters used for internal selection and organization of items.

The lessons we learned from the GCM experience include the benefits of having: a central *fact* entity that helps structuring the search; a number of surrounding *groups of dimensions* capturing organization, biological and experimental conditions to describe the facts; a direct representation of a data structure suitable for conceptually organizing genomic elements and their describing information. a data layout that is easy to learn for first-time users and that helps the answering of practical questions (demonstrated in Section 6.4.7).

We hereby propose the Viral Conceptual Model (VCM), which is influenced by our past experience with human genomes, with the comparable goal of providing a simple means of integration between heterogeneous sources. However, there are significant differences between the two conceptual models. The human DNA sequence is long (3 billion base pairs) and has been understood in terms of *reference genomes* (named h19 and GRCh38) to which all other information is referred, including genetic and epigenetic signals. For human genomes, we had chosen ITEMS as a central entity because this is the basic data unit employed for bioinformatic tertiary analysis (i.e., a small portion of data that provides meaningful information on an individual and can be handled easily in terms of computational effort). Human genomes could also be represented as sequences, but the amount of information would be huge and it would be impossible to compare different individuals, aggregate, and compute interesting properties. Therefore, to represent them we chose the abstraction of genomic regions, collected in *GDM samples* (or *GCM items*, as explained in Section 4.1, where we dedicated a paragraph to granularity). Instead, viruses are many; extracted samples are associated with a host sample of another species. Moreover, their sequences are short (order of thousands of base pairs) and each virus has its own reference sequence, thus we deem nucleotide sequences to be a suitable representation to be used as central information in our new conceptual model. The different choice is driven by the specific use and computational pipelines that can be used on these kinds of data.

With a bird's eye view, the VCM conceptual model is centered on the SEQUENCE entity that describes individual virus sequences; sequences are analyzed from four groups of dimensions:

1. The *Biological* group (including the VIRUS and HOSTSAMPLE tables) is concerned with the virus species characterization and the host organism, including the temporal/spatial information regarding the extraction of the biological sample.
2. The *Technological* group (EXPERIMENTTYPE table) describes the sequencing method.
3. The *Organizational* group (SEQUENCINGPROJECT table) describes the project producing each sequence.
4. The *Analytical* group provides annotations for specific sub-sequences and characterizes the variants in the nucleotide sequence and in the amino acid sequence with respect to reference sequences for the specific virus species. It includes the ANNOTATION, AMINOACIDVARIANT, NUCLEOTIDEVARIANT and VARIANTIMPACT.

We next illustrate the central entity and the four groups of dimensions.

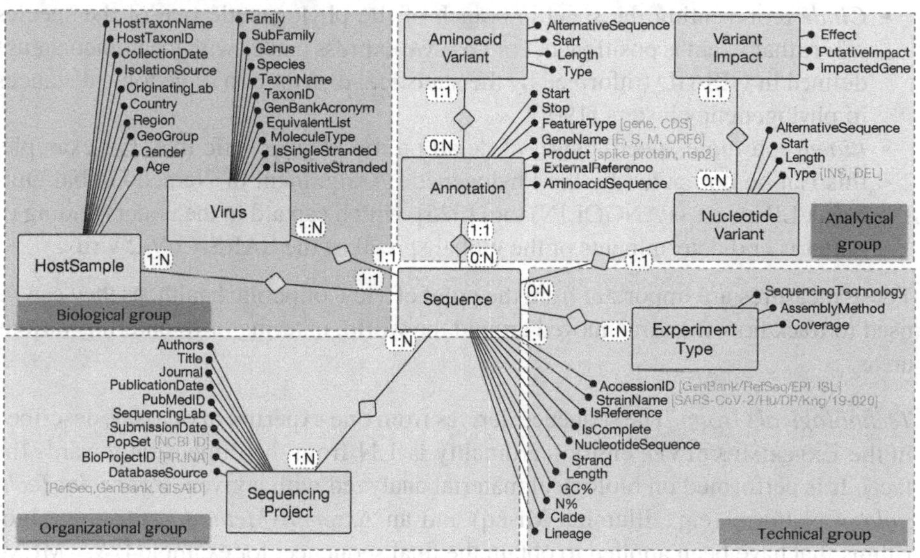

Fig. 9.1 The Viral Conceptual Model: the central fact SEQUENCE is described by four groups of dimensions (biological, technical, organizational, and analytical).

Central Entity. A viral SEQUENCE can regard DNA or RNA; in either case, databases of sequencing data write the sequence as a DNA *NucleotideSequence*: possible characters include guanine (G), adenine (A), cytosine (C), and thymine (T),[2] but

[2] In RNA sequencing databases uracil (U) is replaced with thymine (T).

also eleven "ambiguity" characters associated with all the possible combinations of the four DNA bases [334]. The sequence has a specific *Strand* (positive or negative, strictly dependent on the type of virus—that could be single/double strand), *Length* (ranging from hundreds to millions, depending on the virus), and a percentage of read G and C bases (*GC%*). As the quality of sequences is very relevant to virologists, we also include the percentage of ambiguous bases (i.e., *N%*)[3] to give more complete information on the reliability of the sequencing process. Each sequence is uniquely identified by an *AccessionID*, which is retrieved directly from the source database (GenBank's are usually formed by two capital letters, followed by six digits, GISAID by the string "EPI_ISL_" and six digits) and is used to backtrack the sequence in its original database. Sequences can be complete or partial[4] (as encoded by the Boolean flag *IsComplete*) and they can be a reference sequence (stored in RefSeq) or a regular one (encoded by *IsReference*). Sequences have a corresponding *StrainName* (or isolate) assigned by the sequencing laboratory, somehow hard-coding relevant information (e.g., hCoV-19/Nepal/61/2020 or 2019-nCoV_PH_nCOV_20_026). For genomic epidemiology studies of active viruses (in this particular time, SARS-CoV-2), it is relevant to characterize a strain using other two classifications:

- *Clade* representing the specific branch of the phylogenetic tree of the species where that strain is positioned. For now, we express clades with the nomenclature defined in GISAID (informed by the statistical distribution of genome distances in phylogenetic clusters [190]).
- *Lineage*, a more specific level of detail on the phylogenetic tree. For example, this can be computed by the Phylogenetic Assignment of Named Global Outbreak LINeages (PANGOLIN) tool [325], which can aid in the understanding of patterns and determinants of the global spread of the SARS-CoV-2 virus.

These measures are important from the point of view of public health, as they can be used to track how the virus moved around the world, entering or re-entering various areas.

Technological Group. The sequence derives from one experiment or assay, described in the ExperimentType entity (cardinality is 1:N from the dimension towards the fact). It is performed on biological material analyzed with a given *SequencingTechnology* platform (e.g., Illumina Miseq) and an *AssemblyMethod*, collecting algorithms that have been applied to obtain the final sequence, for example BWA-MEM, to align sequence reads against a large reference genome; BCFtools, to manipulate variant calls; Megahit, to assemble NGS reads. Another technical measure is captured by *Coverage* (e.g., 100x or 77000x), representing the number of unique reads that include a specific nucleotide in the reconstructed sequence.

---

[3] N is the IUPAC character to represent the ambiguity between the four possible bases A, C, T, and G. See `https://genome.ucsc.edu/goldenPath/help/iupac.html`.

[4] At the time of writing, the algorithm used to compute viral sequence completeness is not made available by sources; reasonably, it should include checks on the length of a sequence, the completeness of each protein it codes for, and the number of N's and gaps it contains. Until more precise information is released, we only import the information provided by sources, leaving the field empty when this is not available.

Biological Group. VIRUS captures the relevant information of the analyzed pathogen, summarizing the most important levels of the taxonomy branch it belongs to (viruses are described by complex taxonomies [229]. The most precise definition is given by the pair ⟨ *TaxonName*,*TaxonID* ⟩ (e.g., ⟨ Severe acute respiratory syndrome coronavirus 2, 2697049⟩, according to the NCBI Taxonomy [144]), related to a simpler *GenBankAcronym* (e.g., SARS-CoV-2) and to many comparable forms, contained in the *EquivalentList* (e.g., 2019-nCoV, COVID-19, SARS-CoV2, SARS2, Wuhan coronavirus, Wuhan seafood market pneumonia virus, ...), as these are the names given to the virus at different time points and considered synonyms within the NCBI Taxonomy. This term in the taxonomy is contained in a broader category called *Species* (e.g., Severe acute respiratory syndrome-related coronavirus), which belongs to a *Genus* (e.g., Betacoronavirus), part of a *SubFamily* (e.g., Orthocoronavirinae), finally falling under the most general category of *Family* (e.g., Coronaviridae). Each virus species corresponds to a specific *MoleculeType* (e.g., genomic RNA, viral cRNA, unassigned DNA), which has either a double- or single-stranded structure; in the second case, the strand may be either positive or negative. These possibilities are encoded within the *IsSingleStranded* and *IsPositiveStranded* Boolean variables.

An assay is performed on a tissue extracted from an organism that has hosted the virus; this information is collected in the HOSTSAMPLE entity. The host is defined by the pair ⟨ *HostTaxonName*, *HostTaxonID* ⟩ (e.g., ⟨ Homo Sapiens, 9606 ⟩, according to the NCBI Taxonomy). The sample is extracted on a *CollectionDate*, from an *IsolationSource* that is a specific host tissue (e.g., nasopharyngeal or oropharyngeal swab, lung), in a certain location identified by the quadruple *OriginatingLab* (when available), *Region*, *Country*, and *GeoGroup* (i.e., continent) – for such attributes, ISO standards may be used. In some cases information related to the *Age* and *Gender* of the individual donating the HOSTSAMPLE may also be available. Both dimensions of this group are in 1:N cardinality with the SEQUENCE.

Organizational Group. The entity SEQUENCINGPROJECT describes the management aspects of the production of the sequence. Each sequence is connected to a number of studies, usually represented by a research publication (with *Authors*, *Title*, *Journal*, *PublicationDate* and eventually a *PubMedID* referring to the most important biomedical literature portal[5]). When a study is not available, just the *SequencingLab* (or submitting laboratory) and *SubmissionDate* (different from the date of collection, captured in the HOSTSAMPLE entity) are provided.[6] On rare occasions, a project is associated with a *PopSet* number, which identifies a collection of related sequences derived from population studies (submitted to GenBank), or with a *BioProjectID* (an identifier to the BioProject external database[7]). We also include the name of *DatabaseSource*, denoting the organization that primarily stores the sequence, from which integration pipelines can retrieve the information. The cardinality of the rela-

---

[5] `https://www.ncbi.nlm.nih.gov/pubmed/`

[6] Note that this condition could have been represented with an *is-a* hierarchy where a new entity STUDY specializes the SEQUENCINGPROJECT entity.

[7] `https://www.ncbi.nlm.nih.gov/bioproject/`

tionship between the two involved entities is many-to-many as sequences can be part of multiple projects; conversely, sequencing projects contain various sequences.

Analytical Group. This group of entities allows to store information that is useful during the secondary analysis of genomic sequences. The NucleotideVariant entity contains sub-parts of the main Sequence that differ from the reference sequence of the same virus species. They can be identified just by using the *AlternativeSequence* (i.e., the nucleotides used in the analyzed sequence at position *Start* for an arbitrary *Length*, typically just equal to 1) and a specific *Type*, which can correspond to insertion (INS), deletion (DEL), substitution (SUB) or others. The content of the attributes of this entity is not retrieved from existing databases; instead, it is computed in-house by our procedures. Indeed, we use the well-known dynamic programming algorithm of Needleman-Wunsch [298], which computes the optimal alignment between two sequences. From a technical point of view, we compute the pair-wise alignment of every sequence to the reference sequence of RefSeq (e.g., NC_045512 for SARS-CoV-2); from such alignment, we then extract all insertions, deletions, and substitutions that transform (i.e., edit) the reference sequence into the considered sequence.

Each mutation at the nucleotide sequence level is connected to its own Impact, an entity which contains annotations of the variant computed using SnpEff tool [104]; it calculates the *Effect* that the variant produces on a certain *ImpactedGene* (a variant may, for example, be irrelevant, silent, produce small changes in the transcript, or be deleterious for the transcript), with a measure of the generated *PutativeImpact* (high, moderate, low...).

The Annotation entity contains information on the structure of the full sequence; it defines a number of sub-sequences, each representing a segment (defined by *Start* and *Stop* coordinates) of the original sequence, with a particular *FeatureType* (e.g., gene, protein, coding DNA region, or untranslated region, molecule patterns such as stem-loops and so on), the recognized *GeneName* to which it belongs (e.g., gene "E", gene "S" or open reading frame genes such as "ORF1ab"), the *Product* it concurs to produce (e.g., leader protein, nsp2 protein, RNA-dependent RNA polymerase, membrane glycoprotein, envelope protein...), and eventually related *ExternalReference* when the protein is present in a separate database such as UniProtKB. Additionally, for each Annotation we also store the corresponding *AminoacidSequence* (encoded according to the notation of the International Union of Pure and Applied Chemistry[8]). Example codes are A (Alanine), D (Aspartic Acid), and F (Phenylalanine).

The AminoacidVariant entity contains sub-parts of the *AminoacidSequence* stored in the specific Annotation, which differ from the reference amino acids of the same virus species. These variants are calculated similarly to the NucleotideVariants (note that a comparable approach is used within CoV-GLUE [371]). Also here we include the *AlternativeSequence*, the *Start* position, the *Length*, and a specific *Type* (SUB, INS, DEL...).

[8] https://en.wikipedia.org/wiki/Nucleic_acid_notation#IUPAC_notation

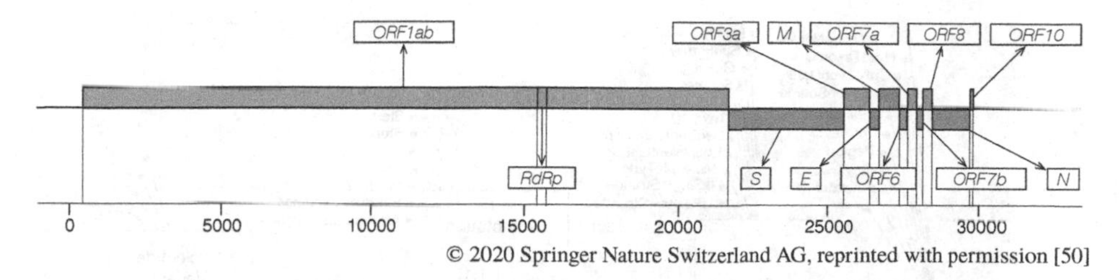

Fig. 9.2 Location of major structural protein-encoding genes (as red boxes: S = Spike glycoprotein, E = Envelope protein, M = Membrane glycoprotein, N = Nucleocapsid phosphoprotein), accessory protein ORFs = Open Reading Frames (as blue boxes), and RNA-dependent RNA polymerase (RdRp) on the sequence of the SARS-CoV-2.

9.2 Answering Complex Biological Queries

In addition to very general questions that can be easily asked through our conceptual model (e.g., retrieve all viruses with given characteristics), in the following, we propose a list of interesting application studies that could be backed by the use of our conceptual model. In particular, they refer to the SARS-CoV-2 virus, as it is receiving most of the attention of the scientific community. Figure 9.2 represents the reference sequence of SARS-CoV-2,[9] highlighting the major structural subsequences that are relevant for the encoding of proteins and other functions. It has 56 regions ANNOTATIONS, of which Figure 9.2 represents only the 11 genes (ORF1ab, S, ORF3a, E, M, ORF6, ORF7a, ORF7b, ORF8, N, ORF10), plus the RNA-dependent RNA polymerase enzyme, with an approximate indication of the corresponding coordinates. We next describe biological queries supported by VCM, from the easy to complex ones, typically suggested by existing studies.

Q1. The most common variants found in SARS-CoV-2 sequences can be selected for US patients; the query can be performed on entire sequences or only on specific genes.

Q2. COVID-19 European patients affected by a SARS-CoV-2 virus can be selected when they have a specific one-base variant on the first gene (ORF1ab), indicated by using the triple ⟨start, reference_allele, alternative_allele⟩. Patients can be distributed according to their country of origin. This conceptual query is illustrated in Figure 9.3, where selected attribute values are specified in red, in place of attribute names in the ER model; values in NUCLEOTIDEVARIANT show one possible example. *Country* is in blue as samples will be distributed according to such field.

Q3. According to [108], E and RdRp genes are highly mutated and thus crucial in diagnosing COVID-19 disease; first-line screening tools of 2019-nCoV should perform an E gene assay, followed by confirmatory testing with the RdRp gene assay. Conceptual queries are concerned with retrieving all sequences with mutations

[9] It represents the positive-sense, single-stranded RNA virus (from 0 to the 29903th base) of NC_045512 RefSeq staff-curated complete sequence (*StrainName* "Wuhan-Hu-1"), collected in China from a "Homo Sapiens" HOSTSAMPLE in December 2019.

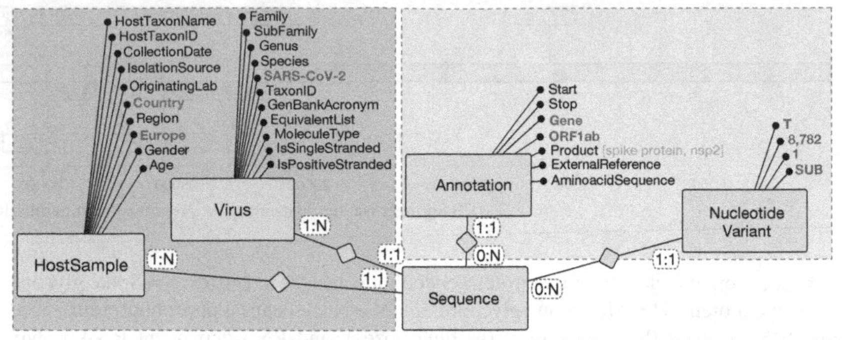

Fig. 9.3 Visual representation of query Q2.

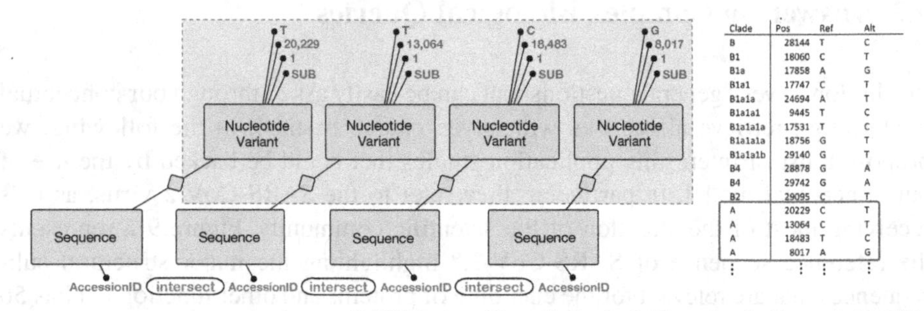

Fig. 9.4 Illustration of the selection predicate for the *A clade* [178], used in query Q4.

within genes E or RdRp and relating them to given hosts, e.g. humans affected in China.

Q4. To inform SARS-CoV-2 vaccine design efforts, it may be needed to track antigenic diversity. Typically, pathogen genetic diversity is categorized into distinct *clades* (i.e., a monophyletic group on a phylogenetic tree). These clades may refer to "subtypes", "genotypes", or "groups", depending on the taxonomic level under investigation. In [178], Gudbjartsson *et al.* use specific sequence variants to define clades/haplogroups (e.g., the *A group* is characterized by the 20,229 and 13,064 nucleotides, originally C mutated to T, by the 18,483 nucleotide T mutated to C, and by the 8,017, from A to G). VCM supports all the information required to replicate the definition of SARS-CoV-2 clades requested in the study. Figure 9.4 illustrates the conjunctive selection of sequences with all four variants corresponding to the *A clade group* defined in [178].

Q5. Morais Junior at al. [289] propose a subdivision of the global SARS-CoV-2 population into sixteen subtypes, defined using "widely shared polymorphisms" identified in nonstructural (nsp3, nsp4, nsp6, nsp12, nsp13, and nsp14) cistrons, structural (spike and nucleocapsid), and accessory (ORF8) genes. VCM supports all the information required to replicate the definition of all such subtypes.

9.3 Related Works

In the variety of types of genomic databases [122], aside from the resources dedicated to humans [48], several ones are devoted to viruses [367, 162], including databases, web servers, and tools.

During the years, several epidemics have triggered the necessity for particular virus-centered resources; see the ones for Influenza (NCBI-IVR [26], IVDB [94], EpiFlu^{TM} [370]), HIV (Stanford HIV drug resistance DB [365], LANL HIV database [10]), Dengue (NCBI-VVR [192]). The conception of these systems has been mostly driven by the urgent need – arisen time by time – for a computational resource (and relative storage) that was able to handle specific data types as rapidly as possible. Works in this field indeed require a considerable effort of data modeling and integration in the background. However this is hardly ever described in publications; instead, the focus is on describing database content and interface functionalities.

From a computer science point of view, conceptual modeling (i.e., the entity-relationship approach [32]) can be used as a driver of the integration process – concerned with providing a global, unified schema for all the data that describe some kind of data within various sources [33]. In the context of viral sequences, as the amount of data is easily manageable, it is common to import data at the integrator site; then, data curation/reprocessing can be performed in a homogeneous way.

Even if the use of conceptual modeling to describe genomics databases dates back to more than 20 years ago [312], we found very few works regarding virus databases related to conceptual data modeling. Among these, [377] considers host information and normalized geographical location and [259] focuses on influenza A viruses. The closest work to ours, described in [371], is a flexible software system for querying virus sequences; it includes a basic conceptual schema for SARS-CoV-2.[11]

In comparison, VCM uses the full power of conceptual modeling to structure metadata and organize data integration and curation; it covers many dimensions and attributes, which are very useful for supporting research queries on virus sequences; it provides an extensible database and associated query system that works seamlessly with any kind of virus, as we will detail in the next two chapters.

---

[10] `http://www.hiv.lanl.gov/`

[11] `http://glue-tools.cvr.gla.ac.uk/images/projectModel.png`

Chapter 10
Integrating Viral Sequence Data

"I have been impressed with the urgency of doing. Knowing is not enough; we must apply. Being willing is not enough; we must do."
— Leonardo da Vinci

The pandemic outbreak of the coronavirus disease COVID-19, caused by the virus species SARS-CoV-2, has created unprecedented attention to the genetic mechanisms of viruses. The sudden outbreak has also shown that the research community is generally unprepared to face pandemic crises in a number of aspects, including well-organized databases and search systems. We respond to such urgent need by means of a novel integrated database collecting and curating virus sequences with their properties. Data is captured, standardized, and organized, so as to facilitate current and future research studies.

In our work, we are driven by the Viral Conceptual Model for virus sequences, described in Chapter 9, which was recently developed by interviewing a variety of experts on the various aspects of virus research (including clinicians, epidemiologists, drug and vaccine developers). The conceptual model is general and applies to any virus. The sequence of the virus is the central information; sequences are analyzed from a *biological* perspective describing the virus species and the host environment, a *technological* perspective describing the sequencing technology, an *organizational* perspective describing the project which was responsible for producing the sequence, and an *analytical* perspective describing properties of the sequence, such as known annotations and variants. Annotations include known genes, coding and untranslated regions, and so on. Variants are extracted by performing data analysis and include both nucleotide variants – with respect to the reference sequence for the specific species – with their impact, and amino acid variants related to the genes.

We built an integration pipeline and corresponding repository – called ViruBase in the following – to feed the content of a novel database, including viral sequences of several viruses (giving priority to viruses causing recent epidemics) enriched by annotations and their mutations with respect to the species-specific reference sequence (established by the research community or – more commonly – by the NCBI RefSeq [321]).

A. Bernasconi: *Model, Integrate, Search... Repeat*, LNBIP 496, pp. 193–200, 2023.
https://doi.org/10.1007/978-3-031-44907-9_10

Chapter organization. In the following, we provide complete descriptions of the content of ViruBase (Section 10.1), the logical schema of the corresponding relational database (Section 10.2), the process used to import data at our site (Section 10.3), the in-house computed content regarding annotations, variants, and their impact (Section 10.4), and finally, the data curation efforts performed during our pipeline (Section 10.5).

10.1 Database Content

Currently, ViruBase includes reference sequences from RefSeq [301] and regular sequences from GenBank [356] of SARS-CoV-2 and SARS-related coronavirus, as well as MERS-CoV, Ebola and Dengue viruses; the pipeline is generic and other virus species will be progressively added next, giving precedence to those species which are most harmful to humans. For what concerns SARS-CoV-2, we also include sequences from COG-UK [335] and NMDC.[1] GenBank and COG-UK data are made publicly available and can be freely downloaded and re-distributed. Special arrangements have been agreed with GISAID [370, 137], resulting in a GISAID-enabled version of ViruBase. Due to constraints imposed by GISAID, the database exposed in this version lacks the original sequences, certain metadata, and nucleotide variants; moreover, GISAID requires their dataset not to be merged with other datasets. Hence, the two versions of ViruBase should be used separately, and a certain amount of integration effort must be carried out by the user. We also reviewed other available sources (GenomeWarehouse and CNGdb) but observed that they do not add substantial value to the integration effort, as most of their sequences overlap with those stored in the four cited sources.

Table 10.1 provides a quantitative description of the current ViruBase content: for each virus, we report the rank, ID, and name from NCBI Taxonomy, the number of sequences included from each source, and the reference sequence. We next provide the average number per sequence of each annotation and nucleotide/amino acid variants computed against the reference sequence. Note that, although GISAID uses a different reference sequence, provided amino acid variants are relative to protein sequences (which are the same as in other sources), hence they can be compared with other variants.

Some content of ViruBase is extracted from the sources and used without changes, some are curated by tailored pipelines, and some are computed in-house (nucleotide and amino acid variants, impact, and quality measures). The current content corresponds to data available at the sources on August 4th, 2020.

We have an overall strategy for content updates, summarised as follows. Data is updated and integrated automatically; updates are performed according to different strategies which depend on the source. For what concerns GenBank, which is the most critical as we consider adding other viral species, updates can be done incrementally,

[1] `http://nmdc.cn/`

						Computed content		
Taxon rank	Taxon ID	Taxon name (or equiv.)	Source	#Seq.	Reference	Avg Annot.	Avg Nuc. Var.	Avg AA Var.
No rank	2697049	SARS CoV-2	GISAID all	76,664	EPI_ISL_402124	-	-	4.8
No rank	2697049	SARS-CoV-2	GISAID only	46,366	EPI_ISL_402124	-	-	4.7
No rank	2697049	SARS-CoV-2	GenBank + RefSeq	13,309	NC_045512.2	28.0	17.6	24.0
No rank	2697049	SARS-CoV-2	COG-UK	38,124	NC_045512.2	28.0	25.5	58.0
No rank	2697049	SARS-CoV-2	NMDC	295	NC_045512.2	28.1	28.8	56.5
Species	694009	SARS virus	GenBank + RefSeq	673	NC_004718.3	14.0	91.6	19.7
Species	1335626	MERS-CoV	GenBank + RefSeq	1,381	NC_019843.3	27.0	104.0	87.4
Species	2010960	Bombali ebolavirus	GenBank + RefSeq	5	NC_039345.1	9.0	126.8	41.6
Species	565995	Bundibugyo ebolavirus	GenBank + RefSeq	22	NC_014373.1	9.0	130.5	40.9
Species	186539	Reston ebolavirus	GenBank + RefSeq	57	NC_004161.1	9.0	126.1	29.4
Species	186540	Sudan ebolavirus	GenBank + RefSeq	40	NC_006432.1	9.0	368.6	51.0
Species	186541	Tai Forest ebolavirus	GenBank + RefSeq	9	NC_014372.1	9.0	4.6	2.8
Species	186538	Zaire ebolavirus	GenBank + RefSeq	2,938	NC_002549.1	9.0	503.7	66.3
Strain	11053	Dengue virus 1	GenBank + RefSeq	11,185	NC_001477.1	15.0	469.7	200.9
Strain	11060	Dengue virus 2	GenBank + RefSeq	8,692	NC_001474.2	15.0	410.0	117.7
Strain	11069	Dengue virus 3	GenBank + RefSeq	5,344	NC_001475.2	15.0	269.3	118.6
Strain	11070	Dengue virus 4	GenBank + RefSeq	2,492	NC_002640.1	15.0	265.1	147.2

Table 10.1 Summary of ViruBase content as of August 4th, 2020. For each taxon name (identified by a taxon ID and rank) and each source, we specify the number of distinct sequences and the reference genome; we also provide the average number of annotations, nucleotide variants, and amino acid variants per sequence. The GISAID-only entry refers to those GISAID sequences that are not also present in the other three sources.

as new *AccessionIds* are generated at each sequence change; thus, it is possible to identify the sequences that should be added or deleted at every update. For what concerns COG-UK, instead, *AccessionIds* are reused; we monitored the sequences at different times and found that a large fraction was changed while keeping the same identifier. Thus, it is preferable to reload the entire database and reprocess it. NMDC sequences are very few and can also be reloaded. Finally, GISAID provides an interchange format that is frequently updated at the source, and also in this case reload is preferred, but this is not computationally demanding as GISAID provides amino acid variants.

The automatic loading procedure currently requires about 2 days, it includes Python scripts to complete the database with indexes, views, and constraints. In the end, we manually check the execution log for errors, and we run testing queries. Finally, we switch the production database with the development one. We will synchronize the automatic management so that update occurs monthly at a fixed date, e.g. the first day of the month.

10.2 Relational Schema

The relational schema of ViruBase is a direct translation of the Viral Conceptual Model shown in Chapter 9. Represented in Figure 10.1, it is inspired to classic data marts [67], with a central fact table describing the SEQUENCE, featuring several char-

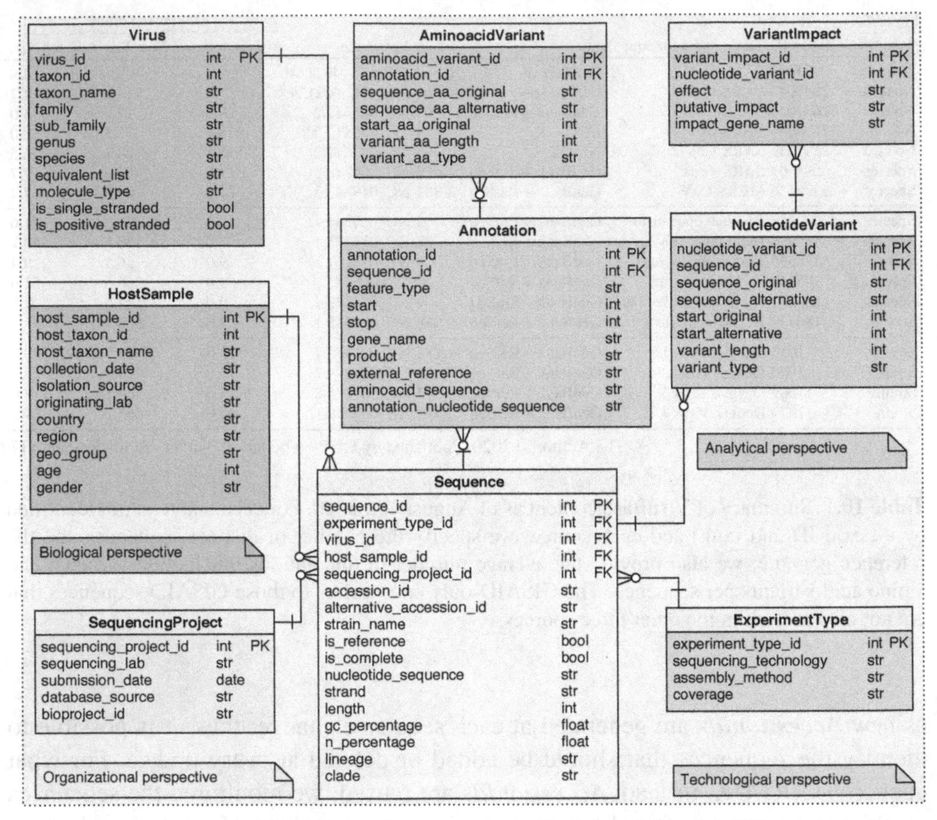

Fig. 10.1 Logical schema of the ViruBase relational database.

acterizing attributes, and then four groups of dimensions, regarding – respectively – biological, technological, organizational and analytical aspects.

All tables have a numerical sequential primary key (PK), conventionally named using the table name and the post-fix "_id", and indicated as PK in Figure 10.1; we indicate with foreign keys (FK) the relationships from a non-key attribute to a primary key attribute of a different table. Note that, being the central table of the schema, SEQUENCE contains four foreign keys fields to connect to `Virus`, `HostSample`, `SequencingProject` and `ExperimentType` tables. Relationships from the SEQUENCE towards VIRUS, HOSTSAMPLE, SEQUENCINGPROJECT and EXPERIMENTTYPE are functional (e.g. one SEQUENCE has one EXPERIMENTTYPE, while an EXPERIMENTTYPE may be the same for multiple SEQUENCE); instead, relationships in the analytical perspective are 1:N (e.g. one SEQUENCE has many ANNOTATIONS, and an ANNOTATION has many AMINOACIDVARIANTS).

The attributes of tables directly interpret the conceptual model attributes; we highlight some slight changes/additions that are convenient for practical use of the database. In SEQUENCE we also store an `alternative_accession_id`, in case the sequence is also present in a second public database. The SEQUENCINGPROJECT table has been reduced to only three attributes, given the limitations of data input from external sources and the cardinality of the relation with the SEQUENCE table has been reduced to one-to-many as this was enough to capture real data. In the ANNOTATION table we store the corresponding pre-computed `nucleotide_sequence` and `aminoacid_sequence`. In NUCLEOTIDEVARIANT characterizing fields include `sequence_original` and `sequence_alternative`, referring respectively to the positions `start_original` and `start_alternative`, `variant_length` and `variant_type` and, consequently, AMINOACIDVARIANT contains `sequence_aa_original`, `sequence_aa_alternative`, `start_aa_original`, `variant_aa_length`, `variant_aa_type`.

We built one additional materialized view to improve performance of relevant queries on the database: NUCLEOTIDEVARIANTANNOTATION joins the two tables ANNOTATION and NUCLEOTIDEVARIANT by directly reporting, for each variant identifier, the type of feature, gene in which it is contained and related protein A number of indexes have been created on critical attributes of the tables that are most used in heavy computations. Specifically, all FK fields have corresponding indexes; moreover, in the tables of nucleotide and amino acid variants, we index original and alternative sequences, their length and types; of annotations, we index start-stop coordinates.

10.3 Data Import

The pipeline used to import the content of the ViruBase database from sources is shown in Figure 10.2. We use different download protocols for each source:

- For NCBI data (including GenBank and RefSeq sequences), we employ the extraction tools available in the E-utilities [354]: the Python APIs allows to retrieve one complex XML file for each sequence ID available in NCBI.
- COG-UK instead provides a single multi-FASTA file on its website;[2] this is associated with a text file for metadata.
- NMDC exposes an FTP server with FASTA files for each sequence, while metadata are captured directly from the HTML description pages.
- GISAID provides us with an export file in JSON format, updated every 15 minutes. The file is produced by GISAID technical team in an *ad-hoc* agreed form for ViruBase.

Automatic pipelines have been implemented to extract metadata and fill the SEQUENCE, VIRUS, HOSTSAMPLE, SEQUENCINGPROJECT, and EXPERIMENTTYPE tables; some attributes require data curation, as next described.

[2] `https://www.cogconsortium.uk/data/`

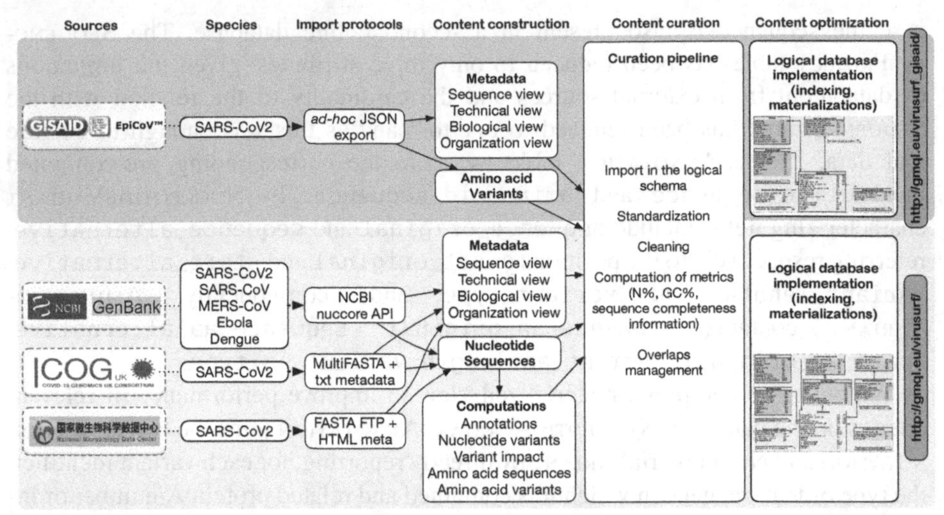

Fig. 10.2 General pipeline of the ViruBase platform. For given sources and species, we use download procedures to construct content, perform data curation, and load the content into two distinct databases, for GISAID and for the other sources, which are schema-compatible (the former is a subset of the latter).

10.4 Annotation and Variant Calling

In order to provide homogeneous information for sequence annotations and variants, we use a unique annotation procedure for GenBank, COG-UK and NMDC; resulting variants for amino acid sequences are consistent with those provided by GISAID. We extract structural annotations, nucleotide and amino acid sequences for each annotated segment, nucleotide variants and their impact, amino acid variants for the proteins, and other information such as the percentage of specific nucleotide bases.

For each virus, we manually select a reference sequence and a set of annotations, comprising coordinates for codifying and structural regions, as well as the amino acid sequences of each protein. Usually, such data are taken from the RefSeq entry for the given virus (e.g., NC_045512 for SARS-CoV-2). For each imported sequence, the pipeline starts by computing the optimal global alignment to the reference by means of the dynamic programming Needleman-Wunsch (NW) algorithm [298]. The time and space complexity of NW is quadratic in the length of the aligned sequences, which often hinders its adoption in genomics, but viral sequences are relatively short, thus we can use NW rather than faster heuristic methods. We configured the algorithm to use an affine gap penalty, so as to favor longer gaps which are very frequent at the ends of sequences.

Once the alignment is computed, all the differences from the reference sequence are collected in the form of variants (substitutions, insertions, or deletions). Using the SnpEff tool [104] we annotate each variant and predict its impact on the codifying regions; indeed, a variant may, for example, be irrelevant (e.g, when the mutated codon codifies for the same amino acid of the original codon), produce small changes, or be deleterious. Based on the alignment result, the sub-sequences corresponding to the reference annotations are identified within the input sequence.

Coding regions are then translated into their equivalent amino acid sequences; the translation takes into consideration annotated ribosomal frameshift events (e.g., within the ORF1ab gene of SARS-CoV-2). When translation fails (e.g., because the nucleotide sequence retrieved from the alignment is empty or its length is not a multiple of 3), we ignore the amino acid product; failures are due to incompleteness and poor quality of the input sequence, further computation of amino acid variants would produce erroneous information. Instead, when an aligned codon contains any IUPAC character ambiguously representing a set of bases,[3] it is translated into the X (unknown) amino acid, which automatically becomes a variant. Note that queries selecting known amino acids are not impacted; unknown amino acids are usually not of interest. Translated amino acid sequences are then aligned with the corresponding amino acid sequences (using NW), annotated with the reference, and amino acid variants are inferred.

Alignment, variant calling, and variant impact algorithms are computationally expensive, so we decided to parallelize this part of the pipeline, taking advantage of Amazon Elastic Compute Cloud (Amazon EC2). We implemented a chunked and parametrized execution modality for distributing the analysis of the sequences associated with each virus to multiple machines so that the total execution time of the process can be divided by the number of available machines.

10.5 Data Curation

Our first curation contribution is to provide a unique schema (the VCM) for different data sources. Each source comes with different terminological choices when describing the different metadata. We have surveyed the terms used in many sources to come up with appropriate reconciling solutions. The mappings between VCM attribute names and those used at the original sources are in the Tables C.1-C.4 reported in Appendix C for space reasons.

Specific value curation efforts have been dedicated to: location information (to properly divide information among region, country, and continent—also using external maps); collection and submission dates (which very often miss day/month/year parts or use different formats); completion of virus and host taxonomy names/identifiers (using NCBI Taxonomy services and a specific module to resolve misspellings); choice of the appropriate reference sequence (cross-checking with several research papers to ascertain that the typical reference sequence used for variant calling is defined); coverage of the sequencing assay (indicated in very heterogeneous ways, possibly with different semantics and measure units). We also compute metrics re-

[3] `https://genome.ucsc.edu/goldenPath/help/iupac.html`

garding the percentage of G and C bases, of unknown bases and the information about sequence completeness. Between the two versions of the database, the one that integrates different sources and the one dedicated to GISAID, we also homogenized the nomenclature used for proteins.

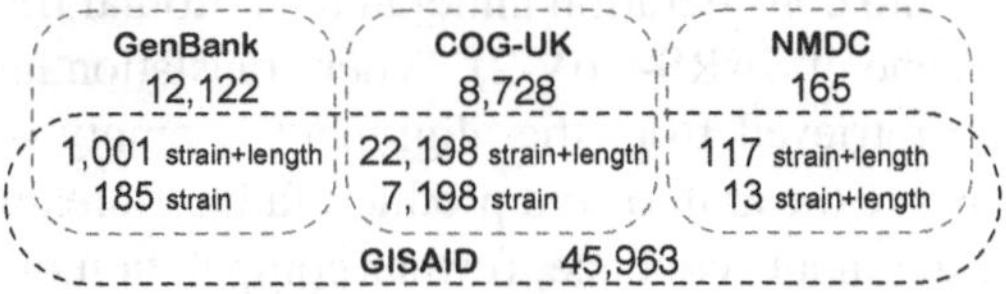

Fig. 10.3 Counts of SARS-CoV-2 overlapping sequences from each source. Overlaps are computed by means of either the strain name, or both strain name and length.

Some sequences are deposited to multiple sources; we detect such redundancy by matching sequences based on their strain name or the pair of strain names and length. Overlaps among sources are illustrated in Figure 10.3. As all overlaps occur between GISAID and the other sources, we store the information about overlaps within the GISAID database to allow the possibility of performing "GISAID only" queries, i.e., restricted to GISAID sequences that are not present in GenBank, COG-UK or NMDC. As a consequence of this duplicate detection procedure, the records of COG-UK, which typically only contain few metadata, can be enriched with 5 additional metadata (*OriginatingLab*, *SubmissionDate*, *SubmissionLab*, *IsolationSource*, *IsComplete*) when they correspond to GISAID records that hold such information.

Chapter 11
Searching Viral Sequence Data

"What we see depends on mainly what we are looking for."
— Sir John Lubbock

The content of ViruBase, a novel integrated and curated database for virus sequences, has been made searchable through a flexible and powerful query interface called ViruSurf, available at `http://gmql.eu/virusurf/`. ViruSurf has a companion system, ViruSurf-GISAID, available at `http://gmql.eu/virusurf_gisaid/` that offers a subset of the functionalities and operates on a GISAID-specific database.[1] ViruSurf can be employed for searching sequences using their describing metadata, but it also exposes a sophisticated search mechanism upon nucleotide and amino acid variants, as well as their genetic impact, where each condition can be combined with others in powerful ways.

While continuous additions are being made to ViruBase (as we will explain in Chapter 12) – by enlarging its scope in terms of included species but also of kind of represented data (requiring changes in the schema) – the ViruSurf interface will be consistently updated, offering more practical functionalities to users.

Chapter Organization. Section 11.1 briefly describes the requirements elicitation process that was conducted before and during the design of ViruSurf. Section 11.2 explains the various modules of the ViruSurf one-page web application; Section 11.3 shows example queries that can be performed on the interface—these are relevant for virology research recent studies; Section 11.4 discusses the usefulness of ViruSurf in the critical times of the COVID-19 pandemic; finally, Section 11.5 reviews concurrent related works.

[1] The two versions were kept separate as a requirement from the GISAID data-sharing agreement, which prevented us to merge together the datasets.

A. Bernasconi: *Model, Integrate, Search... Repeat*, LNBIP 496, pp. 201–212, 2023.
https://doi.org/10.1007/978-3-031-44907-9_11

ID	#Interv.	Title/Expertise Level	Research Area/Specialization	Location	Institution Type
I1	3	Associate Professors/Clinicians + Researcher	Emergency medicine/Hepatology	IT	Hospital
I2	1	Full Professor	Virology	US	University
I3	1	Researcher	Molecular Biology	IT	University
I4	1	R&D Manager	Bioinformatics	IT	Private Company
I5	1	Full Professor	Molecular Microiology	US	University
I6	1	Senior Researcher	Epidemiology	UK	Research Center
I7	2	Researchers	Veterinary Virology/Veterinary Medicine	IT	Research Center
I8	1	Researcher	Bioinformatics	US	Private Company
I9	1	Researcher	Veterinary Virology	IT	Research Center
I10	2	Full Professor + Associate Professor	Medical Genetics	IT	University/Hospital
I11	1	Researcher	Virology	US	University
I12	1	Research Associate	Medical Genetics	BE	University/Hospital
I13	2	Full Professor + Researcher	Molecular Biology	IT	University
I14	1	Full Professor	Molecular Virology	HK	University
I15	1	Full Professor	Computational Biology	SI	University

Table 11.1 Groups of interviewees with their characterization. #Interv. indicates the size of the group; for each group we specify the titles or expertise levels demonstrated by its components, their research areas or specializations, and the location and type of their institution at the time of the interview.

11.1 Requirements Analysis

The ViruSurf development team included six members – holding mixed expertise in Computer Science and Bioinformatics – led by a Principal Investigator, two senior researchers, and three Ph.D. students. Given my background in interviewing experts in human genomics for understanding the requirements of exploratory software (during collaborations with IEO, `www.ieo.it`, and IFOM, `www.ifom.eu`), and in evaluating their usability with empirical studies (see Section 6.4.7), I was responsible for the requirement elicitation design and the direction of the methodological framework.

In total, twenty researchers were interviewed, knowledgeable in various fields, related to the virus from different points of view: the specific mechanisms of its biology, the pathogenesis, and its interaction with human or animal hosts. Interviewees had different levels of expertise and were sourced from different locations from both public and private institutions (as shown in Table 11.1). All participants were interviewed individually except for I7, I10, I13 (in pairs), and I1 (in three).

We conducted lightly structured interviews; the typical setting included an introductory 1.5 hours video call session where the participants were interviewed jointly by the PI, me as the requirements analysis designer, and other two members of our team, supported by slide presentations. All such sessions were approximately divided into three parts of 30 minutes each. First, we explained our interest in the topic, showing the progress of our systems thus far. Then, we made questions to gather the participants' attention toward i) particular areas/questions of interest, or ii) specific functionalities, using a more technical attitude.

In some cases, participants showed particular interest in follow-ups of our research and offered their availability for successive sessions, organized with a more "hands-on" angle. In particular, with I2, I3, I7, I8, and I11 we were able to perform other two sessions of 1 hour each, joined by other members of the development team:

(1) a "mock-up session", using prototypes of the system or slides with simulated user workflows, allowing a "satisfied/non-satisfied" answer from participants, with an open discussion to formulate variations with respect to the proposed design; (2) a "demonstration session" for the implemented features, with the presentation of a small application use case, and finally request for feedback on the compliance with previously formulated requirements and on the usefulness of the functionality in real case scenarios.

11.1.1 Lessons Learnt

Our exercise did not follow any predefined requirement elicitation framework but was of great importance to learn a series of lessons: in spite of our expertise in developing software for genomics, we were lacking a substantial amount of domain knowledge. The work proceeded continuously, in an agile setting that alternated data and software design sessions with interviews with experts in the domain, to understand their requirements in what we call an *Extreme Requirements Elicitation* (ERE) process. We here summarize the main understandings gathered from this experience, characterized by the urgency of the pandemics and the domain knowledge gap with the virology experts, addressed with a "show, don't tell" approach to transfer our need for understanding.

Deal with diversity. Domain-knowledge understanding has been addressed in general [184] and in contexts related to bionformatics [66, 338] or genomics [112, 113]. However, the domain experts involved in COVID-19 research represent a novel scenario: a plethora of different specializations exist (biologists, clinicians, geneticists, virologists – focusing on phylogenesis, epidemiology, veterinary, etc.). It is important to capitalize on the diversity of interviewed people and also on the fact that a similar diversity is present in the many stakeholders who could take advantage of our system. Note that some features may, for example, be useful to both virologists and clinicians: however, every single functionality is described by experts in different ways, and each type of expert must be approached differently.

Diversity characterizes not only disciplines of specialization but also the level of expertise and personal talents (e.g., how to address young researchers and senior professors). For example, during interviews, we observed that young researchers appreciated prototype/mockup-driven presentations (allowing them to gain insight into the details of the user interaction and implementation), while professors appreciated high-level presentations, as they were able to highlight completely new directions from more abstract descriptions.

Investing in short, just-in-time pre-interview meetings. While requirements engineering usually features a long and controlled process of steps – including the interpretation, analysis, modeling, and validation of requirements' completeness and correctness [299] – we argue that, in the specific setting described in this work, exceptions should be allowed. Domain experts have limited time available; they allow it for interviews if they see potential in the proposal, so interaction should be as productive as possible.

For successfully conducting the interviews, we understood that organizing quick preparatory meetings preceding the call was essential to: i) refresh the foundational aspects of the specific research expertise of the person to be interviewed;[2] ii) discuss it within the team; iii) quickly agree on a clear scheme for the interview; and iv) identify/plan roles for each member of the team.

Being use-case driven. When speaking different domain languages, one quite successful approach is that of "show, don't tell" [403] (typical of design thinking mindsets [70]), e.g., to bridge the knowledge gap between computer scientists and virologists. Interactions need to be driven by specific problems and use cases: proposing examples, small prototypes, showcasing alternatives, and mimicking a live interface – by using slides with animations on specific modules and objects – led to promising results of transferring our need of knowing or understanding.

Being curious. Using all possible means to facilitate discovery. While collecting requirements, we were continuously reasoning about *new resources, new use cases, new experts* that could broaden our views; this has produced new interesting data sources and data analysis methods that were not envisioned at the beginning of the project and that were integrated within our data ingestion and curation pipelines. In parallel, we activated continuous monitoring of literature and news of interesting facts, so as to enrich our interviews with every novel aspect that could create stimulating interaction niches (e.g., the emergence of a given important mutation, the publishing of a study with broad societal impact...). Finding *new experts* was essential: in our case, next contacts were often suggested during previous interviews.

Being aware that a second opportunity of interaction is worthwhile. A challenging aspect was to identify the right moment in which developments had made a big enough step for the next interview: a second round of discussion with the stakeholders was used to show progress, ask for feedback, and proceed with following requirements, embracing a similar approach to the one of agile requirements analysis [357].

11.2 Web Interface

The web interface of ViruSurf is composed of 4 sections, numbered in Figure 11.1: (1) the menu bar, for accessing services, documentation, and query utilities; (2) the search interface over metadata attributes; (3) the search interface over annotations and nucleotide/amino acid variants; (4) the result visualization section, showing resulting

[2] Even if, in some cases, prior domain knowledge may introduce negative bias in interviews [184], in this context it is vital that requirements analysts have some (flexible) experience in the domain. This is helpful in view of time scarcity and also because most of the interviewed experts interacted under the implicit assumption that explanations of basic concepts of their discipline were not needed.

sequences with their metadata. Results produced by queries on the metadata search interface (2) are updated to reflect each additional search condition, and counts of matching sequences are dynamically displayed to help users in assessing if query results match their intents. The interface allows to choose multiple values for each attribute at the same time (these are considered in *disjunction*); it enables the interplay between the searches performed within parts (2) and (3), thereby allowing to build of complex queries given as the logical conjunction – of arbitrary length – of filters set in parts (2) and (3).

Menu Bar. The menu bar includes links to the GISAID-specific ViruSurf system, to the GenoSurf system, and pointers to the data curation detail page, to the wiki, to a video compilation, and to a pedagogical survey supporting the user by documenting the aspects of search queries. Below, users can use a button to clear the current set query, and on the right, they can select various "Predefined queries" from a drop-down menu.

Metadata Search. The Metadata search section is organized into four parts: *Virus* and *Host Organism* (from the *biological* dimension), *Technology* and *Organization* (from the respective dimensions). Its functioning is analogous to the one of the "Metadata search" of GenoSurf described in Chapter 6. It includes attributes that are present in most of the sources, described by an information tab that is opened by clicking on blue circles; values can be selected using drop-down menus. At the side of each value, we report the number of items in the repository with that value. The field Metadata search is dynamically compiled to show in a single point which values have been selected in the table below. We allow to choose multiple values in one attribute drop-down list at the same time (these are considered as *alternative*). Values chosen over different attributes, instead, are considered conditions that should coexist in the resulting items.

The user can compose desired queries by entering values from all the drop-down menus; the result is the set of sequences matching all the filters. Note that the special value N/D (Not Defined) indicates the null value, which can also be used for selecting items.[3] For numerical fields (age, length, GC% and N%) the user must specify a range between a minimum and maximum value; in addition, the user can check the N/D flag, thereby including in the result those sequences having the value set to N/D. Similarly, the collection date and submission date have calendar-like drop-down components, supporting a range of dates and the N/D flag.

Variant Search. The Variant search section allows searching sequences based on their nucleotide variants (with their impact) and the amino acid variants. When the user selects the "ADD CONDITION ON AMINO ACIDS" or "ADD CONDITION ON NUCLEOTIDES" buttons, a dedicated panel is opened, with a series of drop-down menus for building search conditions. A user can add multiple search conditions within the same panel; these are considered in disjunction. Once the panel is com-

[3] This corresponds to a simplifying choice made to have a first system up and running soon. In the future we may consider different kinds of nulls [34], differentiating between what is not known and what is not available.

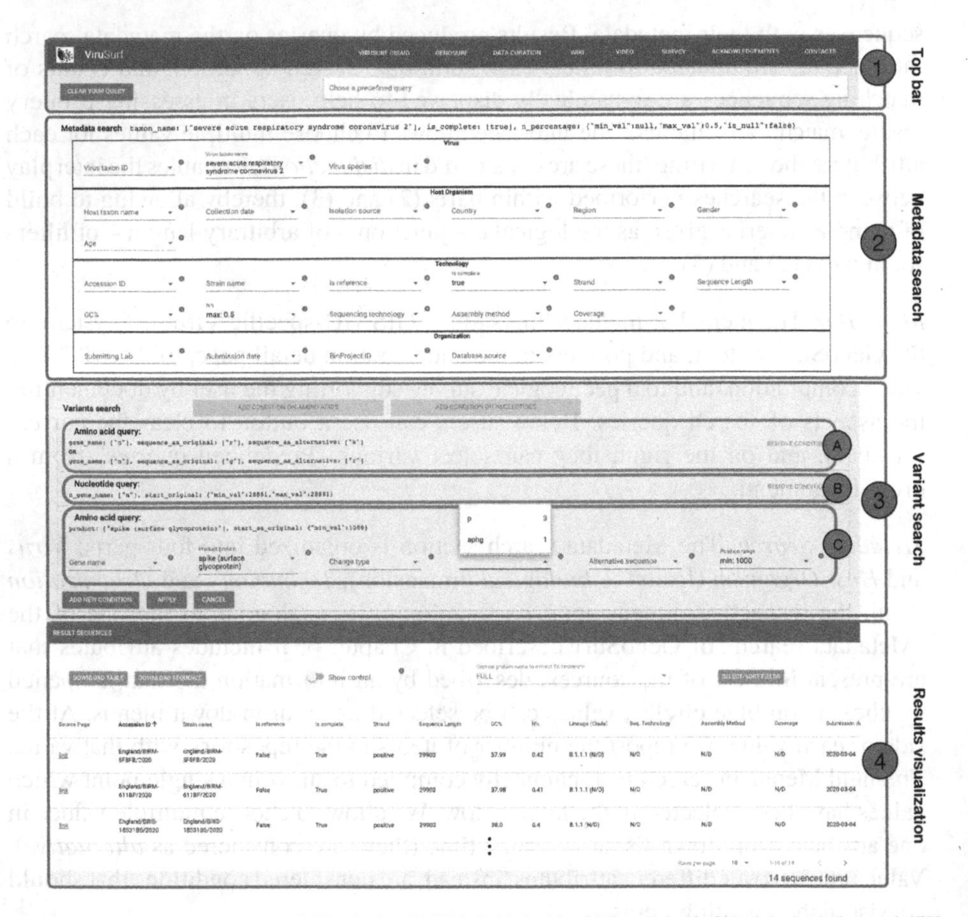

Fig. 11.1 Overview of ViruSurf interface. Part 1 (Top bar) allows you to reset the previously chosen query or select predefined example queries. Queries are composed by using Part 2 (Metadata search) and Part 3 (Variants search). In our example, Part 2 includes three filters on Virus taxon name, Is complete, and N%. Part 3 includes three panels. Panel "A" is a query on amino acid variants, selecting sequences with RK and GR changes in gene N; Panel "B" is a query on nucleotide variants, selecting sequences with a variant at position 28,881. Panels "A" and "B" are closed, they can be removed but not changed. Panel "C" is another query on amino acid variants, currently open; it includes two filters selecting given positions of the Spike protein and visualizes available values for the original amino acid involved in the change. Part 4 shows the Result Visualization. The resulting sequences already reflect the filters of Part 2 and the queries of the closed panels "A" and "B" of Part 3, applied in conjunction. Results can be downloaded, in CSV or FASTA format; they can be selected as either cases (default) or controls (switch), and both the nucleotide and amino acid sequences can be projected on a specific protein; table columns can be omitted and reordered. In the bottom right corner, the number of sequences resulting from the search is visualized (in the Figure we show only three sequences out of 14 sequences found).

pleted, it is registered; registered panels can be then deleted from a query if needed. Variants selected in different panels are intended in conjunction. Heterogeneous variant searches (i.e., on amino acid and nucleotide ones) can only be combined in panels, thus in conjunction.

In the example shown in Figure 11.1 (which represents the construction of the "Predefined query 8", from Pachetti *et al.* [306]) the user is choosing all SARS-CoV-2 sequences that are complete, have a maximum percentage of unknown bases of 0.5%, and have R to K *or* G to K amino acid changes in gene N *and* a nucleotide variant at position 28,881. The filters set up to this point have selected a set of 14 sequences (as indicated at the bottom right of the page) In the represented snapshot, a third variant panel is in the process of being compiled with an amino acid condition that could be added to the two existing ones by pressing "APPLY". This holds a filter on the spike protein and on the position of the variant on the protein (> 1000); the filter on the original amino acid allows to select P (3 sequences available) or APHG (1 sequence). In addition, the user could add other amino acid queries that can be disjoint from the currently open one.

Result Visualization and Download. The result table describes the sequences resulting from the selections of the user. The columns of the table can be ordered/included/excluded from the visualization; the resulting table can be downloaded for further processing. Whenever the user either adds or removes a value in the Metadata search, by clicking on a drop-down menu, the results table is updated; instead, it is updated only when a panel of the Variant search section is complete.

The "Show control" switch allows one to visualize the sequences of the control group, defined by those sequences selected by the Metadata search filters for which: 1) there exist some variants and 2) the variant filters set by the users are not satisfied. This option, suggested to us by virologists, is very sensible for describing the effects of variant analysis.

A user can select from a drop-down menu which sub-part of a nucleotide or amino acid sequence should be visualized; the default returns a "FULL" nucleotide sequence (leaving the amino acid field empty), but with this menu option it is possible to return in the result the specific segment of interest. The whole result table – as it is visualized, inclusive of selected metadata and nucleotide or amino acid sequences – can be downloaded for further analysis as a CSV. Alternatively, the user may download either full or selected sequences by using their accession ID, either as CSV or FASTA files. Note that, most of the time, bioinformaticians specifically require data in these common formats so that they can further develop the analysis by employing off-the-shelf libraries.

GISAID-Specific ViruSurf. ViruSurf presents a version that is specific for data imported from GISAID (available at `http://gmql.eu/virusurf_gisaid/`), as requested by a specific Data Agreement. This interface presents limited functionalities but is nevertheless powerful and allows for combining its results with the ViruSurf main interface. Notable differences are here summarized: 1) After selecting filters, a user must explicitly apply her search by pressing an execution button. 2) Searches may be performed on the full dataset from GISAID or on the specific subset of sequences that are only present on GISAID (button "Apply GISAID specific") – this may result particularly useful when the user wishes to compare or sum up results from the two interfaces (see Q5 in the following for an example). 3) Both drop-down menus and the result table's columns hold the original GISAID attribute name, when available – when this differs from ViruSurf's, the second one is provided in the second position inside parentheses.

11.3 Example Queries

ViruSurf enables simple queries to retrieve sets of sequences that adhere to filters set on metadata such as "How many sequences were collected in China in January 2020?" or "How many sequences were retrieved using either Ion Torrent or Ion Torrent X5Plus sequencing technology and have a maximum of 10% of unknown bases?".

Moreover, by means of more complex search queries over the database, it is possible to help virus research, according to the requirements provided by several domain experts; this is not currently supported by other existing systems, which typically offer very nice visual interfaces reporting results of data analysis but limited search capabilities. We cite some examples inspired by recent research works.[4]

Q1. Artesi *et al.* [17] studied how the failure of the cobas®SARS-CoV-2 (Roche) E-gene assay is associated with a C-to-T transition at position 26340 of the SARS-CoV-2 genome. ViruSurf can be used to quickly retrieve 1,104 SARS-CoV-2 sequences with such mutation (original base C, alternative base T at position 26340 of the virus) in different time periods.

Q2. Khailany *et al.* [223] have proposed a characterization of the novel SARS-CoV-2 genome; they show a list of non-coding mutations detected in SARS-CoV-2 genomes. ViruSurf may be employed to select such mutations one by one (or in sets) and correspondingly check what are the annotated effects of such mutations. For example, the mutation from C to T at the 241st nucleotide is an "intergenic region" and an "upstream gene variant".

Q3. Yi *et al.* [407] proposed to create synthetic single amino acid substitutions in the SARS-CoV and SARS-CoV-2 Receptor-Binding Domains,[5] to either enhance or diminish the specific binding activity of the RDB with the functional receptor human ACE2. ViruSurf may be used to programmatically check if there are any publicly available sequences that naturally present at least some of the changes that bring increased infectivity (i.e., P499, Q493, F486, A475, and L455) or the ones that lead to decreased infectivity (i.e., N501, Q498, E484, T470, K452 and R439) in

[4] Note that counts reported as results of the queries corresponding to the first released version of ViruSurf, on August 4th, 2020.

[5] The Receptor-Binding Domain (RBD) is a key part of the virus located in its spike protein – it allows it to dock to body receptors to gain entry into the host's cells and lead to infection. For this reason, it is believed to be a major target to block viral entry. Both SARS-CoV-2 and SARS-CoV share this mechanism.

the Spike protein. These counts may be studied in combination with space and time information, so as to understand in which kind of host population they are found.

Q4. Tang *et al.* [379] claim that there are two clearly definable major types (S and L) of SARS-CoV-2 in the COVID-19 outbreak, which can be differentiated by transmission rates. The S and L types can be distinguished by two tightly linked variants at positions 8782 (within the ORF1ab gene from T to C) and 28144 (within the ORF8 gene from C to T, resulting in a change from Serine to Leucine at the 84 positions of the ORF8 protein). MacLean *et al.* [264] later discredited the claims made in [379] stating the difficulty in demonstrating the existence or nature of a functional effect of a viral mutation. Anyways, ViruSurf can be used to perform a query to isolate the "S" type, based on the amino acid change in position 84 of gene ORF8, which returns 1,947 sequences in our main database and 5,026 from the GISAID-specific one (by choosing the option to only retrieve sequences that are not present in the other database).

Tang *et al.* also suggest that patients that have the genotype Y (ambiguous base representing either C or T[6]) at both positions 8782 and 28144 (differing from the general trend of having respectively C and T in the two distinct positions) could have been infected with viruses from multiple strains of different types. We perform on ViruSurf a query that searches for sequences that have both 1) a variant with original nucleotide C, alternative one Y, at position 8782, and 2) a variant that has original nucleotide T, alternative Y, in position 28144. We thus extract 5 sequences with such characteristics, which may be useful for downstream studies regarding heteroplasmy[7] of SARS-CoV-2 viruses in SARS-CoV-2 patients.

Q5. A study from Scripps Research, Florida, found that the mutation D614G stabilized the SARS-CoV-2 virus's spike proteins, which emerge from the viral surface. As a result, the viruses with D614G seem to infect a cell more likely than viruses without that mutation; the G genotype was not present in February and was found with low frequency in March; instead, it increased rapidly from April onward. The scientific manuscript by Zhang et al. [416], cited by mass media,[8] has not been peer-reviewed yet, but others on the same matter are [230, 36]).

ViruSurf can be used to illustrate this trend. Let us consider two queries on complete sequences. Sequences with the D614G mutation collected before March 30 are 6,592, against 4,664 without the mutation; sequences with the D614G mutation collected after April 1 are 23,649, against 3,331 without the mutation. In both queries, case/control checks are obtained by using the "Show control" switch, which retrieves – for the population specified by metadata filters – sequences that either have or do not have the chosen variants. This allows us to answer questions such as "In the user-defined population of virus sequences extracted from SARS-CoV-2 in Wuhan from 01-Jan-2020 till 31-Mar-2020, how many did have the variants V1 and V2, and how many had neither V1 nor V2?".

---

[6] `https://genome.ucsc.edu/goldenPath/help/iupac.html`

[7] Heteroplasmy is the presence of more than one type of organellar genome (mitochondrial DNA, plastid DNA, or viral RNA) within a cell or individual.

[8] `https://www.nytimes.com/2020/06/12/science/coronavirus-mutation-genetics-spike.html`

The same queries can be repeated on the GISAID-specific version of ViruSurf. Sequences with the D614G mutation collected before March 30 are 15,034, against 8,821 without the mutation; sequences with the D614G mutation collected after April 1 are 18,421, against 3,369 without the mutation. By summing up the query results from the two non-overlapping databases, we obtain that the sequences with the D614G mutation are 61.6% of those collected before March 30 and 86.3% of those collected after April 1.

Q6. In SARS-CoV-2, the G-T transversion at 26144, which caused an amino acid change in ORF3 protein (G251V), is investigated in Chaw *et al.* [95]. The paper claims that this mutation showed up on 1/22/2020 and rapidly increased its frequency. We can use ViruSurf to find out that GenBank currently provides 3 complete sequences with such mutation collected before 1/22/2020, while GISAID provides other 13 sequences, non-overlapping with GenBank ones.

Q7. Pachetti *et al.* [306] report about a mutation located in SARS-CoV-2 gene N at position 28,881, which is related to a double codon mutation, inducing the substitution of two amino acids, namely 28881 (R to K) and (G to R). We reproduce this on ViruSurf by first looking for complete sequences that have two alternative amino acid changes in gene N and then filtering only the sequences that have a nucleotide variation specifically at position 28881. We extract 45 sequences from ViruSurf (5 GenBank, 40 COG-UK).

11.4 Discussion

Note that ViruSurf increases the possibilities available to users with respect to existing resources: queries Q1-Q7 could not be asked directly to the native sources. Indeed, while these provide fairly advanced filters concerning metadata describing sequences, they do not provide much support for variants: GenBank, COG-UK, and NMDC have no in-house computed variants (either on nucleotides or on amino acids), while GISAID (since the last version was released in December 2020) provides filters on around 40 single or combined amino acid variants. More precisely, not all possible variants can be searched for; no disjunction filters can be built; no nucleotide variant can be requested; no question on variants present in position ranges can be posed; and no query on insertion/deletion length can be asked. ViruSurf offers all these possibilities. Moreover, ViruSurf provides integration between four sources (GISAID, GenBank, and COG-UK, plus a tiny dataset from the Chinese NMDC), with duplicate elimination, thus it provides a good estimate of the sizes of results matching a given query. Note that most COG-UK sequences are already included in GISAID, but as such they are "blinded" to the few degrees of freedom offered by that system. We thus reprocess all UK data and offer full ViruSurf functionality over it. To account for data duplication, we cross-check COG-UK sequences with GISAID ones and flag as `gisaid_only = false` the repeated entries. In this way, counts obtained on ViruSurf and on ViruSurf-GISAID can be seamlessly summed up for a

Mutation	ViruSurf				ViruSurf-GISAID					
	COG-UK	GenBank	NMDC	all	gisaid_only	all	**VS+VSG**	GISAID	NCBI	COG-UK
Spike_A222V	53394	395	0	53789	6526	46054	**60315**	46357	-	-
Spike_D614G	138811	39138	125	178074	130156	240811	**308230**	242562	-	-
Spike_H69-	5740	29	0	5769	2022	4906	**7791**	5346	-	-
Spike_H69-,N501Y	3301	0	0	3301	58	1437	**3359**	1489	-	-
Spike_H69-,N501Y, A570D	3231	0	0	3231	57	1436	**3359**	-	-	-
Spike_S477N	2880	10381	0	13261	12774	15394	**26035**	15417		-
NS3_G251V	4296	488	13	4797	3325	6312	**8122**	6328	-	-
N_S197L	311	258	0	569	1924	2212	**2493**	-	-	-
1 nuc. sub. in [27848,28229]	659	94	2	755	-	-	**755**	-	-	-
15/30-long nuc. del.	2547	1696	19	4262	-	-	**4262**	-	-	-

Table 11.2 Comparison of sequences matched by queries for specific mutations, in our system versus original sources (GISAID, NCBI, COG-UK). VS + VSG represents the sum of sequences matched in ViruSurf and in ViruSurf-GISAID, considering only the ones that are specific to GISAID (i.e., "gisaid_only").

complete representation of all currently available SARS-CoV-2 data. In addition, in the ViruSurf a virus-independent infrastructure; thus in the future it may be used to compare characteristics of different virus species in parts with similar functions; as an example, a virologist may be interested in the number of SARS-CoV-2 sequences that have amino acid variants in the specific range 438-488 of spike protein, as opposed to SARS-CoV sequences that have variants in 425-474 of the same protein in this other species (as these ranges define the RDBs of the two species when considering their genes aligned [236]).

For more tangible evidence of ViruSurf value, we refer to Table 11.2, where we show, for a small number of interesting queries, the counts that can be retrieved from different systems, considering submissions up to December 15th, 2020. As can be observed, only queries for notable amino acid mutations (even combined among them) can be performed on GISAID, while on other systems this is not possible. First, we show the number of sequences matched in ViruSurf, splitting them by source of origin (in order: COG-UK, GenBank, NMDC). Then we show the sequences matched within ViruSurf-GISAID In the two columns dedicated to ViruSurf-GISAID, we provide the number of "gisaid-only" sequences and the number of total sequences matched in the database that we retrieve directly from GISAID. VS+VSG column shows the total amount of sequences that can be retrieved in our system, by summing the column ViruSurf/all with the column ViruSurf-GISAID/gisaid_only. Matches found in our system are consistently greater than what can be found on GISAID

ViruSurf provides a single point of access to curated and integrated data resources about several virus species. Example queries show that ViruSurf is able to replicate research results and to monitor how such results are confirmed over time and within different segments of available viral sequences, in a simple and effective way. The relevance of ViruSurf as a tool for assisting the research community will progressively increase with the growth of available sequences and of the knowledge about viruses (we can see this as a "network effect" applied to the data integration problem).

While today's efforts are concentrated on SARS-CoV-2, ViruSurf can similarly be useful for studying other virus species, such as other Coronavirus species, the Ebola Virus, the Dengue Virus, and MERS-CoV, epidemics which are a current threat to mankind; ViruSurf will also enable faster responses to future threats that could arise from new viruses, informed by the knowledge extracted from existing virus sequences available worldwide.

11.5 Related Works

Resources to visualize viral sequence data (and most importantly SARS-CoV-2 related data) are springing up in these critical times in which the COVID-19 pandemic has not been put under control yet. We briefly mention some systems that are at their early stages and are trying, as ViruSurf, to support viral research and provide insights on sequence variants and their characteristics or distribution:

- The COVID-19 Viral Genome Analysis Pipeline [230] provides an environment to explore the most interesting mutations of SARS-CoV-2, by combining time and geographic information;
- GESS [142] aims to uncover single nucleotide variants' behavior in the virus;
- coronApp [282] serves the double purpose of monitoring worldwide mutations and annotating user-provided mutations of SARS-CoV-2 strains;
- WashU [151] is a web-based portal that provides different ways to visualize viral sequencing data;
- The UCSC SARS-CoV-2 browser [145] extends the functionalities of the classical UCSC genome-browser visualization tool [222] for this virus;
- 2019nCoVR [417], already discussed in Chapter 8, not only provides integrated data but also a visualization-based dashboard for mutation inspection.

Chapter 12
Future Directions of the Viral Sequence Repository

"Science appears calm and triumphant when it is completed; but science in the process of being done is only contradiction and torment, hope and disappointment."
— Pierre Paul Émile Roux, French bacteriologist and developer of the first effective diphtheria treatment

In the previous chapters, we have presented our approach to modeling viral sequences (Chapter 9), building a repository collecting data from different viral data sources (Chapter 10), and exposing its content over a web interface, with enhanced functionalities on variant selection and filtering (Chapter 11). This work has been conducted in the last nine months of 2020, during the spreading of the SARS-CoV-2 epidemic worldwide. After setting the first milestones (the conceptual model, the integrated database, and the web entry point), we are now moving forward, considering the next challenges of this new domain with growing interest. In the close future, we plan consistent additions to our project:[1]

- We are carrying on our process of continuous requirements elicitation, constantly improving our understanding of the domain and aiming at a systematization of our interviews with experts (Section 12.1).
- We aim to progressively extend and consolidate ViruSurf so as to make it a strong resource for supporting viral research (Section 12.2).
- We are producing VirusViz, an interface to provide visualization support to the results generated by ViruSurf (Section 12.3).
- We will add data analysis services that can provide sophisticated use of our big data collection, with a focus on variations monitoring, data summarization, and data mining tailored to specific biological questions (Section 12.4).
- We will consider the connection between the phenotype and genotype of the pathogen-host organism and the viral genome, opening new research directions, in search of dependencies between sequence variants, genome signals, and their impact on clinical outcomes of the disease (Section 12.5).

[1] Note of the author, August 2023. Several of these research threads have been explored after the conclusion of this thesis, leading to the following publications–reported as references in this chapter.

A. Bernasconi: *Model, Integrate, Search... Repeat*, LNBIP 496, pp. 213–225, 2023.
https://doi.org/10.1007/978-3-031-44907-9_12

12.1 Research Agenda

We aim to develop a systematic interviewing method driven by three principles: i) overcoming strong initial differences in the understanding of a given domain; ii) converging towards a solution in a short time; iii) conducting extreme requirements elicitation (ERE) sessions in parallel with design and development, neglecting the typical precedence between phases and enforcing cycles of ERE sessions that instrument design and vice versa. After experimenting with these concepts during the first COVID-19 pandemic wave for ViruSurf, in our agenda, we are planning to produce many significant research additions, including the support for: queries for testing the stability of viral regions (useful for vaccine design), emerging knowledge about the mutations' impact, effective visualization for data analysis, and ad-hoc services for covering the needs of hospitals and other parties who are not interested in publishing their sequences. To support all these directions, the ERE design shall pursue several research questions:

RQ1. *How could we train groups of system designers and developers to manage extreme requirements?* Scenarios that are driven by urgency can be simulated or reproduced by means of training exercises, where urgency is artificially created by simulating a disaster and then instructors play the role of domain experts, also setting artificial goals that represent the need for rushing towards the definition of requirements, with strict submission deadlines and fixed time schedules.

RQ2. *Could it be worth applying the ERE method proposed for ViruSurf – conceived to manage developing knowledge in an interdisciplinary setting – within a known disciplinary context?* The method was created to cover two kinds of emergencies, besides urgency, we were also driven by the need to dynamically create enough knowledge within the group in order to manage the forthcoming interviews one after the other. We should put the ERE method at work for collecting requirements within a known domain, shared between designers and interviewed experts, and compare its performance with a structured and conventional plan of interviews.

RQ3. *How will we diversify our portfolio of ERE use cases in the forthcoming design experiences?* After the lockdown period, we managed to hire new people in the group, including a biologist, a communication designer, and a UI expert. We will test if our performance in dealing with extreme requirements will increase or decrease in the context of this new group.

12.2 ViruSurf Extensions

The Web interface of ViruSurf has been planned with virologists so as to facilitate their routine interaction with sequence data, including provisions for fast extraction of result subsequences in various formats and for easing the comparison of search results with their controls. We briefly discuss short-term improvements that we are already designing for ViruSurf.

Schema Additions. We are adding epitopes, i.e., short amino acid sequences that are recognized by the host immune system antigens – a substance capable of stimulating antibody responses in the organism invaded by the virus; epitopes, described by their location in the sequence, evidence, and type of response, are thus candidate binding sites for *vaccine design.*

We do not dedicate our efforts to epitope design, as this topic is covered by specialists; instead, we capitalize on Immune Epitope Database and Analysis Resource (IEDB) [389], an open resource where we expect epitope sequences will continue to be deposited and curated through literature review. While, as of August 2020, there are only 283 epitopes specifically deposited for SARS-CoV-2, IEDB has several thousand sequences deposited for species that are already loaded in ViruSurf.[2] In the case of a new virus spread, information systems to integrate and elaborate available data acquire enormous relevance. Indeed, wet-laboratory experiments to discover novel biological knowledge or validate hypotheses require, among other resources, a considerable amount of time, that may not be available under the pressure of a pandemic event. Thus, computational methods and frameworks to both prioritize biological experiments and computationally infer new insights are effective resources to speed up the research.

As pointed out by Prof. Limsoon Wong of the National University Singapore, of particular importance are those methods capable of transferring consolidated knowledge from related virus species to the one of interest. A significant example is provided by Grifoni *et al.* [174], a bioinformatics method to infer putative epitopes of SARS-CoV-2 from similar viruses, namely Severe Acute Respiratory Syndrome CoronaVirus 2 (SARS-CoV) and MERS-CoV. An interesting transfer learning could concern the adoption of future COVID-19 vaccines to other existing or future virus species.

We will also add haplotype descriptors (such as clades and lineages); we plan to use several approaches proposed in the literature: GISAID system,[3] PANGOLIN system [325], and the operational classification system described by Chiara *et al.* in [100, 99]. The last one is a new clustering method for viral sequences based upon common variants; the method has the advantage of producing a sequence classification reliably indicating the sequence's country of origin, so as to trace viral outbreaks that may occur within a small group and linking it to their recent travels.

Content Additions and Optimizations. ViruSurf already stores, in addition to about 300K sequences of SARS-CoV-2 collected as of December 2020, also the sequences

[2] Note of the author, August 2023. The EpiViruSurf system [52] has been built after the conclusion of this thesis, contributing a search engine for selecting sequences and epitopes and integrating them for testing epitope stability.

[3] https://www.gisaid.org/references/statements-clarifications/clade-and-lineage-nomenclature-aids-in-genomic-epidemiology-of-active-hcov-19-viruses/

of other Coronaviruses (i.e., SARS-CoV and MERS-CoV), of Ebola and Dengue. We plan to progressively enrich ViruSurf with other virus species.

The loading of ViruSurf sequences and their incremental maintenance is a huge parallel process, as we engage in computing the sequence-specific annotations and both nucleotide and amino acid variants; loading occurs in parallel and takes advantage of Amazon Cloud services. Similarly, the storage of VisuSurf sequences is optimized with classic data warehousing methods (materialized views, database indexes) so as to enable real-time production of query results. Both aspects will be stressed with the schema and content additions.[4]

12.3 Visualization Support: VirusViz

We are building a Web application (available at `http://gmql.eu/virusviz/`) for analyzing viral sequences and visualizing their variants and characteristics. VirusViz (published in [57]) analyses sequences (FASTA files) and their metadata (CSV files) and produces a JSON file that comprises each sequence: alignment to the reference genome using the Needleman-Wunsch algorithm with affine gap penalty; ranked list of 20 most similar sequences from the ViruSurf database using BLASTN;[5] nucleotide variants with associated putative impact computed using SnpEff;[6] amino acid variants; for SARS-CoV-2 sequences, lineage assignment using Pangolin.[7]

VirusViz projects can be created either (i) by submitting FASTA+CSV files (for long processing, the user is notified of completion by email); or (ii) by generating a query on ViruSurf/ViruSurf-GISAID and then invoking VirusViz from the result panel. Projects can be saved in local files, reloaded, and merged.

The Web application has five pages, respectively for defining sub-populations (groups), inspecting each sub-population, visualizing and exporting single sequences, visualizing the distribution of variants of single sub-populations, and comparatively visualizing the variant distributions of different sub-populations. By using VirusViz it is possible to build simple visualizations for detecting and tracing emerging mutation patterns. In particular, we have focused on variant diffusion events that attracted high attention in the press. Next, we show four screens that allow us to compare the variants distributions of different populations, each represented by one track. Each track is a histogram where the X-axis represents positions along the nucleotide or amino acid sequence (of a specific protein) and the Y-axis represents the count of sequences in the selected population that exhibits a mutation in that X position. While comparing tracks, it is the user's responsibility to contextualize

[4] Note of the author, August 2023. The ViruSurf database quickly evolved and contains now more than 7 million sequences.

[5] `https://blast.ncbi.nlm.nih.gov/Blast.cgi`

[6] `https://pcingola.github.io/SnpEff/`

[7] `https://github.com/cov-lineages/pangolin`

different observed behaviors and percentages with respect to the size of the sample represented in a track.

Figure 12.1 illustrates that Italians acquired numerous mutations after traveling during summer; the user can grasp at first sight that the track named "After_Summer" (Sept-Nov) has many more black bars than the ones named "Before_Summer" (Feb-May) or "Mid_Summer" (June-Aug). These groups have been previously prepared by dividing the general population by the metadata collection_date values.

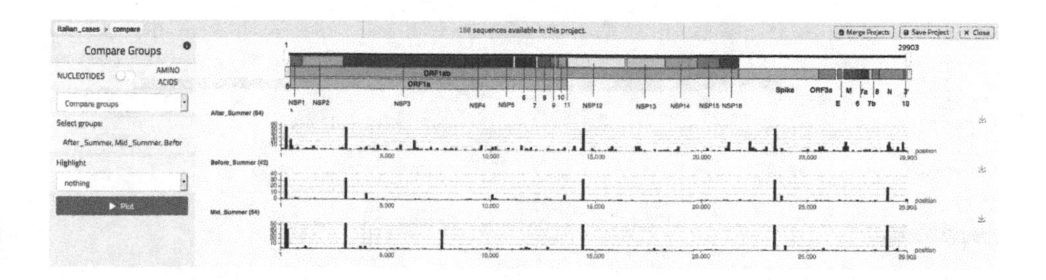

Fig. 12.1 VirusViz screen that compares the variants distribution of a set of Italian sequences before, during, and after summer 2020.

Figure 12.2 shows that a specific Spike mutation, i.e., A222V, had not been observed in Italy before summer, whereas its presence increased considerably after. On the contrary, this mutation was already present in Spain before summer. This may suggest that the specific strains of SARS-CoV-2 featuring this mutation were imported by tourists returning back to their home countries, as observed in [198].

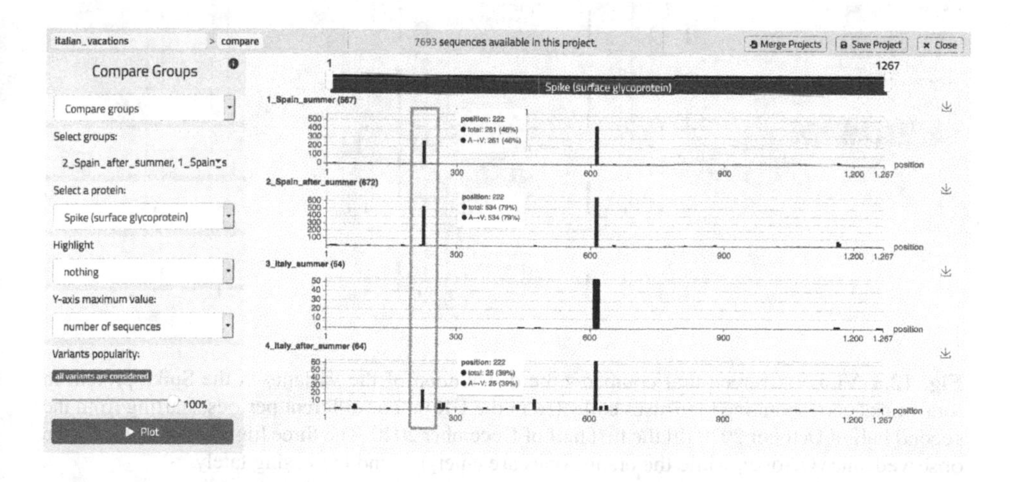

Fig. 12.2 VirusViz screen that compares the distribution of the variants in Spike protein, highlighting the presence of the A222V mutation, with different percentages before and after summer in Italy vs. Spain.

The public debate on possible spillovers of SARS-CoV-2 virus infection from humans to other animals and back has heated up in the last months. Two cases are worth mentioning: the infection in the Netherlands of the minks species *Mustela Lutreola* in June[8] and the infection in Denmark of the minks species *Neovison vison* in November.[9] Figure 12.3 suggests that some viruses found in humans in November acquired frequent mutations from minks, for example in the ORF1ab protein, where two mutations at the amino acids 1244 and 1263 of the NSP3 sub-protein increased notably their presence w.r.t. before November.

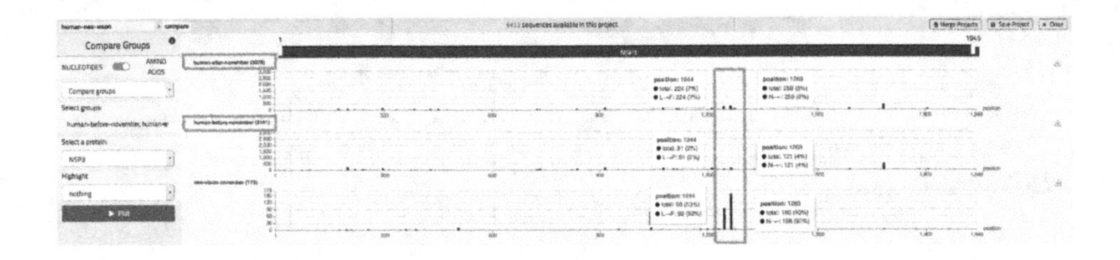

Fig. 12.3 VirusViz screen that compares the distribution of the variants in the NSP3 protein, highlighting the presence of two mutations highly present in minks, while increasing in humans possibly after the spillover.

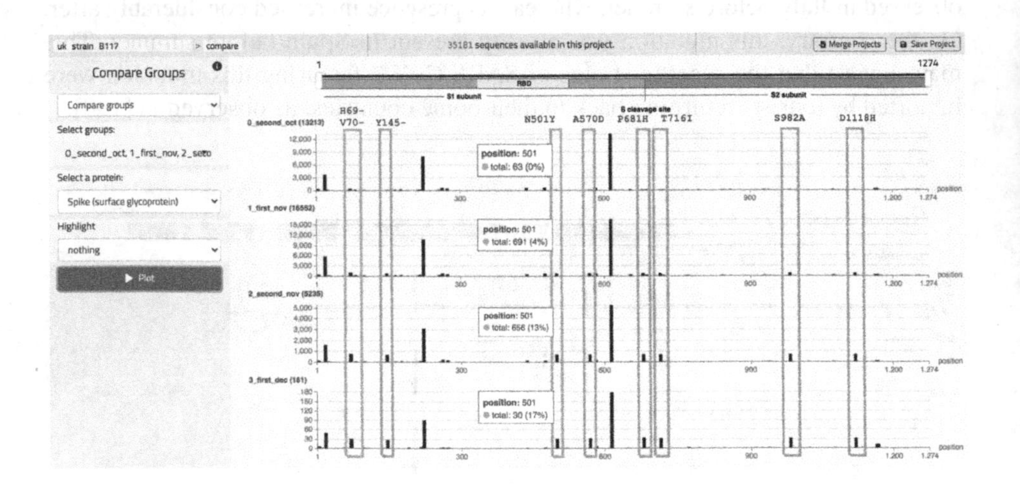

Fig. 12.4 VirusViz screen that compares the distribution of the variants in the Spike protein in four tracks corresponding to viruses collected in the UK in four different periods, starting from the second half of October 2020 till the first half of December 2020. The three highlighted in blue were observed since October, while the orange ones are emerging and increasing lately.

8 https://www.sciencemag.org/news/2020/06/coronavirus-rips-through-dutch-mink-farms-triggering-culls-prevent-human-infections

9 https://www.aa.com.tr/en/europe/mutated-variety-of-coronavirus-found-in-danish-mink/2032390

According to a press release in major mass media,[10] a new strain has started spreading from the UK. Figure 12.4 illustrates the spreading of mutations of this new UK cluster with high transmissibility. H69-, V70-, and N501Y were anticipated by a post on Twitter[11] on December 14th, 2020, while a post on Virological[12] on December 19th, 2020 uncovered several other mutations. We illustrate the distribution of nine such mutations on the Spike protein comparing four time periods since October.

12.4 Active Monitoring of SARS-CoV-2 Variations

As we observed in ViruSurf, SARS-CoV-2 sequences can be connected with their genetic variation (defined by coordinates, size, and site along the genome). These variations may affect not only the genetic interaction of particular genes, but also proteomic aspects (e.g., change or deletion of relevant amino acids due to missense or frameshift mutations) that, in turn, correspond to phenotypic behaviors (i.e., a particular deletion may attenuate the pathological phenotype [409] or the ability of the virus to replicate itself [242]).

At present there is no consolidated knowledge about such impacts of mutations: neither of their role in increasing/decreasing pathogenicity, nor of their correlation with viral transmission. However, a wealth of studies started being published (or just submitted as preprints) targeting particular aspects of the virus regarding, for example, sub-proteins of the long ORF1ab polyprotein [235, 383], ORF3a gene [80], as well as the widely-studied D614G mutation in the spike protein [36, 230]. Some works are claiming particular effects of amino acid variants, such as lower or higher disease severity [5], fatality rate [305], infectivity [350], protein stability [310], sensitivity to convalescent sera or monoclonal antibodies [248], viral transmission [390]. To each of these mutations, we link its effect, a characterization (higher, lower, or null), and the publications that are claiming this result. Surprisingly, we found that in some cases effects provided by papers are contrasting or different in their significance. We are aware that such annotations should only be considered after careful examination of the publications and evaluation of their methods. We are considering assigning them a weight based on the relevance of the publication venue, as in these hectic times attention is given also to results that are still under submission as preprints and low-quality studies could be published under accelerated review processes.

[10] https://www.bbc.com/news/health-55308211

[11] This tweet https://twitter.com/firefoxx66/status/1338533710178775047 was posted by Emma Hodcroft who is part of the Neher's lab https://nextstrain.org/groups/neherlab/ncov/

[12] This post is by Rambaut's group at COG-UK: https://virological.org/t/preliminary-genomic-characterisation-of-an-emergent-sars-cov-2-lineage-in-the-uk-defined-by-a-novel-set-of-spike-mutations/563/5

This collection of mutation effects is growing day by day into what we call a *Knowledge base*, which from semantic annotations on single mutations is extended also to the co-occurrence of mutations (it is already known that particular pairs or triplets of variants are commonly present together in sequences [248]).

This loose integration with papers is so far manually curated, by tagging papers with a set of labels that we deem fundamental and orthogonal. However, we will compare our terminology with the ones used in pre-existing resources, especially community-driven efforts that are systematizing COVID-19-related knowledge (see VODAN [287], CORD-19 [166], and CIDO [194]). We have not considered extracting entities with natural language/ontology-supported techniques as this is a too-hard task to be automatized yet. However, using text mining on COVID-19-related literature could be doable in the future. There have been some efforts reporting encouraging results (see worldwide challenges e.g., CORD-19 [393]) which could be exploited by our work.[13]

We envision an integrative reactive system that may, for instance, put in place daily surveillance of consensus sequences that are published on the sources described earlier. This system could then extract the sequence variants and compare them with a knowledge base of literature-based manually curated information on the impact of the mutations; by such means, it could connect each change in the virus to the implications on its spread, properly contextualized using statistics on the temporal/spatial coordinates of the collected sample and on the variant's representativeness in the observed population.

This same paradigm could then be used to monitor the spread of the virus also in hosts different from humans [124], as it should not be excluded that the pandemic could turn into a panzootic [169].

Some form of a watch of these variations is already put in place by GISAID, but results are not made publicly available; bulletins from the WHO [232] only give a static picture. Online applications – such as the COVID-19 Viral Genome Analysis Pipeline [230], GESS [142], and coronApp [282] – provide environments to explore mutations of SARS-CoV-2. So far, in ViruSurf we only present queries that replicate the results of research papers for demonstration purposes, so that results deposited at given dates can be continuously checked. However, the user still needs to drive the search while we plan to produce an automatic monitoring mechanism that could be used in the future as a means to provide alerts to the general scientific community on the emergence of potentially dangerous mutations in selected regions or in otherwise refined clusters of collected sequences.

In addition, ViruSurf will be equipped with data warehousing tools capable of producing continuous summarization, with expressive power ranging from descriptive

---

[13] Note of the author, August 2023. This thread of research, after the conclusion of this thesis, has contributed different works: CoV2K, an abstract model integrating data and knowledge aspects of SARS-CoV-2 [10, 13] and CoVEffect [363], an interactive web application to extract mutation and variant impact in a semi-automatic way exploiting expert curation and large-language-model-driven prediction.

statistics to complex data mining tasks, concerned for example with the distribution of mutations over relevant locations of the virus.[14]

We will combine database expertise with a dynamic semi-automatic learning approach to build reactive reports on sequence characteristics and distribution; these will be of uttermost importance for clinics, hospitals, or triage centers, providing them with quasi-real-time feedback to respond to relevant clinical questions that will be either automatically generated by internal data mining methods or provided to us by the virus research community: *What are the infection clusters? Where are isolates with specific characteristics coming from? Which are the most common variants? What can we say about co-occurring variants?*[15]

12.5 Integrating Host-Pathogen Information

Together with virus sequences and their metadata, it is critical to integrate also information about the related host phenotype and genotype; this is key to allow supporting the paramount host-pathogen genotype-phenotype analyses and the related pathogenesis [290]. In this section, we briefly mention how these crucial aspects are being investigated.

12.5.1 The Virus Genotype – Host Phenotype Connection

At the time of writing, big integration search engines are only starting to provide clinical information related to hosts of virus sequences. 2019nCoVR [417] has currently 208 clinical records[16] related to specific assemblies (as FASTA files), including information such as the onset date, travel/contact history, clinical symptoms, and tests in a semi-structured format (i.e., attribute-value), where values are free-text and not homogeneous with respect to any dictionary. GISAID is also progressively adding information regarding the "patient status" (e.g., "ICU; Serious", "Hospitalized; Stable", "Released", "Discharged") to its records (in 5,126 out of 75,507 on August 1st, 2020).

So far this kind of effort has not been systematized. Some early findings connecting virus sequences with the human phenotype have been already published, but these include very small datasets (e.g., [246] with only 5 patients, [260] with 9, [65] with 16, and [379] with 103 sequenced SARS-CoV-2 genomes). We are not aware

[14] Note of the author, August 2023. Subsequent work in this direction has been produced in a data-driven approach to identify variants early during their initial growth [53], in ViruClust [103], a tool for fine-tuning clusters of sequences used for fine-grain surveillance, and in VariantHunter [316], an application that monitors the evolution of mutations and indicates possible emerging variants.

[15] Note of the author, August 2023. A study on the co-occurrence of specific mutations and their effect on convergent/divergent evolution has been published in [11].

[16] `https://bigd.big.ac.cn/ncov/clinic`

of big sources comprising linked phenotype data and viral sequences, where the link connects the phenotype (or clinical aspects) of the virus-host organism to the viral genome. There is a compelling need for the combination of phenotypes with virus sequences, so to enable more interesting queries, e.g., concerning the impact of sequence variants. We are confident that in the near future, there will be many more studies like [246, 260, 65, 379]. Additional comprehensive studies, linking the viral sequences of SARS-CoV-2 to the phenotype of patients affected by COVID-19, should be encouraged.

12.5.2 The Host Genotype – Host Phenotype Connection

In addition to investigating the relationship between viral sequences and host conditions, much larger efforts are being conducted for linking the genotype of the human host to the COVID-19 phenotype.

COVID-19 is a multi-organ systemic disease differentially affecting humans, and for understanding it we must primarily focus on the human host; several studies demonstrate the influence of inherited variants or expressed genes. Compared to viral sequences (e.g., of about 30K bases in SARS-CoV-2), the human genome is huge (3 billion bases) and several sequencing technologies support the extraction of different types of information, including variants, gene expression, copy number alterations, epigenomic signals, contact maps, and so on. Genomic information is collected from several open sources, including ENCODE, Roadmap Epigenomics, TCGA, and 1000 Genomes (as described in Chapter 2).

Current COVID-19 research is focusing on understanding which human genes are mostly responsible for clinical severity, e.g., the "genetic variability of the ACE2 receptor is one of the elements modulating virion intake and thus disease severity" [38]. In this direction, an important proposal is the COVID-19 Host Genetics Initiative (HGI),[17] aiming at *bringing together the human genetics community to generate, share, and analyze data to learn the genetic determinants of COVID-19 susceptibility, severity, and outcomes.*

To find important genotype-phenotype correlations, well-characterized phenotypes need to be ascertained in a quantitative and reproducible way [292]; also the activity of sampling from clearly defined case and control groups is fundamental. For this reason, many efforts in the scientific community have been dedicated to harmonizing clinical records of COVID-19 patients. We next illustrate the data dictionary we produced within the context of the COVID-19 HGI;[18] it was contributed by about 50 active participants and is under continuous improvement; the connected

[17] `https://www.covid19hg.org/`

[18] The COVID-19 Host Genetics Initiative Data Dictionary is available at `http://gmql.eu/phenotype/`. The FREEZE 1 version was released on April 16th, 2020, and the FREEZE 2 version on August 16th, 2020.

genotype data is currently being collected and hosted by EGA [150], the European Genome-phenome Archive of EMBL-EBI.[19]

The dictionary is illustrated by the Entity-Relationship diagram in Figure 12.5; the PATIENT phenotype information is collected at admission and during the course of hospitalizations, hosted by a given HOSPITAL.

Attribute groups of patients (grouping about 200 clinical variables that have been progressively consolidated and annotated) describe: DEMOGRAPHY&EXPOSURE, COMORBIDITIES, ADMISSIONSYMPTOMS, RISKFACTORS, HOSPITALIZATIONCOURSE; For ease of visualization, attributes are clustered within attribute groups, indicated with white squares instead of black circles.[20] For instance, COMORBIDITIES include the subgroups *ImmuneSystem, Respiratory, GenitoUrinary, CardioVascular, Neurological, Cancer*; for brevity, these are not further expanded into specific attributes. Each patient is characterized by multiple ENCOUNTERS; attribute groups of encounters describe ENCOUNTERSYMPTOMS, TREATMENTS, and LABORATORYRESULTS. The patient is connected to her human genome information (e.g., single nucleotide variations).

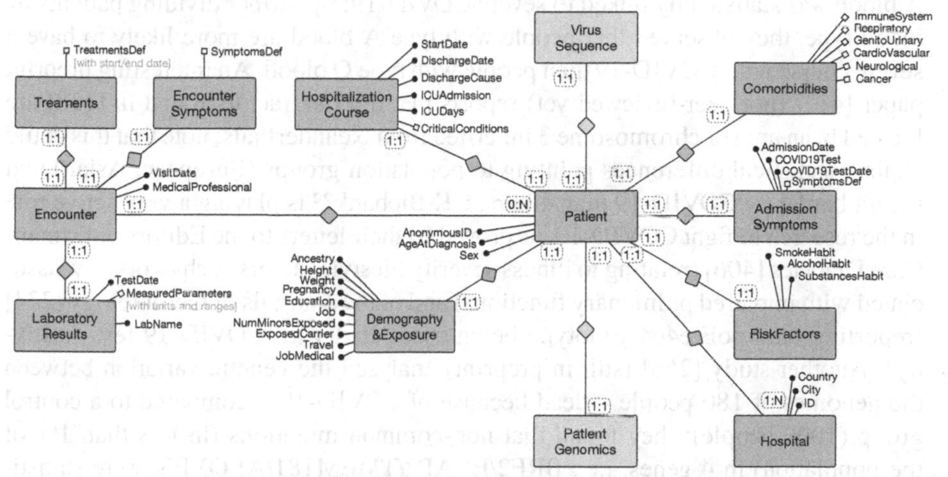

Fig. 12.5 Entity-Relationship diagram of patient phenotype for a viral disease that may lead to hospitalization, such as COVID-19, linked to heterogeneous genomic information and to the sequence of the infecting virus. Patient information is collected at admission and during the hospitalization course, within groups: Demography, RiskFactors, Comorbidities, AdmissionSymptoms, HospitalizationCourse. Each patient has multiple Encounters, with Treatments, Symptoms, and LabResults.

[19] Note of the author, August 2023. The design of the model has been explained in [51]; it has been used in a number of studies of the COVID-19 Host Genetics Initiative, whose initial work is described in [111].

[20] Note that the dictionary representation deviates from a classic Entity-Relationship diagram as some attribute groups would typically deserve the role of an entity; however, this simple format allows easy mapping of the dictionary to questionnaires and implementation by EGA in the form of a spreadsheet.

Researchers can extract the patient phenotype and differentiate cases and controls in a number of ways. For example, one analysis performed within the COVID-19 HGI will discriminate between mild, severe, or critical COVID-19 disease severity, based on a set of EncounterSymptomps and HospitalizationCourse conditions; another analysis will distinguish cases and controls based on Comorbidities and AdmissionSymptoms.

Many other efforts to systematize clinical data collection and harmonization are proposed by international organizations (e.g., WHO [401]), national projects (e.g., AllOfUs [107]), or private companies (e.g., 23AndMe[21]).

In these recent months, a number of studies are concerned with associations of human genetic variants with severe COVID-19 [83]. Genome-Wide Association Studies are being conducted massively, in the attempt to find genetic determinants of COVID-19 severity in specific human genes: a recent study published in the New England Journal of Medicine [138] analyzed the genetic data of almost 2,000 Italian and Spanish patients in Italy and Spain with severe respiratory failure due to COVID-19. The authors of the study found that in the ABO gene, a marker identifying type A blood was statistically linked to severe COVID-19; by further dividing patients by blood type, they observed that people with type A blood are more likely to have a severe course with COVID-19 than people with type O blood. An interesting preprint paper [412] (not peer-reviewed yet) reports that the risk factors found in [138] are located in an area of chromosome 3 inherited from Neanderthals; note that this could explain statistical differences pointing to population groups (European, Asian) that are hit harder by COVID-19 than others. UK Biobank[22] is playing a very active role in the research to fight COVID-19, as proved by their letters to the Editors of Primary Care Diabetes [406] (relating to illness severity lifestyle factors such as obesity, associated with impaired pulmonary functions) and of the Journals of Gerontology [234] (reporting on ApoE e4e4 genotype being associated with COVID-19 test positivity). Another study [258] (still in preprint) analyzed the genetic variation between the genomes of 180 people – dead because of COVID-19 – compared to a control group (1000 people); they found that non-common mutations (in less than 1% of the population) in 4 genes, i.e., BRF2/ERAP2/TMEM181/ALOXE3, were statistically significant in the people who died compared with controls. Other studies are trying to understand the genetic variants associated with the disease by considering sex-related genetic differences [386, 314].

Vision. Note that Figure 12.5 illustrates the possibility of connecting each patient also to the viral sequence of the SARS-CoV-2 virus (of which she is the host organism providing the HostSample in the previously presented Viral Conceptual Model, in Figure 9.1). In general, there is a strong need to connect human genotype, human phenotype, and viral genotype, so as to build a complete and fully encompassing scenario for data analysis. For now, very few works relate the three systems of viral genomics, host genotype, and phenotype in recent literature. An interesting approach to studying the interactome of viruses with their host is proposed in [284]. A number

[21] `https://www.23andme.com/`

[22] `https://www.ukbiobank.ac.uk/`

of works are instead reviewing host-pathogen interactions in SARS and SARS-CoV-2 mainly on experimental animals [352, 93], but we are not aware of similar works on humans.

When the three systems will be connected, we will be able to generate services to hospitals for associating patients or patient groups with reports indicating the most significant information about isolated viruses from a clinical perspective, also considering the patient's genetic characteristics.

Part III
Epilogue

Chapter 13
Conclusions and Vision

"I am still learning." — Michelangelo di Lodovico Buonarroti Simoni

Thousands of new experimental genomic datasets are becoming available every day; in many cases they are produced within the scope of large cooperative efforts, involving a variety of laboratories and world-wide consortia, who are important enablers of biological research as they in general share these datasets openly for public use. Moreover, all experimental datasets leading to publications in genomics must be deposited in public repositories and made available to the research community. Thus, the potential collective amount of available information is huge; these datasets are typically used by biologists for validating or enriching their experiments; their content is documented by metadata.

However, the effective combination of such public sources are hindered by data heterogeneity, as the datasets exhibit a wide variety of notations, protocols, and formats, concerning experimental values and, especially, metadata, which are not homogenized across the sources and are often unstructured and incomplete. Thus, data integration is becoming a fundamental activity, to be performed prior to data analysis and biological knowledge discovery, consisting of subsequent steps of data extraction, normalization, matching, enrichment, reconciliation, and fusion; once applied to heterogeneous data sources, it builds multiple perspectives over the genome, leading to the identification of meaningful relationships that could not be perceived by using incompatible data formats. As a consequence of the underlying data disorganization, access datasets at each distinct repository become problematic, with heterogeneous search interfaces, not interoperable and often with limited capabilities.

Similar issues have been observed in the last months, immediately preceding the submission of this thesis. When COVID-19 hit, the world was unprepared. Scientists had to fill in, under pressure, information about the virus (how it spreads, how much it is contagious, how/when to test for it in human hosts), the disease (which organs are affected, how to care for it – it has been long interpreted as a respiratory disease while now we consider it mostly as a vascular one, with long-term effects largely

A. Bernasconi: *Model, Integrate, Search... Repeat*, LNBIP 496, pp. 229–236, 2023.
https://doi.org/10.1007/978-3-031-44907-9_13

still unknown). The world was flooded with huge amounts of information, often unstructured, from which it is hard to extract and maintain solid knowledge.

13.1 Summary of Thesis Contributions

In the first part of this thesis, we have analyzed and proposed solutions for the domain of Human Genomics Data.

- In Chapter 2 we have described the panorama of genomic actors that play an important role in the production and integration of data, usually providing interfaces to allow domain researchers to download and use such data for analysis; our own system has been already put in perspective of a precisely described dynamic environment that includes consortia, independent initiatives, and academic and research – public or private – institutions.
- In Chapter 3 we have stressed the need for a unifying data model and a structured proposal for metadata organization (the Genomic Conceptual Model), that also allows us to take into consideration the semantics of data descriptions; we have thus proposed a general conceptual model to drive integration of heterogeneous datasets of many genomic signals described by their metadata.
- In Chapter 4 we have described a full integration pipeline that starts from the extraction of controlled partitions of content from origin genomic data sources, transforms formats and cleans metadata keys into a uniform simple model, imports the metadata into a relational representation driven by the Genomic Conceptual Model and connects the stored values to specialized ontological knowledge, which is well-recognized in the genomics and biomedical domains.
- In Chapter 5 we have shown the product realized by the application of the previously overviewed pipeline: the META-BASE repository contains several datasets of important genomic sources, fully interoperable both at the level of region data and of metadata, which are enriched with many levels of ontological expansion. Specific efforts dedicated to the most critical sources are also explained.
- In Chapter 6 we have described the features of semantic search targeted to the specific domain of genomic data; we have then proposed interfaces that allow such augmented search experience. Users can access all the datasets of our repository by choosing metadata from easy-to-use drop-down menus that report dynamically the number of genomic region data files available for given selected characteristics.
- In Chapter 7 we have hinted at future directions initiated by this thesis' work: i) inclusion of new data sources and types; ii) introduction of quality-aware metrics to evaluate our integration and search process; iii) adoption of community-driven interoperability practices; iv) extension of GeCo tools through the support of a conversational agent; v) creation of a repository of best practices of data extraction and analysis in the form of a publicly available marketplace service.

We here summarize the main threats to the wide adoption of the work presented in this first part. These points should be complemented with the content of the 'Theoretical rationale' provided in Section 4.1 for an introduction to our modeling and integration choices, which are to be considered assumptions upon which this thesis builds its contribution.

- *Granularity.* The Genomic Conceptual Model and the integration pipeline include a few attributes, which do not allow for capturing specific aspects of given sources (such as Cancer Genomics clinical metadata in TCGA). However, we point out that GCM purposefully contains a restricted number of metadata, as they are the only ones shared by most data sources. Other source-specific (or area-specific) attributes are instead stored as key-value pairs, easily reachable through the interface search engine. In the future, we do not exclude using other sub-models that represent specific areas of interest connected to the central GCM (performing some sort of area-driven integration).
- *Generality.* As an abstract idea, the model we proposed applies to genomic datasets including processed data belonging to donors of any animal species. However, in practice, it is easily applicable to human genomics, while it needs further tuning and the addition of attributes for other species (as demonstrated by the need for a new conceptual model for viral sequences). In the future, we may consider integration with more general models (as suggested in [159]).
- *Offline process.* The low-level integration performed by our process could be criticized bringing the argument that our application domain includes many resources that change almost on a daily basis, both in their schemata and in their values. Records soon become deprecated or are updated without holding a proper tag/date to inform our integration process. In the community, many works are addressing similar problems with *on-the-fly* approaches, as we hinted in Chapter 7. However, we debate that an offline process like ours holds the considerable advantage of building a stable resource of datasets that, at a certain timestamp, are homogeneous and can be employed seamlessly together for analysis. Our pipelines allow for periodical updates of the repository content that can be done in a modular way, such that failed executions can be repeated only on the specific modules that caused problems (usually due to unexpected changes in the sources).
- *Scalability.* Our integration process requires considerable manual work prior to the inclusion of a new data source (specifically for the phases of download, transformation, cleaning, and mapping). This hinders scalability but only in terms of adding new sources and data types; instead, once the modules are developed, datasets can be updated (potentially including more and more files) very often and changes are only required when sources renovate consistently.
- *Reliability of results.* As briefly discussed in Chapter 4, our pipeline produces a lossless integration. Once a specific partition of interest is defined over the source, our download process, followed by transformation, cleaning, and mapping, does not lose information, while it represents it in a more suitable form for subsequent use.

- *Reliability of values.* We do not address quality problems in the data (as done instead in [307]). This limit will be addressed as a future work of this thesis, currently hinted in Chapter 7, by considering quality checks and improvements that can be performed directly while integrating datasets.

In the second part of this thesis, we have analyzed and proposed solutions for the domain of Viral Sequence Data, produced in a timely manner to quickly contribute to the general call to arms to the scientific community, made necessary by the COVID-19 pandemic:

- In Chapter 8 we have reviewed the data integration efforts required for accessing and searching genome sequences and metadata of SARS-CoV-2, the virus responsible for the COVID-19 disease, which has been deposited into the most important repositories of viral sequences.
- In Chapter 9 we have presented the Viral Conceptual Model, centered on the virus sequence and described from four dimensions: biological (virus type and hosts/sample), analytical (annotations, nucleotide and amino acid variants), organizational (sequencing project) and technical (experimental technology).
- In Chapter 10 we have detailed the integration pipeline that feeds our ViruBase, a large public database of viral sequences and integrated/curated metadata from heterogeneous sources (RefSeq, GenBank, COG-UK, NMDC, and GISAID), which also exposes computed nucleotide and amino acid variants, called from original sequences. Given the current pandemic outbreak, SARS-CoV-2 data are collected from the four sources; but ViruSurf contains other virus species harmful to humans, including SARS-CoV, MERS-CoV, Ebola, and Dengue.
- In Chapter 11 we have shown the web interface realized to enable expressing complex search queries in a simple way; arbitrary queries can freely combine conditions on attributes from the four dimensions, extracting the resulting sequences. Effective search over large and curated sequence data may enable faster responses to future threats that could arise from new viruses.
- In Chapter 12 we have hinted at ongoing and future directions suggested by this second part of the thesis: i) a systematization of our requirement elicitation process with virologists; ii) schema and content extensions of ViruSurf; the provision of visualization support for variants analysis with VirusViz; iii) the design of a monitoring mechanism for notable (groups of) variants, based on a semantic knowledge base that predicates on their effects; iv) the integration of genomic and phenotype information regarding the host of the viral disease.

The main threats to the wide adoption of the presented work are:

- *ER semantics.* We are aware that our ER diagram representation for Viral Data is atypical with respect to conceptual modeling best practices: some properties have been assigned to a host sample, while conceptually they belong to a host species. Authors and publication data are properties of a bibliographic entry rather than of a sequencing project. However, the purpose of the proposed ER model is to facilitate the retrieval of data from practical web interfaces, thus we aim to keep dimensions and schema complexity to a minimum. Our proposal is a starting point that is undergoing changes day by day, as we develop data structures to contain physical data; we already applied many changes following the received suggestions.

- *Lack of visual support.* ViruSurf still lacks data summarization indicators and a graphical representation of the data's most relevant aspects (e.g., the number of variants in sequences by time and space). We are working towards the addition of such features, as indicated in Chapter 12.
- *Lack of connection with relevant information.* At the moment, ViruSurf information on variants can only be connected to coarse-grained information on time and space and some quality measure. More interesting queries concerning the status of the infected patient or her habits, social background, etc. cannot be asked. This limitation comes from the external context: unfortunately linked data to enable such inferences is missing in open source databases.

13.2 Achievements within GeCo Project

The work presented in this thesis has been included in the achievements of the ERC AdG GeCo Project, covering an entire Working Package and setting the basis for future developments. The full GeCo platform, under development since 2015, is fully dedicated to *tertiary* data analysis. It now comprises two main components: the data repository we have described, which integrates heterogeneous genomic and epigenomic data from various public sources and the GenoMetric Query Language, to perform complex queries on such data. A bird's-eye view of GeCo components is presented in Figure 13.1.

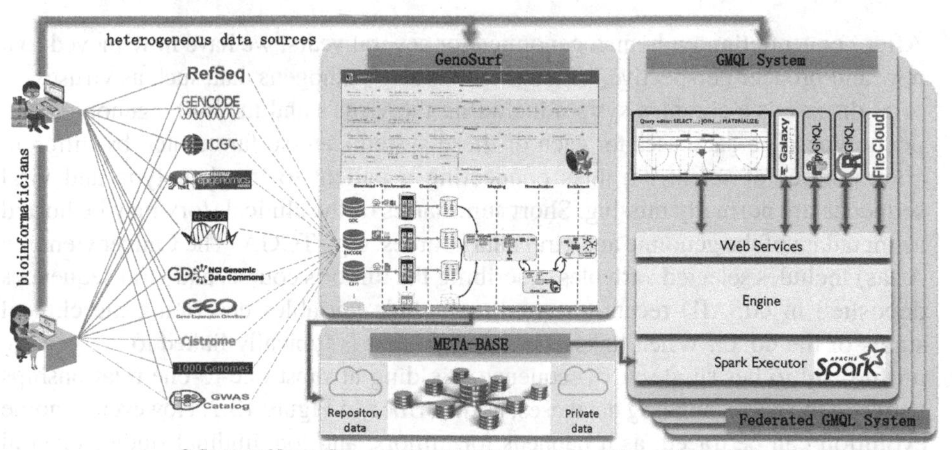

Fig. 13.1 Bird's eye view of the GeCo platform

In discontinuity with the common practice of accessing many different data sources interfaces, with a variety of access modes, download protocols, and terminologies, GenoSurf [76] offers to bioinformaticians (and more in general experts of biology/genomics domain) a unique entry-point to the META-BASE repository (containing datasets organized by using the Genomic Data Model [276], which couples the outcome of a biological experiment with clinical and biological information of the studied sample). A portion of the metadata, which has been identified to be the most valuable and frequently used, has further been modeled using a relational schema, namely the Genomic Conceptual Model [54], and stored in a relational database. Open data are retrieved from different sources by using the data integration pipeline of META-BASE, described in [47]. Our repository can be accessed also by the Genometric Query Language system [275]), so that genomic data can be analyzed using a high-level language for querying genomic datasets; this part is not discussed in this thesis.

As observed in the review Section 2.3, among the integrators currently available in the landscape of genomic players, the GeCo approach is the only one that joins together a broad range of genomic data, which spans from epigenomics to all data types typical of cancer genomics (e.g., mutation, variation, expression, etc.), until annotations, including all the integration steps for data and metadata that follow the genomic secondary analysis. The META-BASE approach is independent of specific sub-branches of genomics and can be applied to a large number of heterogeneous sources by any integration designer, in *a-posteriori* fashion, i.e., without having to follow any guidelines in the preliminary production of metadata.

13.3 Outlook

After concentrating on human genomics for several years, we have now moved to a new and broader perspective, which includes also pathogens data such as viruses. A viral disease is a complex system including the virus's and the host's genotype and phenotype. The databases for each of these systems are so far curated by different communities of scientists; links connecting a patient to its phenotype and viral sequence are normally missing. Short summaries of the clinical story may be hosted as metadata of the genomic and viral data sources, e.g., TCGA (The Cancer Genome Atlas) includes selected variables describing the tumor's course, and viral sequences deposited in GISAID recently started to include variables describing the clinical status of the donor. When links exist, each patient is typically linked to her genetic profile and to her single virus sequence, yielding at most one-to-one relationships between the corresponding databases, as described in Figure 13.2. However, genome evolution can be traced, as it happens for tumors, and longitudinal studies of viral sequences start to emerge for COVID-19, showing how the virus mutates when repeatedly sampled from the same human [341]. Therefore, the clinical course of a patient should be linked to multiple sequencing events of both the human genome and the virus. Possibly, also geopolitical and social data may be linked, exploiting

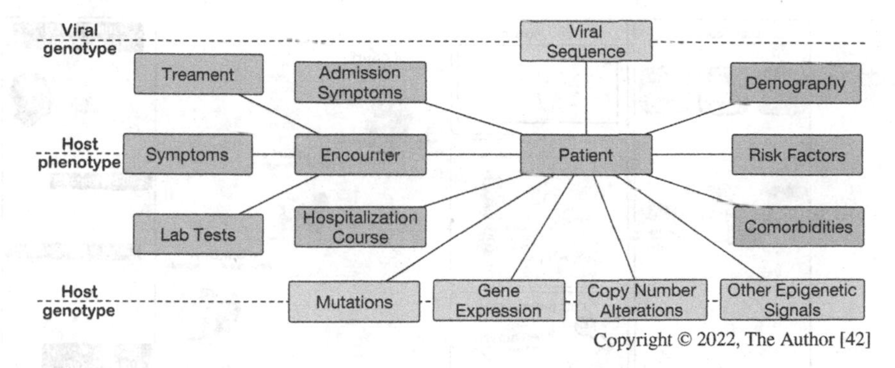

Fig. 13.2 Schema of patient phenotype for a viral disease linked to heterogeneous genomic information and to the sequence of the infecting virus, bridging Part I and Part II of this thesis.

the connection with the information referring to the location where the viral sample was collected.

We are in the unique position of having strong experience with the three systems, as we developed both GenoSurf (Chapter 6) and ViruSurf (Chapter 11), resulting from recent reviews of available data sources in the respective domains (Chapters 2 and 8); within the COVID-19 Host Genetics Initiative,[1] we coordinated about 50 clinicians for cooperatively designing COVID-19's patient phenotype, by curating the format for submitting the clinical/phenotype data, illustrated in Figure 13.2.

We will work on building the missing links between databases, either explicitly collected or learned through sound computational methods. With such connections available, viral mutations could be linked to specific organs or to global severity (e.g., requiring intensive care) and be seen as an aspect of clinical practice.

It is quite possible, according to [263], that infections by emerging viral populations will lead to more severe and/or longer-lasting symptoms, as it has been suggested for other viruses. While some ongoing studies connect human genetics to COVID-19 (e.g., [38]), few studies so far connect COVID-19 to the viral sequence, each with very few patients (e.g., [246, 65]); studies encompassing combinations of variables from the three systems have not been performed yet. We conclude this thesis by proposing a long-term vision: *better linking among databases (and, correspondingly, improved communication between specialists in the various disciplines) will help us in better understanding infectious diseases and empowering a richer precision medicine.* For example, it has been studied that the co-occurrence of certain lab-generated mutations modifies the antigenicity of the SARS-CoV-2 virus, therefore its sensitivity to specific neutralizing monoclonal antibodies [248]. This kind of knowledge, when properly structured and applied in a hospital context dealing with COVID-19 patients, can practically inform the treatment decisions of clinicians on given sets of patients, of which either the clinical profile or the infecting viral strain is known.

[1] https://www.covid19hg.org/

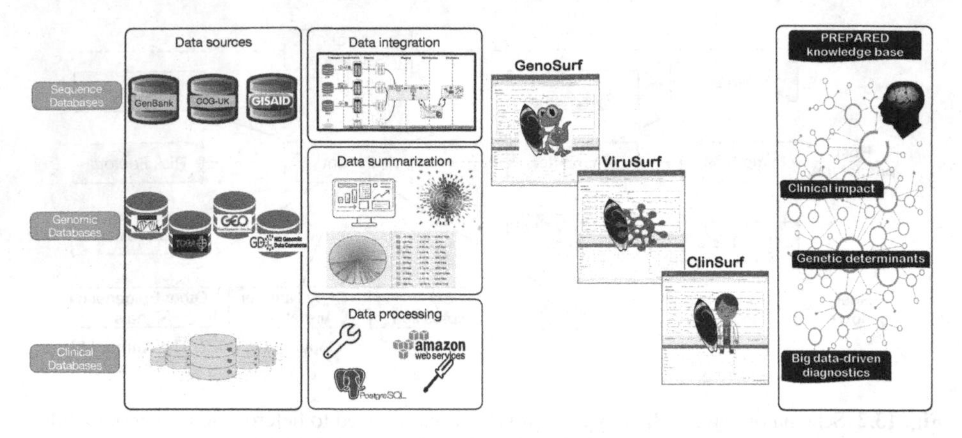

Fig. 13.3 Vision of main data sources and data receiving/processing steps yielding to the project's main results: ViruSurf, GenoSurf, and ClinSurf, and the impact knowledge base.

Figure 13.3 attempts to visualize a possible future architecture: we will need to adapt GenoSurf to become more supportive of viral research, so as to create a COVID-19-specific section of the system; we will use ViruSurf with its enhanced functionalities and we will also produce a large database of clinical cases, inspectable through a web-based interface (provisionally named ClinSurf), allowing access to those clinical cases that are linked to viral sequences and to genomic signals. We will also continuously feed a knowledge base about the impact of these features on COVID-19 phenotype. We can easily link new viral mutations to all deposited sequences, e.g. by searching along time and space. We could then provide big data-driven services, based on a large curated repository of viral sequences so far available, giving real-time insights into the spread of SARS-CoV-2 and how it affects the COVID-19 disease, possibly driving therapeutic strategies.

Appendix A
META-BASE tool configuration

A.1 User Manual

META-BASE, described in Chapter 4 is a software architecture implemented in Scala,[1] available open-source.[2] Assuming that the compiled code is in the executable Metadata-Manager.jar, running the software requires invoking the following line from the command line:

```
java -jar Metadata-Manager.jar <configuration_xml_path>
    <gmql_conf_folder>
```

where:

- `configuration_xml_path` is the location of the configuration file;
- `gmql_conf_folder` contains the path to the folder with corresponding variables to start the GMQLRepository service.

Optional following arguments are:

- `log`: shows the number of executions already performed and stored in the database;
- `log -n`: shows the statistical summary for the n-th run—multiple runs can be requested at the same time by separating them with a comma;
- `retry`: Metadata-Manager tries to download the files whose download has failed in previous runs.

Inside the project, we also provide another small tool, i.e., the Rule Base Generator (described in Section 4.5, which can be run with the following command:

```
java -cp Metadata-Manager.jar it.polimi.genomics.metadata.cleaner.
    RuleBaseGenerator
<transformed_files_directory> <rules_file> <key_files_directory>
```

[1] `https://www.scala-lang.org/`
[2] `https://github.com/DEIB-GECO/Metadata-Manager/`

A. Bernasconi: *Model, Integrate, Search... Repeat*, LNBIP 496, pp. 237–241, 2023.
https://doi.org/10.1007/978-3-031-44907-9

where:

- `transformed_files_directory` is the folder of the transformed files;
- `rules_file` is the file that contains the *Rule Base*;
- `key_files_directory` contains the seen keys, all keys, and unseen keys files.

A.2 Process Configuration

Metadata-Manager is designed to receive a configuration XML file with the needed parameters to prepare all its steps: downloading, transforming, cleaning, mapping, enriching, checking, flattening, and loading. The inputs required to run each step are summarized in Table A.1; they can be run all together in a single sequential workflow or one at a time.

Step	Inputs
`<Downloader>`	Download query parameters List of file filters Dataset name Downloaded files output path
`<Transformer>`	File selection parameters Input format Downloaded files input path Transformed files output path
`<Cleaner>`	Cleaning rules file path Transformed files input path Cleaned files output path
`<Mapper>`	Mapping rules file path (see Section A.3) Cleaned files input path Relational database connection Flag for importing or excluding key-value pairs metadata
`<Enricher>`	Relational database connection Ontology search service Specific API keys for external services calls, when required Preferred ontologies for annotation Match score threshold fort values annotation Depth of considered ancestors/decendants in ontology (e.g., three is the default)
`<Checker>`	Relational database connection Integrity rules file path
`<Flattener>`	Relational database connection Flattened files output path Dataset name Separation characters Prefix for new attributes resulting from flattening

Table A.1 Configurable parts of the framework, which can be changed in the settings.

An XML Schema Definition (XSD) schema file is designed to validate configuration XML files used as input to the tool. The schema is organized in a tree structure, starting from a *root node* where general settings and a list of *sources* are stored; sources represent NGS data providers which provide those genomic data and experimental metadata divided into datasets (examples are ENCODE, TCGA, GDC,

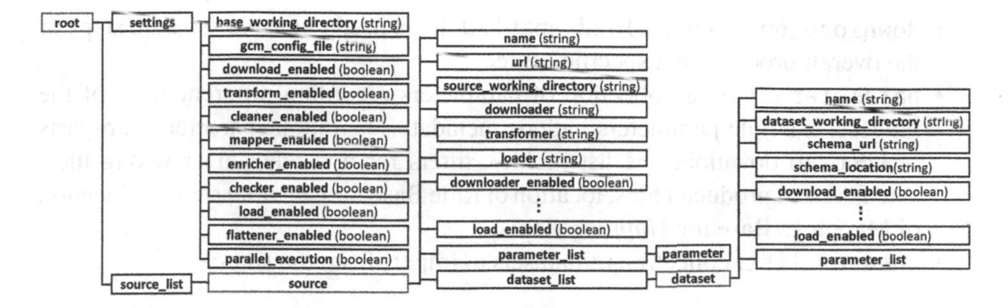

Fig. A.1 First level (root), second level (source), and third level (dataset) of the XML configuration XSD.

and Roadmap Epigenomics). Each source contains, in turn, a list of *datasets*. After processing, each dataset represents a GDM dataset where every sample has a region data file and a corresponding metadata file; moreover, every sample share the same region data schema.

In Figure A.1 we show the details of the three levels of the schema. Note that we allow for enabling/disabling each step of the process at different granularity. E.g., the XML element `downloader_enabled` is present at the level of the entire repositories (i.e., for all sources), at the level of a single source (i.e., for all the datasets from that origin), and at the level of a single dataset (i.e., the smallest organizational unit in this project).

First level elements. `root` contains general settings and a list of sources to import. It contains:

- `settings`:
  - `base_working_directory`: folder used to save downloaded, transformed, cleaned, and flattened files;
  - `download_enabled`, ..., `load_enabled`: flags used to enable separate steps of the overall process;
  - `parallel_execution` flag to enable execution with single-thread processing or multi-thread processing.
- `source_list`: collection of sources to be imported.

Second level elements. `source`: represents an NGS data source, contains basic information for the process:

- `name`: use as an identifier of the source;
- `url`: base endpoint of the source;
- `source_working_directory`: sub-directory for processing the source's files;
- `downloader`, `transformer`, `loader`: indicate the specific classes to be used to perform respectively downloading, transforming, and loading;

- `download_enabled`, ..., `load_enabled`: flags used to enable separate steps of the overall process for a specific source;
- `parameter_list`: a collection of parameters regarding different steps of the process; example parameters include metadata separation characters, sub-parts of URLs to download the list of files, filters for non-relevant or wrong files, extensions of produced files, location of Rule Base for *Cleaning* step and location of Mappings Base for *Mapping* step;
- `dataset_list`: collection of datasets to import from the source.

Third level elements. A `dataset` represents a set of samples that share the same region data schema and the same types of experimental or clinical metadata. It is parameterized by:

- `name`: identifier for the dataset.
- `dataset_working_directory` sub-folder where the download and transformation of the dataset is performed;
- `schema_url` location of the dataset-specific schema file;
- `schema_location`: indicates whether the schema is located in FTP, HTTP, or LOCAL destination;
- `download_enabled`, ..., `load_enabled` flags used to enable separate steps of the overall process for a specific dataset;
- `parameter_list` list of dataset specific parameters. Each parameter defines specific information using the quadruple ⟨`key`, `value`,
- `description`, `type`⟩; example parameters are used to specify the loading name and description for the dataset, the source URL to retrieve the files to be included in the dataset.

Note that newer modules can be added to configurations and different parameters could emerge in the future as the addition of other sources and datasets to the project is ongoing.

A.3 Mapper Configuration

The mappings used in the *Mapper* step are stored in a separate XML file, whose XSD schema is represented in Figure A.2. These create a mapping rule between source cleaned keys (i.e., produced during the *Cleaner* phase) and attributes of the global schema (i.e., the GCM).

The first level contains a number of `table` elements, one for each entity in the GCM, which is identified by the `name` attribute and contains an arbitrary number of `mapping` elements, i.e., a single mapping between a row in the source file (`source_key`) and the corresponding attribute of the GCM table (**global_key**). A mapping can be of different kinds, as specified by the `method`. When the method is not specified, the default choice is a direct copy of the source key into the set GCM attribute value.

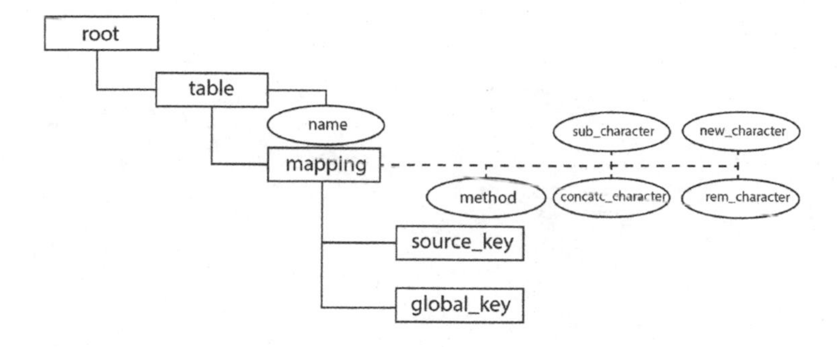

Fig. A.2 Schema of configuration XSD for mapping settings.

Available alternatives include "MANUAL" (i.e., a user-defined fixed value); "CONCAT" (which allows concatenating multiple source key values), "CHECK-PREC" (to define a second choice source key, in case the first corresponds to a null value), "REMOVE" (to remove a string from the value corresponding to the source key), "SUB" (to replace an existing string with an alternative one), "UPPER-CASE/LOWERCASE" (which transforms the case of the input). These methods can be composed and used together on the same mapping by using a dash character. For example, to concatenate uppercase words after, the user can write: `metod = "CONCAT-UPPERCASE"`. Depending on the method choice, other attributes may be required to further define the behavior of the function (`CONCAT_CHARACTER` for concatenation; `SUB_CHARACTER`, `NEW_CHARACTER`, and `REM_CHARACTER` for specifying replacements).

Appendix B
Experimental setup GEO metadata extraction

This short appendix complements the description of the experiment's setup reported in Section 5.3, where we designed three experiments to validate our proposal. Experiments 1 and 2 allow us to compare the performances of the three analyzed models on two different datasets: Cistrome (with input from GEO) and ENCODE (with input both from GEO and ENCODE itself). Experiment 3, instead, tested the performance of the best-proposed model on randomly chosen instances from GEO.

The Transformer library from HuggingFace was used for the GPT-2 model, while the SimpleTransformers library;[1] [2] was used for the RoBERTa model. The LSTM encoder-decoder was built with Tensorflow [2] version 2.1 using the Keras API. For the LSTM model, we performed the tokenization process using the default Keras tokenizer, setting the API parameters to convert all characters into lowercase, using empty space as a word separator, and disabling character-level tokenization. We added a space before and after the following characters: opening/closing parenthesis, dashes, and underscores.[3] We also removed equal signs. For the LSTM models, the resulting vocabulary had a size of 36,107 for Experiment 1 and 17,880 for Experiment 2.

RoBERTa and GPT-2 were trained using a Tesla P100-PCIE-16GB GPU, while the LSTM model was trained on Google Colaboratory[4] with GPU accelerator. Table B.1 lists the configurations for the systems. All models were subject to an early stopping method to avoid over-fitting.

For Experiments 1 and 2, data were split into a training set (80%), validation set (10%), and test set (10%). Some text cleaning and padding processes were adopted: Encoder-Decoder requires input-output pairs that are padded to the maximum length of concatenation of input and output; GPT-2 requires single sentences that are

[1] `https://github.com/huggingface/transformers`

[2] `https://github.com/ThilinaRajapakse/simpletransformers`

[3] Pre-processing was motivated by the fact that important character ngrams often appear in sequences separated by special characters, e.g., "RH_RRE2_14028".

[4] `https://colab.research.google.com/`

A. Bernasconi: *Model, Integrate, Search... Repeat*, LNBIP 496, pp. 243–244, 2023.
https://doi.org/10.1007/978-3-031-44907-9

Model	Batch size	Loss function	Tokenizer	Optimizer	LR	beta_1	beta_2	epsilon
RoBERTa	10	Cross Entropy	BPE	Adam	2e-4	0.9	0.999	1e-6
LSTM	64	Sparse Cross Entropy	keras	Adam	1e-3	0.9	0.999	1e-7
GPT-2	5	Cross Entropy	BPE	Adam	1e-3	0.9	0.999	1e-6

Table B.1 Setup of the three different models for each experiment. BPE = Byte Pair Encoding; LR = learning rate

padded to a maximum length of 500 characters. We excluded sentences exceeding the maximum length.

Appendix C
Mappings of viral sources attributes into ViruSurf

A. Bernasconi: *Model, Integrate, Search... Repeat*, LNBIP 496, pp. 245–249, 2023.
https://doi.org/10.1007/978-3-031-44907-9

ViruSurf	NCBI Nuccore	NCBI Virus HTML
Sequence.AccessionId	INSDSeq_accession-version	Accession
Sequence.AlternativeAccession	×	×
Sequence.StrainName	INSDQualifier_value/isolate	GenBank_Title
Sequence.IsReference	INSDKeyword	×
Sequence.IsComplete	INSDSeq_definition	Nuc_Completeness
Sequence.Strand	✓	×
Sequence.Length	INSDSeq_length	Length
Sequence.GC%	✓	×
Sequence.N%	✓	×
Sequence.Clade	×	×
Sequence.Lineage	×	×
Virus.TaxonomyID	✓	×
Virus.TaxonomyName	INSDSeq_organism	GenBank_Title
Virus.Species	✓	Species
Virus.Genus	✓	Genus
Virus.Subfamily	✓	×
Virus.Family	✓	Family
Virus.EquivalentList	✓	×
Virus.MoleculeType	INSDSeq_moltype	×
Virus.SingleStranded	INSDSeq_strandedness	×
Virus.PositiveStranded	✓	×
HostSample.TaxonomyID	✓	×
HostSample.TaxonomyName	INSDQualifier_value/host	Host
HostSample.CollectionDate	INSDQualifier_value/collection_date	Collection_Date
HostSample.IsolationSource	INSDQualifier_value/isolation_source	Isolation_Source
HostSample.Originating Lab	×	×
HostSample.GeoGroup	✓	×
HostSample.Country	INSDQualifier_value/country	Geo_Location
HostSample.Region	INSDQualifier_value/country	Geo_Location
HostSample.Age	INSDQualifier_value/host	×
HostSample.Gender	INSDQualifier_value/host	×
ExperimentType.SequencingTechnology	INSDSeq_comment/Sequencing Technology	×
ExperimentType.AssemblyMethod	INSDSeq_comment/Assembly Method	×
ExperimentType.Coverage	INSDSeq_comment/Coverage	×
SequencingProject.SubmissionDate	INSDReference_title/Direct Submission	×
SequencingProject.SequencingLab	INSDReference_title/Direct Submission	×
SequencingProject.DatabaseSource	✓	Sequence_Type
SequencingProject.BioprojectId	INSDXref_id/BioProject	×

Table C.1 Mappings between variables of ViruSurf and variables of NCBI direct sources (API Nuccore `https://www.ncbi.nlm.nih.gov/books/NBK25497/` and HTML interface `https://www.ncbi.nlm.nih.gov/labs/virus/vssi/#/virus`). For NCBI we specify both the terminology used in the xml/json export files and the one used in their interfaces. In each cell we specify the mapped field, or ✓if that field is calculated by our pipelines (for example most information about the Virus are computed using the NCBI Taxonomy services), or × when that information is not available. Sometimes we report the double values (e.g., INSDCQualifier/host), this notation indicates that the same information is available under two different elements.

ViruSurf	COG-UK metadata csv
Sequence.AccessionId	sequence_name
Sequence.AlternativeAccession	×
Sequence.StrainName	sequence_name
Sequence.IsReference	✓
Sequence.IsComplete	✓
Sequence.Strand	✓
Sequence.Length	✓
Sequence.GC%	✓
Sequence.N%	✓
Sequence.Clade	×
Sequence.Lineage	lineage
Virus.TaxonomyID	✓
Virus.TaxonomyName	✓
Virus.Species	✓
Virus.Genus	✓
Virus.Subfamily	✓
Virus.Family	✓
Virus.EquivalentList	✓
Virus.MoleculeType	✓
Virus.SingleStranded	✓
Virus.PositiveStranded	✓
HostSample.TaxonomyID	✓
HostSample.TaxonomyName	✓
HostSample.CollectionDate	sample_date
HostSample.IsolationSource	×
HostSample.Originating Lab	×
HostSample.GeoGroup	✓
HostSample.Country	country
HostSample.Region	adm1
HostSample.Age	×
HostSample.Gender	×
ExperimentType.SequencingTechnology	×
ExperimentType.AssemblyMethod	×
ExperimentType.Coverage	×
SequencingProject.SubmissionDate	×
SequencingProject.SequencingLab	×
SequencingProject.DatabaseSource	✓
SequencingProject.BioprojectId	×

Table C.2 Mappings between variables of ViruSurf and of COG-UK metadata file, provided in CSV format on their web page (`https://www.cogconsortium.uk/data/`). The notation used is the same as for Table C.1

ViruSurf	NMDC json	NMDC HTML
Sequence.AccessionId	seqName	NMDC Accession
Sequence.AlternativeAccession	gisaid	Gisa Id
Sequence.StrainName	isolate	Isolation Strain
Sequence.IsReference	✓	×
Sequence.IsComplete	✓	×
Sequence.Strand	✓	×
Sequence.Length	glength/✓	Length/✓
Sequence.GC%	✓	×
Sequence.N%	✓	×
Sequence.Clade	×	×
Sequence.Lineage	×	×
Virus.TaxonomyID	✓	×
Virus.TaxonomyName	spciesname	Organism
Virus.Species	✓	×
Virus.Genus	✓	×
Virus.Subfamily	✓	×
Virus.Family	✓	×
Virus.EquivalentList	✓	×
Virus.MoleculeType	✓	×
Virus.SingleStranded	✓	×
Virus.PositiveStranded	✓	×
HostSample.TaxonomyID	✓	×
HostSample.TaxonomyName	host	Host
HostSample.CollectionDate	collectionDateFormat	Collection Date
HostSample.IsolationSource	isolationSource	Isolate Name
HostSample.Originating Lab	samplingPlace	Sampling Place
HostSample.GeoGroup	✓/Country	Country
HostSample.Country	country	Country
HostSample.Region	country	Country
HostSample.Age	×	×
HostSample.Gender	×	×
ExperimentType.SequencingTechnology	sequencingMethods	Sequencing Methods
ExperimentType.AssemblyMethod	jointMethods	Joint Methods
ExperimentType.Coverage	×	×
SequencingProject.SubmissionDate	submitDateFormat	×
SequencingProject.SequencingLab	dept	Organization
SequencingProject.DatabaseSource	✓	×
SequencingProject.BioprojectId	×	×

Table C.3 Mappings between variables of ViruSurf and of NMDC metadata file, provided in JSON and HTML format on their Web page (http://nmdc.cn/nCov/globalgenesequence/detail/, followed by the sequence id, e.g., "NMDC60013002-07"). The notation used is the same as for Table C.1. Sometimes we report the double values (e.g., glength/✓), this notation indicates that when information is missing in the input file, it is computed by our pipeline.

ViruSurf	EpiCoV export json	EpiCoV HTML
Sequence.AccessionId	covv_accession_id	Accession ID
Sequence.AlternativeAccession	×	×
Sequence.StrainName	covv_virus_name	Virus name
Sequence.IsReference	is_reference	×
Sequence.IsComplete	is_complete	Complete
Sequence.Strand	covv_strand	×
Sequence.Length	sequence_length	Length
Sequence.GC%	gc_content	×
Sequence.N%	n_content	×
Sequence.Clade	covv_clade	Lineage (GISAID Clade)
Sequence.Lineage	covv_lineage	Lineage (GISAID Clade)
Virus.TaxonomyID	✓	✓
Virus.TaxonomyName	✓	×
Virus.Species	✓	×
Virus.Genus	covv_type	Type
Virus.Subfamily	✓	×
Virus.Family	✓	×
Virus.EquivalentList	✓	×
Virus.MoleculeType	✓	×
Virus.SingleStranded	✓	×
Virus.PositiveStranded	✓	×
HostSample.TaxonomyID	✓	×
HostSample.TaxonomyName	covv_host	Host
HostSample.CollectionDate	covv_collection_date	Collection date
HostSample.IsolationSource	covv_specimen	Specimen source
HostSample.Originating Lab	covv_orig_lab	Originating lab, Address
HostSample.GeoGroup	covv_location	
HostSample.Country	covv_location	
HostSample.Region	covv_location	Location
HostSample.Age	×	Patient age
HostSample.Gender	×	Gender
ExperimentType.SequencingTechnology	×	Sequencing Technology
ExperimentType.AssemblyMethod	×	Assembly method
ExperimentType.Coverage	×	Coverage
SequencingProject.SubmissionDate	covv_subm_date	Submission date
SequencingProject.SequencingLab	covv_subm_lab	Submitting lab, Address
SequencingProject.DatabaseSource	✓	×
SequencingProject.BioprojectId	×	×

Table C.4 Mappings between variables of ViruSurf and of GISAID metadata file, provided in JSON (as an export file prepared for ViruSurf ad-hoc) and HTML format on GISAID portal, via authorized login. The notation used is the same as for Table C.1

References

1. Prognosis: A $100 Genome Within Reach, Illumina CEO Asks If World Is Ready, By Kristen V Brown, Bloomberg. https://www.bloomberg.com/news/articles/2019-02-27/a-100-genome-within-reach-illumina-ceo-asks-if-world-is-ready. Accessed: 2021-01-04.
2. M. Abadi, A. Agarwal, P. Barham, E. Brevdo, Z. Chen, et al. Tensorflow: Large-scale machine learning on heterogeneous distributed systems. *arXiv*, 2016. https://arxiv.org/abs/1603.04467.
3. M. Abu-Elmagd, M. Assidi, H.-J. Schulten, A. Dallol, P. N. Pushparaj, et al. Individualized medicine enabled by genomics in Saudi Arabia. *BMC medical genomics*, 8(1):S3, 2015.
4. D. Adams, L. Altucci, S. E. Antonarakis, J. Ballesteros, S. Beck, et al. BLUEPRINT to decode the epigenetic signature written in blood. *Nature biotechnology*, 30(3):224, 2012.
5. P. Aiewsakun, P. Wongtrakoongate, Y. Thawornwattana, S. Hongeng, and A. Thitithanyanont. SARS-CoV-2 genetic variations associated with COVID-19 severity. *medRxiv*, 2020.
6. J. Akoka and I. Comyn-Wattiau. Entity-relationship and object-oriented model automatic clustering. *Data & Knowledge Engineering*, 20(2):87–117, 1996.
7. S. Akther, E. Bezrucenkovas, B. Sulkow, C. Panlasigui, L. Li, et al. CoV Genome Tracker: tracing genomic footprints of Covid-19 pandemic. *bioRxiv*, 2020. https://doi.org/10.1101/2020.04.10.036343.
8. H. Alani, C. Brewster, and N. Shadbolt. Ranking ontologies with AKTiveRank. In *International Semantic Web Conference*, pages 1–15. Springer, 2006.
9. F. Albrecht, M. List, C. Bock, and T. Lengauer. DeepBlue epigenomic data server: programmatic data retrieval and analysis of epigenome region sets. *Nucleic Acids Research*, 44(W1):W581–W586, 2016.
10. T. Alfonsi, R. Al Khalaf, S. Ceri, and A. Bernasconi. CoV2K: a knowledge base of SARS-CoV-2 variant impacts. In S. Cherfi, A. Perini, and S. Nurcan, editors, *International Conference on Research Challenges in Information Science*, pages 274–282, Cham, 2021. Springer International Publishing.
11. R. Al Khalaf, A. Bernasconi, P. Pinoli, and S. Ceri. Analysis of co-occurring and mutually exclusive amino acid changes and detection of convergent and divergent evolution events in SARS-CoV-2. *Computational and Structural Biotechnology Journal*, 20:4238–4250, 2022.
12. T. Alfonsi, A. Bernasconi, A. Canakoglu, and M. Masseroli. Genomic data integration and user-defined sample-set extraction for population variant analysis. *BMC Bioinformatics*, 23(1):401, 2022.
13. T. Alfonsi, R. Al Khalaf, S. Ceri, and A. Bernasconi. CoV2K model, a comprehensive representation of SARS-CoV-2 knowledge and data interplay. *Scientific Data*, 9:260, 2022.
14. C. Alkan, P. Kavak, M. Somel, O. Gokcumen, S. Ugurlu, et al. Whole genome sequencing of Turkish genomes reveals functional private alleles and impact of genetic interactions with Europe, Asia and Africa. *BMC genomics*, 15(1):963, 2014.
15. C. Amid, B. T. Alako, V. Balavenkataraman Kadhirvelu, T. Burdett, J. Burgin, et al. The European Nucleotide Archive in 2019. *Nucleic Acids Research*, 48(D1):D70–D76, 2020.
16. R. Angles and C. Gutierrez. Survey of graph database models. *ACM Computing Surveys (CSUR)*, 40(1):1, 2008.
17. M. Artesi, S. Bontems, P. Göbbels, M. Franckh, P. Maes, et al. A Recurrent Mutation at Position 26340 of SARS-CoV-2 Is Associated with Failure of the E Gene Quantitative Reverse Transcription-PCR Utilized in a Commercial Dual-Target Diagnostic Assay. *Journal of Clinical Microbiology*, 58(10), 2020.
18. P. Artimo et al. ExPASy: SIB bioinformatics resource portal. *Nucleic Acids Res.*, 40(W1):W597–W603, 2012.
19. M. Ashburner, C. A. Ball, J. A. Blake, D. Botstein, H. Butler, et al. Gene ontology: tool for the unification of biology. *Nature genetics*, 25(1):25–29, 2000.

A. Bernasconi: *Model, Integrate, Search... Repeat*, LNBIP 496, pp. 251–270, 2023.
https://doi.org/10.1007/978-3-031-44907-9

20. N. Askham, D. Cook, M. Doyle, H. Fereday, M. Gibson, et al. The six primary dimensions for data quality assessment. *DAMA UK Working Group*, pages 432–435, 2013. https://damauk.wildapricot.org/resources/Documents/DAMA%20UK%20DQ%20Dimensions%20White%20Paper2020.pdf, Accessed: 2021-01-04.
21. A. Athar, A. Füllgrabe, N. George, H. Iqbal, L. Huerta, et al. ArrayExpress update–from bulk to single-cell expression data. *Nucleic Acids Research*, 47(D1):D711–D715, 2018.
22. S. Babcock, L. G. Cowell, J. Beverley, and B. Smith. The Infectious Disease Ontology in the Age of COVID-19. In *OSF Preprints*, pages 1–33. Center for Open Science, 2020. https://doi.org/10.31219/osf.io/az6u5.
23. V. Babenko, B. Brunk, J. Crabtree, S. Diskin, S. Fischer, et al. GUS the genomics unified schema a platform for genomics databases. http://www.gusdb.org/. Accessed: 2021-01-04.
24. C. O. Back, S. Debois, and T. Slaats. Towards an empirical evaluation of imperative and declarative process mining. In C. Woo, J. Lu, Z. Li, T. Ling, G. Li, and M. Lee, editors, *Advances in Conceptual Modeling*, pages 191–198, Cham, 2018. Springer International Publishing.
25. A. Bandrowski, R. Brinkman, M. Brochhausen, M. H. Brush, B. Bug, et al. The ontology for biomedical investigations. *PloS one*, 11(4), 2016.
26. Y. Bao, P. Bolotov, D. Dernovoy, B. Kiryutin, L. Zaslavsky, et al. The influenza virus resource at the National Center for Biotechnology Information. *Journal of virology*, 82(2):596–601, 2008.
27. A.-L. Barabási, N. Gulbahce, and J. Loscalzo. Network medicine: a network-based approach to human disease. *Nature reviews genetics*, 12(1):56, 2011.
28. T. Barrett, K. Clark, R. Gevorgyan, V. Gorelenkov, E. Gribov, et al. BioProject and BioSample databases at NCBI: facilitating capture and organization of metadata. *Nucleic Acids Research*, 40(D1):D57–D63, 2011.
29. T. Barrett, T. O. Suzek, D. B. Troup, S. E. Wilhite, W.-C. Ngau, et al. NCBI GEO: mining millions of expression profiles–database and tools. *Nucleic Acids Research*, 33(suppl_1):D562–D566, 2005.
30. T. Barrett, D. B. Troup, S. E. Wilhite, P. Ledoux, D. Rudnev, et al. NCBI GEO: archive for high-throughput functional genomic data. *Nucleic Acids Research*, 37(suppl_1):D885–D890, 2008.
31. T. Barrett, S. E. Wilhite, P. Ledoux, C. Evangelista, I. F. Kim, et al. NCBI GEO: archive for functional genomics data sets–update. *Nucleic Acids Research*, 41(D1):D991–D995, 2012.
32. C. Batini, S. Ceri, and S. B. Navathe. *Conceptual database design: an Entity-relationship approach*. Benjamin-Cummings Publishing Co., Inc., 1991.
33. C. Batini, M. Lenzerini, and S. B. Navathe. A comparative analysis of methodologies for database schema integration. *ACM computing surveys (CSUR)*, 18(4):323–364, 1986.
34. C. Batini and M. Scannapieco. *Data and Information Quality: Dimensions, Principles and Techniques*. Springer Publishing Company, Incorporated, 2016.
35. G. D. Battista, P. Eades, R. Tamassia, and I. G. Tollis. *Graph drawing: algorithms for the visualization of graphs*. Prentice Hall PTR, 1998.
36. M. Becerra-Flores and T. Cardozo. SARS-CoV-2 viral spike G614 mutation exhibits higher case fatality rate. *International Journal of Clinical Practice*, 74(8):e13525, 2020.
37. S. M. Bello, M. Shimoyama, E. Mitraka, S. J. Laulederkind, C. L. Smith, et al. Disease Ontology: improving and unifying disease annotations across species. *Disease models & mechanisms*, 11(3):dmm032839, 2018.
38. E. Benetti, R. Tita, O. Spiga, A. Ciolfi, G. Birolo, et al. ACE2 gene variants may underlie interindividual variability and susceptibility to COVID-19 in the Italian population. *European Journal of Human Genetics*, 28:1602–1614, 2020.
39. A. Bernasconi. Using Metadata for Locating Genomic Datasets on a Global Scale. In A. Cuzzocrea, F. Bonchi, and D. Gunopulos, editors, *International Workshop on Data and Text Mining in Biomedical Informatics*, volume 2482 of *CEUR Workshop Proceedings*, 2018.
40. A. Bernasconi. Data quality-aware genomic data integration. *Computer Methods and Programs in Biomedicine Update*, 1:100009, 2021.

41. A. Bernasconi. Extreme requirements elicitation: lessons learnt from the COVID-19 case study. In F. B. Aydemir, C. Gralha, M. Daneva, *et al.*, editors, In: *Joint Proceedings of REFSQ 2021 Workshops, OpenRE, Poster and Tools Track, and Doctoral Symposium*, volume 2857 of *CEUR Workshop Proceedings*, 2021.
42. A. Bernasconi. Model, Integrate, Search... Repeat: A Sound Approach to Building Integrated Repositories of Genomic Data In *Special Topics in Information Technology*, pages 89–99. Springer, 2022.
43. A. Bernasconi, A. Canakoglu, and S. Ceri. Exploiting Conceptual Modeling for Searching Genomic Metadata: A Quantitative and Qualitative Empirical Study. In G. Guizzardi, F. Gailly, and R. Suzana Pitangueira Maciel, editors, *Advances in Conceptual Modeling*, pages 83–94, Cham, 2019. Springer International Publishing.
44. A. Bernasconi, A. Canakoglu, and S. Ceri. From a Conceptual Model to a Knowledge Graph for Genomic Datasets. In A. H. F. Laender, B. Pernici, E.-P. Lim, and J. P. M. de Oliveira, editors, *Conceptual Modeling*, pages 352–360, Cham, 2019. Springer International Publishing.
45. A. Bernasconi, A. Canakoglu, A. Colombo, and S. Ceri. Ontology-Driven Metadata Enrichment for Genomic Datasets. In C. J. O. Baker, A. Waagmeester, A. Splendiani, O. D. Beyan, and M. S. Marshall, editors, *International Conference on Semantic Web Applications and Tools for Life Sciences*, volume 2275 of *CEUR Workshop Proceedings*, 2018.
46. A. Bernasconi, A. Canakoglu, and F. Comolli. Processing genome-wide association studies within a repository of heterogeneous genomic datasets. *BMC Genomic Data*, 24:13, 2023.
47. A. Bernasconi, A. Canakoglu, M. Masseroli, and S. Ceri. META-BASE: a novel architecture for large-scale genomic metadata integration. *IEEE/ACM Transactions on Computational Biology and Bioinformatics*, 19(1):543–557, 2022.
48. A. Bernasconi, A. Canakoglu, M. Masseroli, and S. Ceri. The road towards data integration in human genomics: players, steps and interactions. *Briefings in Bioinformatics*, 22(1):30–44, 2021.
49. A. Bernasconi, A. Canakoglu, M. Masseroli, P. Pinoli, and S. Ceri. A review on viral data sources and search systems for perspective mitigation of COVID-19. *Briefings in Bioinformatics*, 22(2):664–675, 2021.
50. A. Bernasconi, A. Canakoglu, P. Pinoli, and S. Ceri. Empowering Virus Sequence Research Through Conceptual Modeling. In G. Dobbie, U. Frank, G. Kappel, S. W. Liddle, and H. C. Mayr, editors, *Conceptual Modeling*, pages 388–402, Cham, 2020. Springer International Publishing.
51. A. Bernasconi, and S. Ceri. Interoperability of COVID-19 clinical phenotype data with host and viral genetics data. *BioMed*, 2(1):69–81, 2022.
52. A. Bernasconi, L. Cilibrasi, R. Al Khalaf, T. Alfonsi, S. Ceri, P. Pinoli, and A. Canakoglu. EpiSurf: metadata-driven search server for analyzing amino acid changes within epitopes of SARS-CoV-2 and other viral species. *Database*, 2021.
53. A. Bernasconi, L. Mari, R. Casagrandi, and S. Ceri. Data-driven analysis of amino acid change dynamics timely reveals SARS-CoV-2 variant emergence. *Scientific Reports*, 11:21068, 2021.
54. A. Bernasconi, S. Ceri, A. Campi, and M. Masseroli. Conceptual Modeling for Genomics: Building an Integrated Repository of Open Data. In H. C. Mayr, G. Guizzardi, H. Ma, and O. Pastor, editors, *Conceptual Modeling*, pages 325–339, Cham, 2017. Springer International Publishing.
55. A. Bernasconi, A. García S, S. Ceri and O. Pastor. A comprehensive approach for the conceptual modeling of genomic data. In J. Ralyté, S. Chakravarthy, M. Mohania, M.A. Jeusfeld, and K. Karlapalem, editors, *Conceptual Modeling*, pages 194–208, Cham, 2022. Springer International Publishing.
56. A. Bernasconi, A. García S, S. Ceri and O. Pastor. PoliViews: A comprehensive and modular approach to the conceptual modeling of genomic data. *Data & Knowledge Engineering*, 147:102201, 2023.
57. A. Bernasconi, A. Gulino, T. Alfonsi, A. Canakoglu, P. Pinoli, A. Sandionigi, and S. Ceri. VirusViz: comparative analysis and effective visualization of viral nucleotide and amino acid variants. *Nucleic Acids Research*, 49(15):e90, 2021.

58. M. Bersanelli, E. Mosca, D. Remondini, E. Giampieri, C. Sala, et al. Methods for the integration of multi-omics data: mathematical aspects. *BMC bioinformatics*, 17(2):S15, 2016.
59. J. Beverley, S. Babcock, G. Carvalho, L. Cowell, S. Duesing, et al. Coordinating Coronavirus Research: The COVID-19 Infectious Disease Ontology. *OSF Preprints*, 2020. https://doi.org/10.17605/OSF.IO/7EJ4H.
60. National Genomic Data Initiatives: A Worldwide Update. https://www.bio-itworld.com/2019/08/12/national-genomic-data-initiatives-worldwide-update.aspx. Accessed: 2021-01-04.
61. A. Black, D. R. MacCannell, T. R. Sibley, and T. Bedford. Ten recommendations for supporting open pathogen genomic analysis in public health. *Nature Medicine*, 26:832–841, 2020.
62. M. Blum, P.-E. Cholley, V. Malysheva, S. Nicaise, J. Moehlin, et al. A comprehensive resource for retrieving, visualizing, and integrating functional genomics data. *Life science alliance*, 3(1), 2020.
63. O. Bodenreider. The unified medical language system (UMLS): integrating biomedical terminology. *Nucleic Acids Research*, 32(suppl_1):D267–D270, 2004.
64. O. Bodenreider. Biomedical ontologies in action: role in knowledge management, data integration and decision support. *Yearbook of Medical Informatics*, page 67, 2008.
65. M. M. Böhmer, U. Buchholz, V. M. Corman, M. Hoch, K. Katz, et al. Investigation of a COVID-19 outbreak in Germany resulting from a single travel-associated primary case: a case series. *The Lancet Infectious Diseases*, pages 920–928, 2020.
66. D. Bolchini, A. Finkelstein, V. Perrone, and S. Nagl. Better bioinformatics through usability analysis. *Bioinformatics*, 25(3):406–412, 2009.
67. A. Bonifati, F. Cattaneo, S. Ceri, A. Fuggetta, and S. Paraboschi. Designing Data Marts for Data Warehouses. *ACM Transactions on Software Engineering and Methodology*, 10(4):452–483, 2001.
68. V. Bonnici et al. Comprehensive reconstruction and visualization of non-coding regulatory networks in human. *Frontiers in bioengineering and biotechnology*, 2:69, 2014.
69. E. Bornberg-Bauer and N. W. Paton. Conceptual data modelling for bioinformatics. *Briefings in Bioinformatics*, 3(2):166–180, 2002.
70. T. Brown. Design thinking. *Harvard business review*, 86(6):84–92,141, 2008.
71. P. Buitelaar, T. Eigner, and T. Declerck. OntoSelect: A dynamic ontology library with support for ontology selection. In *In Proceedings of the Demo Session at the International Semantic Web Conference*, 2004.
72. D. Bujold, D. A. de Lima Morais, C. Gauthier, C. Côté, M. Caron, et al. The international human epigenome consortium data portal. *Cell systems*, 3(5):496–499, 2016.
73. P. Buneman, S. B. Davidson, K. Hart, C. Overton, and L. Wong. A Data Transformation System for Biological Data Sources. In *In Proceedings of 21st International Conference on Very Large Data Bases*, pages 158–169, 1995.
74. A. Buniello, J. A. L. MacArthur, M. Cerezo, L. W. Harris, J. Hayhurst, et al. The NHGRI-EBI GWAS Catalog of published genome-wide association studies, targeted arrays and summary statistics 2019. *Nucleic Acids Research*, 47(D1):D1005–D1012, 2018.
75. Genome Canada. https://www.genomecanada.ca/. Accessed: 2021-01-04.
76. A. Canakoglu, A. Bernasconi, A. Colombo, M. Masseroli, and S. Ceri. GenoSurf: metadata driven semantic search system for integrated genomic datasets. *Database*, 2019. baz132.
77. A. Canakoglu, P. Pinoli, A. Bernasconi, T. Alfonsi, D. P. Melidis, and S. Ceri. ViruSurf: an integrated database to investigate viral sequences. *Nucleic Acids Research*, 49(D1):D817–D824, 2021.
78. A. Canakoglu, P. Pinoli, A. Gulino, L. Nanni, M. Masseroli, and S. Ceri. Federated sharing and processing of genomic datasets for tertiary data analysis. *Briefings in Bioinformatics*, 2020. https://doi.org/10.1093/bib/bbaa091.
79. G. Cannizzaro, M. Leone, A. Bernasconi, A. Canakoglu, and M. J. Carman. Automated Integration of Genomic Metadata with Sequence-to-Sequence Models. In *Proceedings of the 24th European Conference on Machine Learning and Principles and Practice of Knowledge Discovery in Databases (ECML PKDD 2020)*, Part V, pp. 187-203, Springer, 2021.

80. M. R. Capobianchi, M. Rueca, F. Messina, E. Giombini, F. Carletti, et al. Molecular characterization of SARS-CoV-2 from the first case of COVID-19 in Italy. *Clinical Microbiology and Infection*, 2020.
81. E. Cappelli, F. Cumbo, A. Bernasconi, A. Canakoglu, S. Ceri, et al. OpenGDC: Unifying, Modeling, Integrating Cancer Genomic Data and Clinical Metadata. *Applied Sciences*, 10(18):6367, 2020.
82. E. Cappelli, G. Felici, and E. Weitschek. Combining DNA methylation and RNA sequencing data of cancer for supervised knowledge extraction. *BioData mining*, 11(1):22, 2018.
83. M. E. Carter-Timofte, S. E. Jørgensen, M. R. Freytag, M. M. Thomsen, N.-S. Brinck Andersen, et al. Deciphering the Role of Host Genetics in Susceptibility to Severe COVID-19. *Frontiers in immunology*, 11:1606, 2020.
84. F. Casati, S. Ceri, B. Pernici, and G. Pozzi. Conceptual modeling of workflows. In *International Conference on Conceptual Modeling*, pages 341–354. Springer, 1995.
85. M. Caulfield, J. Davies, M. Dennys, L. Elbahy, T. Fowler, et al. The National Genomics Research and Healthcare Knowledgebase. https://figshare.com/articles/GenomicEnglandProtocol_pdf/4530893/5, 2017. Accessed: 2021-01-04.
86. Linking Sequence Accessions. https://github.com/CDCgov/SARS-CoV-2_Sequencing/#linking-sequence-accessions. Accessed: 2021-01-04.
87. Recommended formatting and criteria for sample metadata. https://github.com/CDCgov/SARS-CoV-2_Sequencing/#recommended-formatting-and-criteria-for-sample-metadata. Accessed: 2021-01-04.
88. E. Cerami, J. Gao, U. Dogrusoz, B. E. Gross, S. O. Sumer, et al. The cBio Cancer Genomics Portal: An Open Platform for Exploring Multidimensional Cancer Genomics Data. *Cancer discovery*, 2(5):401, 2012.
89. S. Ceri, A. Bernasconi, A. Canakoglu, A. Gulino, A. Kaitoua, et al. Overview of GeCo: A project for exploring and integrating signals from the genome. In *International Conference on Data Analytics and Management in Data Intensive Domains*, pages 46–57. Springer, 2017.
90. S. Ceri, P. Fraternali, and M. Matera. Conceptual modeling of data-intensive Web applications. *IEEE Internet Computing*, 6(4):20–30, 2002.
91. S. Ceri, G. Gottlob, L. Tanca, et al. What you always wanted to know about Datalog(and never dared to ask). *IEEE Transactions on Knowledge and Data Engineering*, 1(1):146–166, 1989.
92. S. Ceri, and P. Pinoli Data science for genomic data management: challenges, resources, experiences. *SN Computer Science*, 1:5, 2020.
93. J. F.-W. Chan, A. J. Zhang, S. Yuan, V. K.-M. Poon, C. C.-S. Chan, et al. Simulation of the Clinical and Pathological Manifestations of Coronavirus Disease 2019 (COVID-19) in a Golden Syrian Hamster Model: Implications for Disease Pathogenesis and Transmissibility. *Clinical Infectious Diseases*, 71(9):2428–2446, 2020.
94. S. Chang, J. Zhang, X. Liao, X. Zhu, D. Wang, et al. Influenza Virus Database (IVDB): an integrated information resource and analysis platform for influenza virus research. *Nucleic Acids Research*, 35(suppl_1):D376–D380, 2007.
95. S.-M. Chaw, J.-H. Tai, S.-L. Chen, C.-H. Hsieh, S.-Y. Chang, et al. The origin and underlying driving forces of the SARS-CoV-2 outbreak. *Journal of biomedical science*, 27(1):1–12, 2020.
96. G. Chen, J. C. Ramírez, N. Deng, X. Qiu, C. Wu, et al. Restructured GEO: restructuring Gene Expression Omnibus metadata for genome dynamics analysis. *Database*, 2019, 2019. bay145.
97. J. Y. Chen and J. V. Carlis. Genomic data modeling. *Information Systems*, 28(4):287–310, 2003.
98. P. P.-S. Chen. The entity-relationship model—toward a unified view of data. *ACM Transactions on Database Systems (TODS)*, 1(1):9–36, 1976.
99. M. Chiara, D. S. Horner, C. Gissi, and G. Pesole. Comparative genomics provides an operational classification system and reveals early emergence and biased spatio-temporal

distribution of SARS-CoV-2. *bioRxiv*, 2020. `https://doi.org/10.1101/2020.06.26.172924`.

100. M. Chiara, F. Zambelli, M. A. Tangaro, P. Mandreoli, D. S. Horner, and G. Pesole. CorGAT: a tool for the functional annotation of SARS-CoV-2 genomes. *Bioinformatics*, 2020. https://doi.org/10.1093/bioinformatics/btaa1047.
101. Y. S. Cho, H. Kim, H.-M. Kim, S. Jho, J. Jun, et al. An ethnically relevant consensus Korean reference genome is a step towards personal reference genomes. *Nature communications*, 7:13637, 2016.
102. P. Christen. *Data Matching: Concepts and Techniques for Record Linkage, Entity Resolution, and Duplicate Detection*. Springer Publishing Company, Incorporated, 2012.
103. L. Cilibrasi, P. Pinoli, A. Bernasconi, A. Canakoglu, M. Chiara, and S. Ceri. ViruClust: direct comparison of SARS-CoV-2 genomes and genetic variants in space and time. *Bioinformatics*, 38(7):1988–1994, 2022.
104. P. Cingolani, A. Platts, L. L. Wang, M. Coon, T. Nguyen, et al. A program for annotating and predicting the effects of single nucleotide polymorphisms, SnpEff: SNPs in the genome of Drosophila melanogaster strain w1118; iso-2; iso-3. *Fly*, 6(2):80–92, 2012.
105. L. Clarke, S. Fairley, X. Zheng-Bradley, I. Streeter, E. Perry, et al. The international Genome sample resource (IGSR): A worldwide collection of genome variation incorporating the 1000 Genomes Project data. *Nucleic Acids Research*, 45(D1):D854–D859, 2016.
106. CNGBdb; China National GeneBank DataBase. `https://doi.org/10.25504/FAIRsharing.9btRvC`. Accessed: 2021-01-04.
107. F. S. Collins and H. Varmus. A new initiative on precision medicine. *New England journal of medicine*, 372(9):793–795, 2015.
108. V. M. Corman, O. Landt, M. Kaiser, R. Molenkamp, A. Meijer, et al. Detection of 2019 novel coronavirus (2019-nCoV) by real-time RT-PCR. *Eurosurveillance*, 25(3), 2020.
109. M. Cornell, N. W. Paton, C. Hedeler, P. Kirby, D. Delneri, et al. GIMS: an integrated data storage and analysis environment for genomic and functional data. *Yeast*, 20(15):1291–1306, 2003.
110. M. Courtot, L. Cherubin, A. Faulconbridge, D. Vaughan, M. Green, et al. BioSamples database: an updated sample metadata hub. *Nucleic Acids Research*, 47(D1):D1172–D1178, 2018.
111. COVID-19 Host Genetics Initiative Mapping the human genetic architecture of COVID-19. *Nature*, 600(7889):472–477, 2021.
112. P. Crovari, F. Catania, P. Pinoli, P. Roytburg, A. Salzar, et al. Ok, dna! a conversational interface to explore genomic data. In *Proceedings of the 2nd Conference on Conversational User Interfaces*, pages 1–3, 2020.
113. P. Crovari, S. Pidò, and F. Garzotto. Towards an ontology for tertiary bioinformatics research process. In G. Grossmann and S. Ram, editors, *Advances in Conceptual Modeling*, pages 82–91, Cham, 2020. Springer International Publishing.
114. P. Crovari, S. Pidò, P. Pinoli, A. Bernasconi, A. Canakoglu, F. Garzotto, and S. Ceri. GeCoAgent: a conversational agent for empowering genomic data extraction and analysis *ACM Transactions on Computing for Healthcare (HEALTH)*, 3(1):1–29, 2021.
115. F. Cumbo, G. Fiscon, S. Ceri, M. Masseroli, and E. Weitschek. TCGA2BED: extracting, extending, integrating, and querying the cancer genome atlas. *BMC bioinformatics*, 18(1):6, 2017.
116. D. Cyranoski. China embraces precision medicine on a massive scale. *Nature News*, 529(7584):9, 2016.
117. F. K. Dankar, A. Ptitsyn, and S. K. Dankar. The development of large-scale de-identified biomedical databases in the age of genomics–principles and challenges. *Human genomics*, 12(1):19, 2018.
118. S. B. Davidson, J. Crabtree, B. P. Brunk, J. Schug, V. Tannen, et al. K2/Kleisli and GUS: Experiments in integrated access to genomic data sources. *IBM Systems Journal*, 40(2):512–531, 2001.
119. S. B. Davidson, C. Overton, V. Tannen, and L. Wong. BioKleisli: A digital library for biomedical researchers. *International Journal on Digital Libraries*, 1(1):36–53, 1997.

120. C. A. Davis, B. C. Hitz, C. A. Sloan, E. T. Chan, J. M. Davidson, et al. The Encyclopedia of DNA elements (ENCODE): data portal update. *Nucleic Acids Research*, 46(D1):D794–D801, 2017.
121. S. de Coronado, L. W. Wright, G. Fragoso, M. W. Haber, E. A. Hahn-Dantona, et al. The NCI Thesaurus quality assurance life cycle. *Journal of biomedical informatics*, 42(3):530–539, 2009.
122. E. De Francesco, G. Di Santo, L. Palopoli, and S. E. Rombo. A summary of genomic databases: overview and discussion. In *Biomedical Data and Applications*, pages 37–54. Springer, 2009.
123. N. De Maio. Issues with SARS-CoV-2 sequencing data. `https://virological.org/t/issues-with-sars-cov-2-sequencing-data/473`. Accessed: 2021-01-04.
124. N. Decaro and A. Lorusso. Novel human coronavirus (SARS-CoV-2): A lesson from animal coronaviruses. *Veterinary Microbiology*, page 108693, 2020.
125. Genome Denmark. `http://www.genomedenmark.dk/english/`. Accessed: 2021-01-04.
126. J. Devlin, M.-W. Chang, K. Lee, and K. Toutanova. BERT: Pre-training of deep bidirectional transformers for language understanding. *arXiv*, 2018. `https://arxiv.org/abs/1810.04805`.
127. H.-H. Do and E. Rahm. Flexible integration of molecular-biological annotation data: The GenMapper approach. In *International Conference on Extending Database Technology*, pages 811–822. Springer, 2004.
128. X. Dong et al. Knowledge vault: A web-scale approach to probabilistic knowledge fusion. In *Proceedings of the 20th ACM SIGKDD international conference on Knowledge discovery and data mining*, pages 601–610. ACM, 2014.
129. X. L. Dong and F. Naumann. Data fusion: resolving data conflicts for integration. *Proceedings of the VLDB Endowment*, 2(2):1654–1655, 2009.
130. R. Dréos, G. Ambrosini, R. Groux, R. C. Périer, and P. Bucher. MGA repository: a curated data resource for ChIP-seq and other genome annotated data. *Nucleic Acids Research*, 46(D1):D175–D180, 2017.
131. R. Edgar, M. Domrachev, and A. E. Lash. Gene Expression Omnibus: NCBI gene expression and hybridization array data repository. *Nucleic Acids Research*, 30(1):207–210, 2002.
132. Microarray standards at last. *Nature*, 419(323), 2002.
133. The E-ellow Submarine. `https://onehealth.ifas.ufl.edu/activities/circular-health-program/eellow-submarine/`. Accessed: 2021-01-04.
134. A. L. Egyedi, M. J. O'Connor, M. M. Romero, D. Willrett, J. Hardi, et al. Embracing Semantic Technology for Better Metadata Authoring in Biomedicine. In *International Conference on Semantic Web Applications and Tools for Life Sciences*, CEUR Workshop Proceedings, 2017.
135. K. Eilbeck, S. E. Lewis, C. J. Mungall, M. Yandell, L. Stein, et al. The Sequence Ontology: a tool for the unification of genome annotations. *Genome biology*, 6(5):R44, 2005.
136. H. El-Ghalayini, M. Odeh, R. McClatchey, and D. Arnold. Deriving Conceptual Data Models from Domain Ontologies for Bioinformatics. In *2006 2nd International Conference on Information Communication Technologies*, volume 2, pages 3562–3567, April 2006.
137. S. Elbe and G. Buckland-Merrett. Data, disease and diplomacy: GISAID's innovative contribution to global health. *Global Challenges*, 1(1):33–46, 2017.
138. D. Ellinghaus, F. Degenhardt, L. Bujanda, M. Buti, A. Albillos, et al. Genomewide association study of severe Covid-19 with respiratory failure. *New England Journal of Medicine*, 2020.
139. S. E. Ellis, L. Collado-Torres, A. Jaffe, and J. T. Leek. Improving the value of public RNA-seq expression data by phenotype prediction. *Nucleic Acids Research*, 46(9):e54–e54, 2018.
140. A. Even and G. Shankaranarayanan. Value-Driven Data Quality Assessment. In *ICIQ*, 2005.
141. A. Even and G. Shankaranarayanan. Utility cost perspectives in data quality management. *Journal of Computer Information Systems*, 50(2):127–135, 2009.
142. S. Fang, K. Li, J. Shen, S. Liu, J. Liu, et al. GESS: a database of global evaluation of SARS-CoV-2/hCoV-19 sequences. *Nucleic Acids Research*, 49(D1):D706–D714, 10 2020.
143. E. Fast, B. Chen, J. Mendelsohn, J. Bassen, and M. S. Bernstein. Iris: A conversational agent for complex tasks. In *Proceedings of the 2018 CHI Conference on Human Factors in Computing Systems*, pages 1–12, 2018.

144. S. Federhen. The NCBI taxonomy database. *Nucleic Acids Research*, 40(D1):D136–D143, 2012.
145. J. D. Fernandes, A. S. Hinrichs, H. Clawson, J. N. Gonzalez, B. T. Lee, et al. The UCSC SARS-CoV-2 Genome Browser. *Nature Genetics*, 52(10):991–998, 2020.
146. J. D. Fernández, M. Lenzerini, M. Masseroli, F. Venco, and S. Ceri. Ontology-Based Search of Genomic Metadata. *IEEE/ACM Transactions on Computational Biology and Bioinformatics*, 13(2):233–247, 2016.
147. J. M. Fernández, V. de la Torre, D. Richardson, R. Royo, M. Puiggròs, et al. The BLUEPRINT data analysis portal. *Cell systems*, 3(5):491–495, 2016.
148. A. M. M. Ferrandis, O. P. López, and G. Guizzardi. Applying the principles of an ontology-based approach to a conceptual schema of human genome. In *International Conference on Conceptual Modeling*, pages 471–478. Springer, 2013.
149. FinnGen Research Project. `https://www.finngen.fi/`. Accessed: 2021-01-04.
150. P. Flicek and E. Birney. The European Genotype Archive: Background and Implementation [White paper], 2007. `https://www.ebi.ac.uk/ega/sites/ebi.ac.uk.ega/files/documents/ega_whitepaper.pdf`.
151. J. A. Flynn, D. Purushotham, M. N. K. Choudhary, X. Zhuo, C. Fan, et al. Exploring the coronavirus pandemic with the WashU Virus Genome Browser. *Nature Genetics*, 2020.
152. N. Francis, A. Green, P. Guagliardo, L. Libkin, T. Lindaaker, et al. Cypher: An Evolving Query Language for Property Graphs. In *Proceedings of the 2018 International Conference on Management of Data*, SIGMOD '18, page 1433–1445, New York, NY, USA, 2018. Association for Computing Machinery.
153. A. Frankish, M. Diekhans, A.-M. Ferreira, R. Johnson, I. Jungreis, et al. GENCODE reference annotation for the human and mouse genomes. *Nucleic Acids Research*, 47(D1):D766–D773, 2018.
154. K. W. Fung and O. Bodenreider. Knowledge representation and ontologies. In *Clinical Research Informatics*, pages 313–339. Springer, 2019.
155. I. Gabdank, E. T. Chan, J. M. Davidson, J. A. Hilton, C. A. Davis, et al. Prevention of data duplication for high throughput sequencing repositories. *Database*, 2018, 2018. bay008.
156. E. Galeota, K. Kishore, and M. Pelizzola. Ontology-driven integrative analysis of omics data through Onassis. *Scientific Reports*, 10(1):1–9, 2020.
157. E. R. Gansner, Y. Koren, and S. C. North. Topological Fisheye Views for Visualizing Large Graphs. *IEEE Transactions on Visualization and Computer Graphics*, 11(4):457–468, July 2005.
158. G. F. Gao, J. S. Parker, S. M. Reynolds, T. C. Silva, L.-B. Wang, et al. Before and after: Comparison of legacy and harmonized TCGA genomic data commons' data. *Cell systems*, 9(1):24–34, 2019.
159. A. García S. and J. C. Casamayor. Towards the generation of a species-independent conceptual schema of the genome. In G. Grossmann and S. Ram, editors, *Advances in Conceptual Modeling*, pages 61–70, Cham, 2020. Springer International Publishing.
160. G. Gawriljuk et al. A scalable approach to incrementally building knowledge graphs. In *International Conference on Theory and Practice of Digital Libraries*, pages 188–199. Springer, 2016.
161. J. M. Gaziano, J. Concato, M. Brophy, L. Fiore, S. Pyarajan, et al. Million Veteran Program: a mega-biobank to study genetic influences on health and disease. *Journal of clinical epidemiology*, 70:214–223, 2016.
162. A. George and G. J. J. Viroinformatics: Databases and Tools. In *Proceedings of 11th National Science Symposium on Recent Trends in Science and Technology*, volume 2019, pages 117–126. Rajkot, Gujarat, India: Christ Publications, 2019.
163. L. Getoor and A. Machanavajjhala. Entity resolution: theory, practice & open challenges. *Proceedings of the VLDB Endowment*, 5(12):2018–2019, 2012.
164. M. Ghandi, F. W. Huang, J. Jané-Valbuena, G. V. Kryukov, C. C. Lo, et al. Next-generation characterization of the Cancer Cell Line Encyclopedia. *Nature*, 569(7757):503, 2019.
165. C. B. Giles, C. A. Brown, M. Ripperger, Z. Dennis, X. Roopnarinesingh, et al. ALE: Automated Label Extraction from GEO metadata. *BMC Bioinformatics*, 18(14):509, 2017.

166. O. Giles, R. Huntley, A. Karlsson, J. Lomax, and J. Malone. Reference ontology and database annotation of the COVID-19 Open Research Dataset (CORD-19). *bioRxiv*, 2020. `https://doi.org/10.1101/2020.10.04.325266`.
167. V. Gligorijević and N. Pržulj. Methods for biological data integration: perspectives and challenges. *Journal of the Royal Society Interface*, 12(112):20150571, 2015.
168. M. J. Goldman, B. Craft, M. Hastie, K. Repečka, F. McDade, et al. Visualizing and interpreting cancer genomics data via the Xena platform. *Nature Biotechnology*, 38(6):675–678, 2020.
169. R. Gollakner and I. Capua. Is COVID-19 the first pandemic that evolves into a panzootic? *Veterinaria Italiana*, 56(1):11–12, 2020.
170. D. Gomez-Cabrero, I. Abugessaisa, D. Maier, A. Teschendorff, M. Merkenschlager, et al. Data integration in the era of omics: current and future challenges. *BMC Systems Biology*, 8(Suppl 2):I1, 2014.
171. R. S. Gonçalves and M. A. Musen. The variable quality of metadata about biological samples used in biomedical experiments. *Scientific data*, 6:190021, 2019.
172. M. Graves et al. Graph database systems. *IEEE Engineering in Medicine and Biology Magazine*, 14(6):737–745, 1995.
173. A. J. Gray, C. A. Goble, and R. Jimenez. Bioschemas: From Potato Salad to Protein Annotation. In *International Semantic Web Conference (Posters, Demos & Industry Tracks)*, 2017.
174. A. Grifoni, J. Sidney, Y. Zhang, R. H. Scheuermann, B. Peters, and A. Sette. A Sequence Homology and Bioinformatic Approach Can Predict Candidate Targets for Immune Responses to SARS-CoV-2. *Cell Host & Microbe*, 27(4):671–680.e2, 2020.
175. J. Grosjean, T. Merabti, B. Dahamna, I. Kergourlay, B. Thirion, et al. Health multi-terminology portal: a semantic added-value for patient safety. *Stud Health Technol Inform*, 166(66):129–138, 2011.
176. J. Grosjean, L. Soualmia, K. Bouarech, C. Jonquet, and S. Darmoni. An approach to compare bio-ontologies portals. *Studies in health technology and informatics*, 205:1008, 2014.
177. R. L. Grossman, A. P. Heath, V. Ferretti, H. E. Varmus, D. R. Lowy, et al. Toward a shared vision for cancer genomic data. *New England Journal of Medicine*, 375(12):1109–1112, 2016.
178. D. F. Gudbjartsson, A. Helgason, H. Jonsson, O. T. Magnusson, P. Melsted, et al. Spread of SARS-CoV-2 in the Icelandic Population. *New England Journal of Medicine*, 2020.
179. É. Guerin, G. Marquet, A. Burgun, O. Loréal, L. Berti-Équille, et al. Integrating and warehousing liver gene expression data and related biomedical resources in GEDAW. In *International Workshop on Data Integration in the Life Sciences*, pages 158–174. Springer, 2005.
180. A. Gulino, A. Kaitoua, and S. Ceri. Optimal Binning for Genomics. *IEEE Transactions on Computers*, 68(1):125–138, 2019.
181. Z. Guo, B. Tzvetkova, J. M. Bassik, T. Bodziak, B. M. Wojnar, et al. RNASeqMetaDB: a database and web server for navigating metadata of publicly available mouse RNA-Seq datasets. *Bioinformatics*, 31(24):4038–4040, 2015.
182. A. Gupta, W. Bug, L. Marenco, X. Qian, C. Condit, et al. Federated access to heterogeneous information resources in the Neuroscience Information Framework (NIF). *Neuroinformatics*, 6(3):205–217, 2008.
183. M. Gwinn, D. MacCannell, and G. L. Armstrong. Next-generation sequencing of infectious pathogens. *JAMA*, 321(9):893–894, 2019.
184. I. Hadar, P. Soffer, and K. Kenzi. The role of domain knowledge in requirements elicitation via interviews: an exploratory study. *Requirements Engineering*, 19(2):143–159, 2014.
185. J. Hadfield, C. Megill, S. M. Bell, J. Huddleston, B. Potter, et al. Nextstrain: real-time tracking of pathogen evolution. *Bioinformatics*, 34(23):4121–4123, 2018.
186. D. Hadley, J. Pan, O. El-Sayed, J. Aljabban, I. Aljabban, et al. Precision annotation of digital samples in NCBI's gene expression omnibus. *Scientific data*, 4:170125, 2017.
187. S. Haider, B. Ballester, D. Smedley, J. Zhang, P. M. Rice, and A. Kasprzyk. BioMart Central Portal – unified access to biological data. *Nucleic Acids Research*, 37(Web-Server-Issue):23–27, 2009.

188. H. Hakonarson, J. R. Gulcher, and K. Stefansson. deCODE genetics, Inc. *Pharmacogenomics*, 4(2):209–215, 2003.
189. J. Hammer and M. Schneider. The GenAlg project: developing a new integrating data model, language, and tool for managing and querying genomic information. *ACM SIGMOD Record*, 33(2):45–50, 2004.
190. A. X. Han, E. Parker, F. Scholer, S. Maurer-Stroh, and C. A. Russell. Phylogenetic Clustering by Linear Integer Programming (PhyCLIP). *Molecular Biology and Evolution*, 36(7):1580–1595, 03 2019.
191. J. Hastings, G. Owen, A. Dekker, M. Ennis, N. Kale, et al. ChEBI in 2016: Improved services and an expanding collection of metabolites. *Nucleic Acids Research*, 44(D1):D1214–D1219, 2016.
192. E. L. Hatcher, S. A. Zhdanov, Y. Bao, O. Blinkova, E. P. Nawrocki, et al. Virus Variation Resource–improved response to emergent viral outbreaks. *Nucleic Acids Research*, 45(D1):D482–D490, 2017.
193. Y. He, Y. Liu, and B. Zhao. OGG: a biological ontology for representing genes and genomes in specific organisms. In W. R. Hogan, S. Arabandi, and M. Brochhausen, editors, *Proceedings of the 5th International Conference on Biomedical Ontologies (ICBO 2014), , Houston, TX.*, pages 13–20, 2014.
194. Y. He, H. Yu, E. Ong, Y. Wang, Y. Liu, et al. CIDO, a community-based ontology for coronavirus disease knowledge and data integration, sharing, and analysis. *Scientific Data*, 7(1):1–5, 2020.
195. T. Hernandez and S. Kambhampati. Integration of Biological Sources: Current Systems and Challenges Ahead. *SIGMOD Rec.*, 33(3):51–60, Sept. 2004.
196. B. C. Hitz, L. D. Rowe, N. R. Podduturi, D. I. Glick, U. K. Baymuradov, et al. SnoVault and encodeD: A novel object-based storage system and applications to ENCODE metadata. *PloS one*, 12(4):e0175310, 2017.
197. K. A. Hoadley, C. Yau, T. Hinoue, D. M. Wolf, A. J. Lazar, et al. Cell-of-origin patterns dominate the molecular classification of 10,000 tumors from 33 types of cancer. *Cell*, 173(2):291–304, 2018.
198. E. B. Hodcroft, M. Zuber, S. Nadeau, I. Comas, F. G. Candelas, et al. Emergence and spread of a sars-cov-2 variant through europe in the summer of 2020. *medRxiv*, 2020. `https://doi.org/10.1101/2020.10.25.20219063`.
199. E. L. Hong, C. A. Sloan, E. T. Chan, J. M. Davidson, V. S. Malladi, et al. Principles of metadata organization at the ENCODE data coordination center. *Database*, 2016, 2016. baw001.
200. O. Horlova, A. Kaitoua, V. Markl, and S. Ceri. Multi-Dimensional Genomic Data Management for Region-Preserving Operations. In *2019 IEEE 35th International Conference on Data Engineering (ICDE)*, pages 1166–1177, 2019.
201. C.-C. Huang and Z. Lu. Community challenges in biomedical text mining over 10 years: success, failure and the future. *Briefings in Bioinformatics*, 17(1):132–144, 2016.
202. S. Huang, K. Chaudhary, and L. X. Garmire. More is better: recent progress in multi-omics data integration methods. *Frontiers in genetics*, 8:84, 2017.
203. T. Hubbard, D. Barker, E. Birney, G. Cameron, Y. Chen, et al. The Ensembl genome database project. *Nucleic Acids Research*, 30(1):38–41, 2002.
204. W. Huber, V. J. Carey, R. Gentleman, S. Anders, M. Carlson, et al. Orchestrating high-throughput genomic analysis with Bioconductor. *Nature methods*, 12(2):115–121, 2015.
205. C. Hulo, E. De Castro, P. Masson, L. Bougueleret, A. Bairoch, et al. ViralZone: a knowledge resource to understand virus diversity. *Nucleic Acids Research*, 39(suppl_1):D576–D582, 2011.
206. M. Idrees, M. Khan, et al. A review: conceptual data models for biological domain. *JAPS, Journal of Animal and Plant Sciences*, 25(2):337–345, 2015.
207. How next-generation sequencing can help identify and track SARS-CoV-2. `https://www.nature.com/articles/d42473-020-00120-0`. Accessed: 2021-01-04.
208. J. Ison, H. Ienasescu, P. Chmura, E. Rydza, H. Menager, et al. The bio.tools registry of software tools and data resources for the life sciences. *Genome biology*, 20(1):1–4, 2019.

209. J. Ison, M. Kalaš, I. Jonassen, D. Bolser, M. Uludag, et al. EDAM: an ontology of bioinformatics operations, types of data and identifiers, topics and formats. *Bioinformatics*, 29(10):1325–1332, 2013.
210. F. Ji, R. Elmasri, Y. Zhang, B. Ritesh, and Z. Raja. Incorporating concepts for bioinformatics data modeling into EER models. In *The 3rd ACS/IEEE International Conference on Computer Systems and Applications*, pages 189–192. IEEE, 2005.
211. R. J. L. John, N. Potti, and J. M. Patel. Ava: From Data to Insights Through Conversations. In *CIDR 2017, 8th Biennial Conference on Innovative Data Systems Research, Chaminade, CA, USA, January 8-11, 2017, Online Proceedings*, 2017.
212. C. Jonquet, M. A. Musen, and N. Shah. A system for ontology-based annotation of biomedical data. In *International Workshop on Data Integration in The Life Sciences*, pages 144–152. Springer, 2008.
213. C. Jonquet, M. A. Musen, and N. H. Shah. Building a biomedical ontology recommender web service. In *Journal of biomedical semantics*, volume 1, page S1. Springer, 2010.
214. C. Jonquet, N. H. Shah, and M. A. Musen. The open biomedical annotator. *Summit on translational bioinformatics*, 2009:56, 2009.
215. S. Jupp, T. Burdett, C. Leroy, and H. E. Parkinson. A new Ontology Lookup Service at EMBL-EBI. In *International Conference on Semantic Web Applications and Tools for Life Sciences*, CEUR Workshop Proceedings, pages 118–119, 2015.
216. A. Kaitoua, A. Gulino, M. Masseroli, P. Pinoli, and S. Ceri. Scalable Genomic Data Management System on the Cloud. In *2017 International Conference on High Performance Computing Simulation (HPCS)*, pages 58–63. IEEE, 2017.
217. J. Kans. Entrez direct: E-utilities on the UNIX command line. In *Entrez Programming Utilities Help [Internet]*. National Center for Biotechnology Information (US), 2020. `https://www.ncbi.nlm.nih.gov/books/NBK179288/`.
218. K. J. Karczewski, L. C. Francioli, G. Tiao, B. B. Cummings, J. Alföldi, et al. The mutational constraint spectrum quantified from variation in 141,456 humans. *Nature*, 581(7809):434–443, 2020.
219. K. J. Karczewski, B. Weisburd, B. Thomas, M. Solomonson, D. M. Ruderfer, et al. The ExAC browser: displaying reference data information from over 60 000 exomes. *Nucleic Acids Research*, 45(D1):D840–D845, 2017.
220. C. M. Keet. Biological data and conceptual modelling methods. *Journal of Conceptual Modeling*, 29(1):1–14, 2003.
221. C. M. Keet. A taxonomy of types of granularity. In *2006 IEEE International Conference on Granular Computing*, pages 106–111. IEEE, 2006.
222. W. J. Kent, C. W. Sugnet, T. S. Furey, K. M. Roskin, T. H. Pringle, et al. The human genome browser at UCSC. *Genome research*, 12(6):996–1006, 2002.
223. R. A. Khailany, M. Safdar, and M. Ozaslan. Genomic characterization of a novel sars-cov-2. *Gene Reports*, 19:100682, 2020.
224. K. A. Khan and P. Cheung. Presence of mismatches between diagnostic PCR assays and coronavirus SARS-CoV-2 genome. *Royal Society Open Science*, 7(6):200636, 2020.
225. R. Kimball and M. Ross. *The data warehouse toolkit: the complete guide to dimensional modeling*. John Wiley & Sons, 2011.
226. Y. Kodama, J. Mashima, T. Kosuge, E. Kaminuma, O. Ogasawara, et al. DNA Data Bank of Japan: 30th Anniversary. *Nucleic Acids Research*, 46(D1):D30–D35, 2017.
227. Y. Kodama, M. Shumway, and R. Leinonen. The Sequence Read Archive: explosive growth of sequencing data. *Nucleic Acids Research*, 40(D1):D54–D56, 2011.
228. S. Köhler, L. Carmody, N. Vasilevsky, J. O. B. Jacobsen, D. Danis, et al. Expansion of the Human Phenotype Ontology (HPO) knowledge base and resources. *Nucleic Acids Research*, 47(D1):D1018–D1027, 2019.
229. E. V. Koonin, V. V. Dolja, M. Krupovic, A. Varsani, Y. I. Wolf, et al. Global organization and proposed megataxonomy of the virus world. *Microbiology and Molecular Biology Reviews*, 84(2), 2020.

230. B. Korber, W. M. Fischer, S. Gnanakaran, H. Yoon, J. Theiler, et al. Tracking changes in SARS-CoV-2 Spike: evidence that D614G increases infectivity of the COVID-19 virus. *Cell*, 2020.
231. N. V. Kovalevskaya, C. Whicher, T. D. Richardson, C. Smith, J. Grajciarova, et al. DNAdigest and repositive: connecting the world of genomic data. *PLoS Biology*, 14(3):e1002418, 2016.
232. T. Koyama, D. Platt, and L. Parida. Variant analysis of SARS-CoV-2 genomes. *Bulletin of the World Health Organization*, 98(7):495–504, 2020.
233. A. Kundaje, W. Meuleman, J. Ernst, M. Bilenky, A. Yen, et al. Integrative analysis of 111 reference human epigenomes. *Nature*, 518(7539):317–330, 2015.
234. C.-L. Kuo, L. C. Pilling, J. L. Atkins, J. A. H. Masoli, J. Delgado, et al. ApoE e4e4 Genotype and Mortality With COVID-19 in UK Biobank. *The Journals of Gerontology: Series A*, 07 2020. glaa169.
235. S. Laha, J. Chakraborty, S. Das, S. K. Manna, S. Biswas, and R. Chatterjee. Characterizations of sars-cov-2 mutational profile, spike protein stability and viral transmission. *Infection, Genetics and Evolution*, 85:104445, 2020.
236. J. Lan, J. Ge, J. Yu, S. Shan, H. Zhou, et al. Structure of the SARS-CoV-2 spike receptor-binding domain bound to the ACE2 receptor. *Nature*, 581(7807):215–220, 2020.
237. S. G. Landt, G. K. Marinov, A. Kundaje, P. Kheradpour, F. Pauli, et al. ChIP-seq guidelines and practices of the ENCODE and modENCODE consortia. *Genome research*, 22(9):1813–1831, 2012.
238. V. Lapatas, M. Stefanidakis, R. C. Jimenez, A. Via, and M. V. Schneider. Data integration in biological research: an overview. *Journal of Biological Research-Thessaloniki*, 22(1):9, 2015.
239. I. Lappalainen, J. Almeida-King, V. Kumanduri, A. Senf, J. D. Spalding, et al. The European Genome-phenome Archive of human data consented for biomedical research. *Nature genetics*, 47(7):692–695, 2015.
240. W. Lathe, J. Williams, M. Mangan, and D. Karolchik. Genomic data resources: challenges and promises. *Nature Education*, 1(3):2, 2008.
241. J. W. Lau, E. Lehnert, A. Sethi, R. Malhotra, G. Kaushik, et al. The Cancer Genomics Cloud: collaborative, reproducible, and democratized–a new paradigm in large-scale computational research. *Cancer research*, 77(21):e3–e6, 2017.
242. S.-Y. Lau, P. Wang, B. W.-Y. Mok, A. J. Zhang, H. Chu, et al. Attenuated SARS-CoV-2 variants with deletions at the S1/S2 junction. *Emerging microbes & infections*, 9(1):837–842, 2020.
243. L. Leitsalu and A. Metspalu. From Biobanking to Precision Medicine: The Estonian Experience. In *Genomic and precision medicine*, pages 119–129. Elsevier, 2017.
244. M. Lek, K. J. Karczewski, E. V. Minikel, K. E. Samocha, E. Banks, et al. Analysis of protein-coding genetic variation in 60,706 humans. *Nature*, 536(7616):285, 2016.
245. M. Lenzerini. Data Integration: A Theoretical Perspective. In *Proceedings of the Twenty-First ACM SIGMOD-SIGACT-SIGART Symposium on Principles of Database Systems*, PODS '02, page 233–246. Association for Computing Machinery, 2002.
246. F.-X. Lescure, L. Bouadma, D. Nguyen, M. Parisey, P.-H. Wicky, et al. Clinical and virological data of the first cases of COVID-19 in Europe: a case series. *The Lancet Infectious Diseases*, 2020.
247. J. Li, C.-S. Tseng, A. Federico, F. Ivankovic, Y.-S. Huang, et al. SFMetaDB: a comprehensive annotation of mouse RNA splicing factor RNA-Seq datasets. *Database*, 2017, 2017. bax071.
248. Q. Li, J. Wu, J. Nie, L. Zhang, H. Hao, et al. The impact of mutations in SARS-CoV-2 spike on viral infectivity and antigenicity. *Cell*, 182(5):1284–1294, 2020.
249. Z. Li, J. Li, and P. Yu. GEOMetaCuration: a web-based application for accurate manual curation of Gene Expression Omnibus metadata. *Database*, 2018, 2018. bay019.
250. B. Liu, K. Liu, H. Zhang, L. Zhang, Y. Bian, and L. Huang. CoV-Seq, a New Tool for SARS-CoV-2 Genome Analysis and Visualization: Development and Usability Study. *J Med Internet Res*, 22(10):e22299, Oct 2020.
251. Y. Liu, W. K. Chan, Z. Wang, J. Hur, J. Xie, et al. Ontological and bioinformatic analysis of anti-coronavirus drugs and their Implication for drug repurposing against COVID-19. *Preprints*, 2020. https://doi.org/10.20944/preprints202003.0413.v1.

252. Y. Liu, M. Ott, N. Goyal, J. Du, M. Joshi, et al. Roberta: A robustly optimized bert pretraining approach. *arXiv*, 2019. `https://arxiv.org/abs/1907.11692`.
253. M. Lizio, J. Harshbarger, H. Shimoji, J. Severin, T. Kasukawa, et al. Gateways to the FANTOM5 promoter level mammalian expression atlas. *Genome biology*, 16(1):22, 2015.
254. C.-C. Lo, M. Shakya, K. Davenport, M. Flynn, J. Gans, et al. EDGE COVID-19: A Web Platform to generate submission-ready genomes for SARS-CoV-2 sequencing efforts. *arXiv*, 2020. `https://arxiv.org/abs/2006.08058`.
255. J. Lonsdale, J. Thomas, M. Salvatore, R. Phillips, E. Lo, et al. The genotype-tissue expression (GTEx) project. *Nature genetics*, 45(6):580, 2013.
256. A. E. Loraine, I. C. Blakley, S. Jagadeesan, J. Harper, G. Miller, and N. Firon. Analysis and visualization of RNA-Seq expression data using RStudio, Bioconductor, and Integrated Genome Browser. In *Plant Functional Genomics*, pages 481–501. Springer, 2015.
257. B. Louie, P. Mork, F. Martin-Sanchez, A. Halevy, and P. Tarczy-Hornoch. Data integration and genomic medicine. *Journal of biomedical informatics*, 40(1):5–16, 2007.
258. C. Lu, R. Gam, A. P. Pandurangan, and J. Gough. Genetic risk factors for death with SARS-CoV-2 from the UK Biobank. *medRxiv*, 2020. `https://doi.org/10.1101/2020.07.01.20144592`.
259. G. Lu, K. Buyyani, N. Goty, R. Donis, and Z. Chen. Influenza A virus informatics: genotype-centered database and genotype annotation. In *Second International Multi-Symposiums on Computer and Computational Sciences (IMSCCS 2007)*, pages 76–83. IEEE, 2007.
260. R. Lu, X. Zhao, J. Li, P. Niu, B. Yang, et al. Genomic characterisation and epidemiology of 2019 novel coronavirus: implications for virus origins and receptor binding. *The Lancet*, 395(10224):565–574, 2020.
261. R. Lukyananko, J. Parsons, and B. M. Samuel. Artifact sampling in experimental conceptual modeling research. In C. Woo, J. Lu, Z. Li, T. Ling, G. Li, and M. Lee, editors, *Advances in Conceptual Modeling*, pages 199–205, Cham, 2018. Springer International Publishing.
262. M.-T. Luong, H. Pham, and C. D. Manning. Effective approaches to attention-based neural machine translation. *arXiv*, 2015. `https://arxiv.org/abs/1508.04025`.
263. K. A. Lythgoe, M. Hall, L. Ferretti, M. de Cesare, G. MacIntyre-Cockett, et al. Shared SARS-CoV-2 diversity suggests localised transmission of minority variants. *bioRxiv*, 2020. `https://doi.org/10.1101/2020.05.28.118992`.
264. O. A. MacLean, R. J. Orton, J. B. Singer, and D. L. Robertson. No evidence for distinct types in the evolution of SARS-CoV-2. *Virus Evolution*, 6(1), 05 2020. veaa034.
265. D. Maglott, J. Ostell, K. D. Pruitt, and T. Tatusova. Entrez Gene: gene-centered information at NCBI. *Nucleic Acids Research*, 39(suppl_1):D52–D57, 2010.
266. G. Maiga. A Flexible Biomedical Ontology Selection Tool. *Strengthening the Role of ICT in Development*, pages 171–189, 2009.
267. V. S. Malladi, D. T. Erickson, N. R. Podduturi, L. D. Rowe, E. T. Chan, et al. Ontology application and use at the ENCODE DCC. *Database*, 2015, 2015. bav010.
268. J. Malone, E. Holloway, T. Adamusiak, M. Kapushesky, J. Zheng, et al. Modeling sample variables with an Experimental Factor Ontology. *Bioinformatics*, 26(8):1112–1118, 2010.
269. J. Malone, R. Stevens, S. Jupp, T. Hancocks, H. Parkinson, and C. Brooksbank. Ten simple rules for selecting a bio-ontology. *PLoS Computational Biology*, 12(2):e1004743, 2016.
270. M. Martínez-Romero, C. Jonquet, M. J. O'Connor, J. Graybeal, A. Pazos, and M. A. Musen. NCBO Ontology Recommender 2.0: an enhanced approach for biomedical ontology recommendation. *Journal of biomedical semantics*, 8(1):21, 2017.
271. M. Martínez-Romero, M. J. O'Connor, A. L. Egyedi, D. Willrett, J. Hardi, et al. Using association rule mining and ontologies to generate metadata recommendations from multiple biomedical databases. *Database*, 2019, 2019. baz059.
272. M. Martínez-Romero, M. J. O'Connor, R. D. Shankar, M. Panahiazar, D. Willrett, et al. Fast and accurate metadata authoring using ontology-based recommendations. In *AMIA Annual Symposium Proceedings*, volume 2017, page 1272. American Medical Informatics Association, 2017.

273. M. Martínez-Romero, J. M. Vázquez-Naya, J. Pereira, and A. Pazos. BiOSS: A system for biomedical ontology selection. *Computer methods and programs in biomedicine*, 114(1):125–140, 2014.
274. M. Masseroli, A. Canakoglu, and S. Ceri. Integration and Querying of Genomic and Proteomic Semantic Annotations for Biomedical Knowledge Extraction. *IEEE/ACM Transactions on Computational Biology and Bioinformatics*, 13(2):209–219, 2016.
275. M. Masseroli, A. Canakoglu, P. Pinoli, A. Kaitoua, A. Gulino, et al. Processing of big heterogeneous genomic datasets for tertiary analysis of Next Generation Sequencing data. *Bioinformatics*, 35(5):729–736, 08 2018.
276. M. Masseroli, A. Kaitoua, P. Pinoli, and S. Ceri. Modeling and interoperability of heterogeneous genomic big data for integrative processing and querying. *Methods*, 111:3–11, 2016.
277. M. Masseroli, P. Pinoli, F. Venco, A. Kaitoua, V. Jalili, et al. GenoMetric Query Language: a novel approach to large-scale genomic data management. *Bioinformatics*, 31(12):1881–1888, 2015.
278. R. McNaughton and H. Yamada. Regular expressions and state graphs for automata. *IEEE Transactions on Electronic Computers*, (1):39–47, 1960.
279. C. Médigue, F. Rechenmann, A. Danchin, and A. Viari. Imagene: an integrated computer environment for sequence annotation and analysis. *Bioinformatics (Oxford, England)*, 15(1):2–15, 1999.
280. T. F. Meehan, A. M. Masci, A. Abdulla, L. G. Cowell, J. A. Blake, et al. Logical development of the cell ontology. *BMC bioinformatics*, 12(1):6, 2011.
281. S. Mei, Q. Qin, Q. Wu, H. Sun, R. Zheng, et al. Cistrome Data Browser: a data portal for chip-seq and chromatin accessibility data in human and mouse. *Nucleic Acids Research*, 45(D1):D658–D662, 2016.
282. D. Mercatelli, L. Triboli, E. Fornasari, F. Ray, and F. M. Giorgi. Coronapp: A web application to annotate and monitor SARS-CoV-2 mutations. *Journal of Medical Virology*, pages 1–8, 2020.
283. A. Messina, A. Fiannaca, L. La Paglia, M. La Rosa, and A. Urso. BioGraph: a web application and a graph database for querying and analyzing bioinformatics resources. *BMC systems biology*, 12(5):98, 2018.
284. F. Messina, E. Giombini, C. Agrati, F. Vairo, T. Ascoli Bartoli, et al. COVID-19: viral–host interactome analyzed by network based-approach model to study pathogenesis of SARS-CoV-2 infection. *Journal of Translational Medicine*, 18(1):233, 2020.
285. T. Mihara, Y. Nishimura, Y. Shimizu, H. Nishiyama, G. Yoshikawa, et al. Linking virus genomes with host taxonomy. *Viruses*, 8(3):66, 2016.
286. A. Møller. dk.brics.automaton – Finite-State Automata and Regular Expressions for Java. `http://www.brics.dk/automaton/`, 2017. Accessed: 2021-01-04.
287. B. Mons. The vodan in: support of a fair-based infrastructure for covid-19. *European Journal of Human Genetics*, 28(6):724–727, 2020.
288. G. H. Moore. *Zermelo's axiom of choice: its origins, development, and influence*. Dover Publications, 1982.
289. I. J. Morais, R. C. Polveiro, G. M. Souza, D. I. Bortolin, F. T. Sassaki, and A. T. M. Lima. The global population of sars-cov-2 is composed of six major subtypes. *Scientific reports*, 10(1):1–9, 2020.
290. L. Mousavizadeh and S. Ghasemi. Genotype and phenotype of COVID-19: Their roles in pathogenesis. *Journal of Microbiology, Immunology and Infection*, 2020.
291. C. J. Mungall, C. Torniai, G. V. Gkoutos, S. E. Lewis, and M. A. Haendel. Uberon, an integrative multi-species anatomy ontology. *Genome biology*, 13(1):R5, 2012.
292. M. F. Murray, E. E. Kenny, M. D. Ritchie, D. J. Rader, A. E. Bale, et al. COVID-19 outcomes and the human genome. *Genetics in Medicine*, pages 1–3, 2020.
293. M. A. Musen, C. A. Bean, K.-H. Cheung, M. Dumontier, K. A. Durante, et al. The center for expanded data annotation and retrieval. *Journal of the American Medical Informatics Association*, 22(6):1148–1152, 2015.

294. M. A. Musen, S.-A. Sansone, K.-H. Cheung, S. H. Kleinstein, M. Crafts, et al. CEDAR: Semantic Web Technology to Support Open Science. In *Companion Proceedings of the The Web Conference 2018*, pages 427–428. International World Wide Web Conferences Steering Committee, 2018.
295. L. Nanni, P. Pinoli, A. Canakoglu, and S. Ceri. PyGMQL: scalable data extraction and analysis for heterogeneous genomic datasets. *BMC Bioinformatics*, 20(1):560, 2019.
296. Novel Coronavirus (COVID-19) Overview. `https://nanoporetech.com/covid-19/overview`. Accessed: 2021-01-04.
297. F. Naumann, A. Bilke, J. Bleiholder, and M. Weis. Data fusion in three steps: Resolving inconsistencies at schema-, tuple-, and value-level. In *In Bulletin of the Technical Committee on Data Engineering*, pages 21–31, 2006.
298. S. B. Needleman and C. D. Wunsch. A general method applicable to the search for similarities in the amino acid sequence of two proteins. *Journal of molecular biology*, 48(3):443–453, 1970.
299. B. Nuseibeh and S. Easterbrook. Requirements engineering: a roadmap. In *Proceedings of the Conference on the Future of Software Engineering*, pages 35–46, 2000.
300. T. Okayama, T. Tamura, T. Gojobori, Y. Tateno, K. Ikeo, et al. Formal design and implementation of an improved DDBJ DNA database with a new schema and object-oriented library. *Bioinformatics (Oxford, England)*, 14(6):472–478, 1998.
301. N. A. O'Leary, M. W. Wright, J. R. Brister, S. Ciufo, D. Haddad, et al. Reference sequence (RefSeq) database at NCBI: current status, taxonomic expansion, and functional annotation. *Nucleic Acids Research*, 44(D1):D733–D745, 2016.
302. A. Olivé. *Conceptual modeling of information systems*. Springer Science & Business Media, 2007.
303. D. Oliveira, A. S. Butt, A. Haller, D. Rebholz-Schuhmann, and R. Sahay. Where to search top-K biomedical ontologies? *Briefings in Bioinformatics*, 20(4):1477–1491, 2019.
304. M. Ostaszewski, A. Mazein, M. E. Gillespie, I. Kuperstein, A. Niarakis, et al. COVID-19 Disease Map, building a computational repository of SARS-CoV-2 virus-host interaction mechanisms. *Scientific data*, 7(1):1–4, 2020.
305. A. Oulas, M. Zanti, M. Tomazou, M. Zachariou, G. Minadakis, et al. Generalized linear models provide a measure of virulence for specific mutations in SARS-CoV-2 strains. *bioRxiv*, 2020. `https://doi.org/10.1101/2020.08.17.253484`.
306. M. Pachetti, B. Marini, F. Benedetti, F. Giudici, E. Mauro, et al. Emerging SARS-CoV-2 mutation hot spots include a novel RNA-dependent-RNA polymerase variant. *Journal of Translational Medicine*, 18(1):1–9, 2020.
307. A. L. Palacio, Ó. P. López, and J. C. C. Ródenas. A method to identify relevant genome data: conceptual modeling for the medicine of precision. In *International Conference on Conceptual Modeling*, pages 597–609. Springer, 2018.
308. P. Pareja-Tobes et al. Bio4j: a high-performance cloud-enabled graph-based data platform. *BioRxiv*, page 016758, 2015. `https://doi.org/10.1101/016758`.
309. Y. M. Park, S. Squizzato, N. Buso, T. Gur, and R. Lopez. The EBI search engine: EBI search as a service—making biological data accessible for all. *Nucleic Acids Research*, 45(W1):W545–W549, 2017.
310. M. S. A. Parvez, M. M. Rahman, M. N. Morshed, D. Rahman, S. Anwar, and M. J. Hosen. Genetic analysis of SARS-CoV-2 isolates collected from Bangladesh: Insights into the origin, mutational spectrum and possible pathomechanism. *Computational Biology and Chemistry*, page 107413, 2020.
311. Ó. Pastor, A. P. León, J. F. R. Reyes, A. S. García, and J. C. R. Casamayor. Using conceptual modeling to improve genome data management. *Briefings in Bioinformatics*, 2020. `https://doi.org/10.1093/bib/bbaa100`.
312. N. W. Paton, S. A. Khan, A. Hayes, F. Moussouni, A. Brass, et al. Conceptual modelling of genomic information. *Bioinformatics*, 16(6):548–557, 2000.
313. H. Pearson. Human genome done and dusted. *Nature*, 2003.
314. C. Penna, V. Mercurio, C. G. Tocchetti, and P. Pagliaro. Sex-related differences in COVID-19 lethality. *British Journal of Pharmacology*, pages 1–11, 2020.

315. B. E. Pickett, E. L. Sadat, Y. Zhang, J. M. Noronha, R. B. Squires, et al. ViPR: an open bioinformatics database and analysis resource for virology research. *Nucleic Acids Research*, 40(D1):D593–D598, 2012.
316. P. Pinoli, A. Canakoglu, S. Ceri, M. Chiara, E. Ferrandi, L. Minotti, A. Bernasconi. VariantHunter: a method and tool for fast detection of emerging SARS-CoV-2 variants. *Database*, 2023.
317. P. Pinoli, E. Stamoulakatou, A.-P. Nguyen, M. Rodríguez Martínez, and S. Ceri. Pan-cancer analysis of somatic mutations and epigenetic alterations in insulated neighbourhood boundaries. *PloS one*, 15(1):e0227180, 2020.
318. L. Posch, M. Panahiazar, M. Dumontier, and O. Gevaert. Predicting structured metadata from unstructured metadata. *Database*, 2016, 2016. baw080.
319. Promoting best practice in nucleotide sequence data sharing. *Scientific Data*, 7(1):152, 2020.
320. How Blockchain Companies Are Helping Us Protect Our Genomic Data. `https://www.labiotech.eu/features/blockchain-control-genomic-data/`. Accessed: 2021-01-04.
321. K. Pruitt, T. Murphy, G. Brown, and M. Murphy. RefSeq Frequently Asked Questions (FAQ). *RefSeq Help [Internet]*, 2010. `https://www.ncbi.nlm.nih.gov/books/NBK50679/`.
322. Qatar Genome. `https://qatargenome.org.qa/`. Accessed: 2021-01-04.
323. M. O. Rabin and D. Scott. Finite automata and their decision problems. *IBM Journal of Research and Development*, 3(2):114–125, 1959.
324. A. Radford, J. Wu, R. Child, D. Luan, D. Amodei, and I. Sutskever. Language Models are Unsupervised Multitask Learners. *OpenAI Blog*, 1(8):9, 2019.
325. A. Rambaut, E. C. Holmes, Á. O'Toole, V. Hill, J. T. McCrone, et al. A dynamic nomenclature proposal for SARS-CoV-2 lineages to assist genomic epidemiology. *Nature Microbiology*, 5(1):1403–1407, 2020.
326. G. Rambold, P. Yilmaz, J. Harjes, S. Klaster, V. Sanz, et al. Meta-omics data and collection objects (MOD-CO): a conceptual schema and data model for processing sample data in meta-omics research. *Database*, 2019, 2019. baz002.
327. F. Rechenmann. Data modeling: the key to biological data integration. *EMBnet. journal*, 18(B):59–60, 2012.
328. T. C. Redman and A. Blanton. *Data quality for the information age*. Artech House, Inc., 1996.
329. 1000 Genomes Project Consortium. A global reference for human genetic variation. *Nature*, 526(7571):68, 2015.
330. ENCODE Project Consortium. An integrated encyclopedia of DNA elements in the human genome. *Nature*, 489(7414):57–74, 2012.
331. Gene Ontology Consortium. The gene ontology resource: 20 years and still GOing strong. *Nucleic Acids Research*, 47(D1):D330–D338, 2019.
332. National Genomics Data Center Members and Partners. Database resources of the national genomics data center in 2020. *Nucleic Acids Research*, 48(D1):D24–D33, 2020.
333. National Human Genome Research Institute. The cost of sequencing a human genome. `https://www.genome.gov/27565109/the-cost-of-sequencing-a-human-genome/`. Accessed: 2021-01-04.
334. Nomenclature Committee of the International Union of Biochemistry (NC-IUB). Nomenclature for Incompletely Specified Bases in Nucleic Acid Sequences: Recommendations 1984. *Proceedings of the National Academy of Sciences of the United States of America*, 83(1):4–8, 1986.
335. The COVID-19 Genomics UK (COG-UK) consortium. An integrated national scale SARS-CoV-2 genomic surveillance network. *The Lancet Microbe*, 1(3):E99–E100, 2020.
336. S. M. Reynolds, M. Miller, P. Lee, K. Leinonen, S. M. Paquette, et al. The ISB Cancer Genomics Cloud: a flexible cloud-based platform for cancer genomics research. *Cancer research*, 77(21):e7–e10, 2017.
337. D. J. Rigden and X. M. Fernández. The 27th annual Nucleic Acids Research database issue and molecular biology database collection. *Nucleic Acids Research*, 48(D1):D1–D8, 2020.

338. W. Ritzel Paixão-Côrtes, V. Stangherlin Machado Paixão-Côrtes, C. Ellwanger, and O. Norberto de Souza. Development and Usability Evaluation of a Prototype Conversational Interface for Biological Information Retrieval via Bioinformatics. In S. Yamamoto and H. Mori, editors, *Human Interface and the Management of Information. Visual Information and Knowledge Management*, pages 575–593, Cham, 2019. Springer International Publishing.
339. I. Rodchenkov et al. Pathway Commons 2019 Update: integration, analysis and exploration of pathway data. *Nucleic Acids Research*, 48(D1):D489–D497, 2020.
340. J. F. R. Román, Ó. Pastor, J. C. Casamayor, and F. Valverde. Applying conceptual modeling to better understand the human genome. In *International Conference on Conceptual Modeling*, pages 404–412. Springer, 2016.
341. R. Rose, D. J. Nolan, S. Moot, A. Feehan, S. Cross, et al. Intra-host site-specific polymorphisms of SARS-CoV-2 is consistent across multiple samples and methodologies. *medRxiv*, 2020. https://doi.org/10.1101/2020.04.24.20078691.
342. M. Sabou, V. Lopez, E. Motta, and V. Uren. Ontology selection: ontology evaluation on the real Semantic Web. In *15th International World Wide Web Conference (WWW 2006)*, 2006.
343. S. A. Samarajiwa, I. Olan, and D. Bihary. Challenges and Cases of Genomic Data Integration Across Technologies and Biological Scales. In *Advanced Data Analytics in Health*, pages 201–216. Springer, 2018.
344. H. Samimi, J. Tešić, and A. H. H. Ngu. Patient Centric Data Integration for Improved Diagnosis and Risk Prediction. In *Heterogeneous Data Management, Polystores, and Analytics for Healthcare*, pages 185–195. Springer, 2019.
345. D. Sanchez-Cisneros and F. A. Gali. UEM-UC3M: An Ontology-based named entity recognition system for biomedical texts. In *Proceedings of the Seventh International Workshop on Semantic Evaluation (SemEval 2013)*, pages 622–627, 2013.
346. S.-A. Sansone, A. Gonzalez-Beltran, P. Rocca-Serra, G. Alter, J. S. Grethe, et al. DATS, the data tag suite to enable discoverability of datasets. *Scientific Data*, 4:170059, 2017.
347. S.-A. Sansone, P. McQuilton, P. Rocca-Serra, A. Gonzalez-Beltran, M. Izzo, et al. FAIRsharing as a community approach to standards, repositories and policies. *Nature biotechnology*, 37(4):358–367, 2019.
348. S.-A. Sansone, P. Rocca-Serra, M. Brandizi, A. Brazma, D. Field, et al. The first RSBI (ISA-TAB) workshop:"can a simple format work for complex studies?". *OMICS A Journal of Integrative Biology*, 12(2):143–149, 2008.
349. S.-A. Sansone, P. Rocca-Serra, D. Field, E. Maguire, C. Taylor, et al. Toward interoperable bioscience data. *Nature genetics*, 44(2):121, 2012.
350. J. C. Santos and G. A. Passos. The high infectivity of SARS-CoV-2 B.1.1.7 is associated with increased interaction force between Spike-ACE2 caused by the viral N501Y mutation. *bioRxiv*, 2021. https://doi.org/10.1101/2020.12.29.424708.
351. U. Sarkans, M. Gostev, A. Athar, E. Behrangi, O. Melnichuk, et al. The BioStudies database–one stop shop for all data supporting a life sciences study. *Nucleic Acids Research*, 46(D1):D1266–D1270, 2017.
352. J. Sarkar and R. Guha. Infectivity, virulence, pathogenicity, host-pathogen interactions of SARS and SARS-CoV-2 in experimental animals: a systematic review. *Veterinary Research Communications*, 2020.
353. R. Satija, J. A. Farrell, D. Gennert, A. F. Schier, and A. Regev. Spatial reconstruction of single-cell gene expression data. *Nature biotechnology*, 33(5):495, 2015.
354. E. Sayers. The E-utilities in-depth: parameters, syntax and more. *Entrez Programming Utilities Help [Internet]*, 2009. https://www.ncbi.nlm.nih.gov/books/NBK25499/.
355. E. W. Sayers, R. Agarwala, E. E. Bolton, J. R. Brister, K. Canese, et al. Database Resources of the National Center for Biotechnology Information. *Nucleic Acids Research*, 47(Database issue):D23, 2019.
356. E. W. Sayers, M. Cavanaugh, K. Clark, J. Ostell, K. D. Pruitt, and I. Karsch-Mizrachi. GenBank. *Nucleic Acids Research*, 47(D1):D94–D99, 2019.
357. E.-M. Schön, J. Thomaschewski, and M. J. Escalona. Agile Requirements Engineering: A systematic literature review. *Computer Standards & Interfaces*, 49:79–91, 2017.

358. L. M. Schriml, M. Chuvochina, N. Davies, E. A. Eloe-Fadrosh, R. D. Finn, et al. COVID-19 pandemic reveals the peril of ignoring metadata standards. *Scientific data*, 7(1):1–4, 2020.
359. L. M. Schriml, E. Mitraka, J. Munro, B. Tauber, M. Schor, et al. Human Disease Ontology 2018 update: classification, content and workflow expansion. *Nucleic Acids Research*, 47(D1):D955–D962, 2019.
360. N. J. Schurch, P. Schofield, M. Gierliński, C. Cole, A. Sherstnev, et al. How many biological replicates are needed in an RNA-seq experiment and which differential expression tool should you use? *Rna*, 22(6):839–851, 2016.
361. S. C. Schuster. Next-generation sequencing transforms today's biology. *Nature methods*, 5(1):16, 2007.
362. G. Serna Garcia, M. Leone, A. Bernasconi, and M.J. Carman. GeMI: interactive interface for transformer-based Genomic Metadata Integration. *Database*, 2022.
363. G. Serna Garcia, R. Al Khalaf, F. Invernici, S. Ceri, and A. Bernasconi. CoVEffect: interactive system for mining the effects of SARS-CoV-2 mutations and variants based on deep learning. *GigaScience*, 12, 2023.
364. M. Settino, A. Bernasconi, G. Ceddia, G. Agapito, M. Masseroli, and M. Cannataro. Using GMQL-Web for Querying, Downloading and Integrating Public with Private Genomic Datasets. In *Proceedings of the 10th ACM International Conference on Bioinformatics, Computational Biology and Health Informatics*, BCB '19, page 688–693, New York, NY, USA, 2019. Association for Computing Machinery.
365. R. W. Shafer. Rationale and uses of a public HIV drug-resistance database. *The Journal of infectious diseases*, 194(Supplement_1):S51–S58, 2006.
366. N. H. Shah, C. Jonquet, A. P. Chiang, A. J. Butte, R. Chen, and M. A. Musen. Ontology-driven indexing of public datasets for translational bioinformatics. *BMC Bioinformatics*, 10(2):S1, 2009.
367. D. Sharma, P. Priyadarshini, and S. Vrati. Unraveling the web of viroinformatics: computational tools and databases in virus research. *Journal of virology*, 89(3):1489–1501, 2015.
368. L. Shen, D. Maglinte, D. Ostrow, U. Pandey, M. Bootwalla, et al. Children's Hospital Los Angeles COVID-19 Analysis Research Database (CARD)-A Resource for Rapid SARS-CoV-2 Genome Identification Using Interactive Online Phylogenetic Tools. *bioRxiv*, 2020. https://doi.org/10.1101/2020.05.11.089763.
369. R. Shen, A. B. Olshen, and M. Ladanyi. Integrative clustering of multiple genomic data types using a joint latent variable model with application to breast and lung cancer subtype analysis. *Bioinformatics*, 25(22):2906–2912, 2009.
370. Y. Shu and J. McCauley. GISAID: Global initiative on sharing all influenza data–from vision to reality. *Eurosurveillance*, 22(13), 2017.
371. J. Singer, R. Gifford, M. Cotten, and D. Robertson. CoV-GLUE: A Web Application for Tracking SARS-CoV-2 Genomic Variation. *Preprints*, 2020. https://doi.org/10.20944/preprints202006.0225.v1.
372. D. Smedley, S. Haider, S. Durinck, L. Pandini, P. Provero, et al. The BioMart community portal: an innovative alternative to large, centralized data repositories. *Nucleic Acids Research*, 43(W1):W589–W598, 04 2015.
373. B. Smith, M. Ashburner, C. Rosse, J. Bard, W. Bug, et al. The OBO Foundry: coordinated evolution of ontologies to support biomedical data integration. *Nature biotechnology*, 25(11):1251–1255, 2007.
374. M. Stano, G. Beke, and L. Klucar. viruSITE—integrated database for viral genomics. *Database*, 2016, 2016. baw162.
375. Z. Stark, L. Dolman, T. A. Manolio, B. Ozenberger, S. L. Hill, et al. Integrating genomics into healthcare: a global responsibility. *The American Journal of Human Genetics*, 104(1):13–20, 2019.
376. Z. D. Stephens, S. Y. Lee, F. Faghri, R. H. Campbell, C. Zhai, et al. Big Data: Astronomical or Genomical? *PLOS Biology*, 13(7):1–11, 07 2015.
377. T. Tahsin, D. Weissenbacher, D. Jones-Shargani, D. Magee, M. Vaiente, et al. Named entity linking of geospatial and host metadata in GenBank for advancing biomedical research. *Database*, 2017, 2017. bax093.

378. W. Tang, Z. Lu, and I. S. Dhillon. Clustering with multiple graphs. In *2009 Ninth IEEE International Conference on Data Mining*, pages 1016–1021. IEEE, 2009.
379. X. Tang, C. Wu, X. Li, Y. Song, X. Yao, et al. On the origin and continuing evolution of SARS-CoV-2. *National Science Review*, 7(6):1012–1023, 03 2020.
380. J. G. Tate, S. Bamford, H. C. Jubb, Z. Sondka, D. M. Beare, et al. COSMIC: the catalogue of somatic mutations in cancer. *Nucleic Acids Research*, 47(D1):D941–D947, 2018.
381. S. F. Terry. The Global Alliance for Genomics & Health. *Genetic testing and molecular biomarkers*, 18(6):375–376, 2014.
382. K. Thompson. Programming techniques: Regular expression search algorithm. *Communications of the ACM*, 11(6):419–422, 1968.
383. Y. Toyoshima, K. Nemoto, S. Matsumoto, Y. Nakamura, and K. Kiyotani. SARS-CoV-2 genomic variations associated with mortality rate of COVID-19. *Journal of human genetics*, 65(12):1075–1082, 2020.
384. C. Trapnell, B. A. Williams, G. Pertea, A. Mortazavi, G. Kwan, et al. Transcript assembly and quantification by RNA-Seq reveals unannotated transcripts and isoform switching during cell differentiation. *Nature biotechnology*, 28(5):511, 2010.
385. K. A. Tryka, L. Hao, A. Sturcke, Y. Jin, Z. Y. Wang, et al. NCBI's Database of Genotypes and Phenotypes: dbGaP. *Nucleic Acids Research*, 42(D1):D975–D979, 2014.
386. C. I. van der Made, A. Simons, J. Schuurs-Hoeijmakers, G. van den Heuvel, T. Mantere, et al. Presence of Genetic Variants Among Young Men With Severe COVID-19. *JAMA*, 324(7):663–673, 08 2020.
387. A. Vaswani, N. Shazeer, N. Parmar, J. Uszkoreit, L. Jones, et al. Attention is all you need. In *Advances in neural information processing systems*, pages 5998–6008, 2017.
388. M. Verdonck, F. Gailly, R. Pergl, G. Guizzardi, B. Martins, and O. Pastor. Comparing traditional conceptual modeling with ontology-driven conceptual modeling: An empirical study. *Information Systems*, 81:92–103, 2019.
389. R. Vita, S. Mahajan, J. A. Overton, S. K. Dhanda, S. Martini, et al. The immune epitope database (IEDB): 2018 update. *Nucleic Acids Research*, 47(D1):D339–D343, 2019.
390. E. Volz, V. Hill, J. T. McCrone, A. Price, D. Jorgensen, et al. Evaluating the effects of SARS-CoV-2 Spike mutation D614G on transmissibility and pathogenicity. *Cell*, 2020.
391. L. Wang, F. Chen, X. Guo, L. You, X. Yang, et al. VirusDIP: Virus Data Integration Platform. *bioRxiv*, 2020. `https://doi.org/10.1101/2020.06.08.139451`.
392. L. Wang, A. Zhang, and M. Ramanathan. BioStar models of clinical and genomic data for biomedical data warehouse design. *International Journal of Bioinformatics Research and Applications*, 1(1):63–80, 2005.
393. L. L. Wang, K. Lo, Y. Chandrasekhar, R. Reas, J. Yang, et al. Cord-19: The covid-19 open research dataset. *arXiv*, 2020. `https://arxiv.org/abs/2004.10706`.
394. R. Y. Wang and D. M. Strong. Beyond accuracy: What data quality means to data consumers. *Journal of management information systems*, 12(4):5–33, 1996.
395. Z. Wang, A. Lachmann, and A. Ma'ayan. Mining data and metadata from the gene expression omnibus. *Biophysical reviews*, 11(1):103–110, 2019.
396. Z. Wang, C. D. Monteiro, K. M. Jagodnik, N. F. Fernandez, G. W. Gundersen, et al. Extraction and analysis of signatures from the Gene Expression Omnibus by the crowd. *Nature communications*, 7(1):1–11, 2016.
397. J. N. Weinstein, E. A. Collisson, G. B. Mills, K. R. M. Shaw, B. A. Ozenberger, et al. The cancer genome atlas pan-cancer analysis project. *Nature genetics*, 45(10):1113–1120, 2013.
398. E. Weitschek, F. Cumbo, E. Cappelli, and G. Felici. Genomic data integration: A case study on next generation sequencing of cancer. In *2016 27th International Workshop on Database and Expert Systems Applications (DEXA)*, pages 49–53. IEEE, 2016.
399. P. L. Whetzel, N. F. Noy, N. H. Shah, P. R. Alexander, C. Nyulas, et al. BioPortal: enhanced functionality via new Web services from the National Center for Biomedical Ontology to access and use ontologies in software applications. *Nucleic Acids Research*, 39(suppl_2):W541–W545, 2011.

400. WHO's code of conduct for open and timely sharing of pathogen genetic sequence data during outbreaks of infectious disease. https://www.who.int/blueprint/what/norms-standards/GSDDraftCodeConduct_forpublicconsultation-v1.pdf. Accessed: 2021-01-04.
401. Revised case report form for Confirmed Novel Coronavirus COVID-19 (report to WHO within 48 hours of case identification). https://www.who.int/docs/default-source/coronaviruse/2019-covid-crf-v6.pdf. Accessed: 2021-01-04.
402. M. D. Wilkinson, M. Dumontier, I. J. Aalbersberg, G. Appleton, M. Axton, et al. The FAIR Guiding Principles for scientific data management and stewardship. *Scientific Data*, 3:160018, mar 2016.
403. B. Willers. *Show, Don't Tell*, pages 3–22. Springer London, London, 2015.
404. Y. Yang, J. Fear, J. Hu, I. Haecker, L. Zhou, et al. Leveraging biological replicates to improve analysis in ChIP-seq experiments. *Computational and structural biotechnology journal*, 9(13):e201401002, 2014.
405. B. Yates, B. Braschi, K. A. Gray, R. L. Seal, S. Tweedie, and E. A. Bruford. Genenames.org: the HGNC and VGNC resources in 2017. *Nucleic Acids Research*, 45(D1):D619–D625, 10 2016.
406. T. Yates, C. Razieh, F. Zaccardi, M. J. Davies, and K. Khunti. Obesity and risk of COVID-19: analysis of UK biobank. *Primary Care Diabetes*, sep 2020.
407. C. Yi, X. Sun, J. Ye, L. Ding, M. Liu, et al. Key residues of the receptor binding motif in the spike protein of SARS-CoV-2 that interact with ACE2 and neutralizing antibodies. *Cellular & Molecular Immunology*, 17(6):621–630, 2020.
408. B. H. Yoon et al. Use of graph database for the integration of heterogeneous biological data. *Genomics & informatics*, 15(1):19, 2017.
409. B. E. Young, S.-W. Fong, Y.-H. Chan, T.-M. Mak, L. W. Ang, et al. Effects of a major deletion in the SARS-CoV-2 genome on the severity of infection and the inflammatory response: an observational cohort study. *The Lancet*, 396(10251):603–611, 2020.
410. H. Yu, L. Li, H.-h. Huang, Y. Wang, Y. Liu, et al. Ontology-based systematic classification and analysis of coronaviruses, hosts, and host-coronavirus interactions towards deep understanding of COVID-19. *arXiv*, 2020. https://arxiv.org/abs/2006.00639.
411. A. Zaveri, W. Hu, and M. Dumontier. MetaCrowd: crowdsourcing biomedical metadata quality assessment. *Human Computation*, 6(1):98–112, 2019.
412. H. Zeberg and S. Pääbo. The major genetic risk factor for severe COVID-19 is inherited from Neandertals. *bioRxiv*, 2020. https://doi.org/10.1101/2020.07.03.186296.
413. D. R. Zerbino, P. Achuthan, W. Akanni, M. R. Amode, D. Barrell, et al. Ensembl 2018. *Nucleic Acids Research*, 46(D1):D754–D761, 2017.
414. H. Zhang, T. Li, and Y. Wang. Design of an empirical study for evaluating an automatic layout tool. In C. Woo, J. Lu, Z. Li, T. Ling, G. Li, and M. Lee, editors, *Advances in Conceptual Modeling*, pages 206–211, Cham, 2018. Springer International Publishing.
415. J. Zhang, R. Bajari, D. Andric, F. Gerthoffert, A. Lepsa, et al. The international cancer genome consortium data portal. *Nature biotechnology*, 37(4):367–369, 2019.
416. L. Zhang, C. B. Jackson, H. Mou, A. Ojha, E. S. Rangarajan, et al. The D614G mutation in the SARS-CoV-2 spike protein reduces S1 shedding and increases infectivity. *bioRxiv*, 2020. https://doi.org/10.1101/2020.06.12.148726.
417. W.-M. Zhao, S.-H. Song, M.-L. Chen, D. Zou, L.-N. Ma, et al. The 2019 novel coronavirus resource. *Yi chuan= Hereditas*, 42(2):212–221, 2020.
418. R. Zheng, C. Wan, S. Mei, Q. Qin, Q. Wu, et al. Cistrome Data Browser: expanded datasets and new tools for gene regulatory analysis. *Nucleic Acids Research*, 47(D1):D729–D735, 2018.
419. Y. Zhu, S. Davis, R. Stephens, P. S. Meltzer, and Y. Chen. GEOmetadb: powerful alternative search engine for the gene expression omnibus. *Bioinformatics*, 24(23):2798–2800, 2008.
420. M. Zitnik, F. Nguyen, B. Wang, J. Leskovec, A. Goldenberg, and M. M. Hoffman. Machine learning for integrating data in biology and medicine: Principles, practice, and opportunities. *Information Fusion*, 50:71–91, 2019.

Printed in the United States
by Baker & Taylor Publisher Services